Johann Buxtorf, Impresario of Hebrew and Jewish Books

Johann Buxtorf (1564–1629) pursued the study of Hebrew and Aramaic and the writings and rituals of Jewish tradition through a long and productive life as a professor at the University of Basel and an impresario of Jewish texts. The focus of this work is on Buxtorf's scholarly practices – for example, the ways in which he read and made excerpts from a wide variety of Jewish texts, recycled them in his polemical *Jüden schul* (1603), a treatise on the customs and ceremonies of Ashkenazic Jews, and surveyed them in pioneering if incomplete bibliographies. Using a wide variety of unpublished sources, including letters to Buxtorf from Jewish print professionals, this study follows him into the Basel printing houses that produced books in Hebrew, Aramaic, and Yiddish, and reconstructs how he worked there as both an editor and a censor – even rewriting certain Jewish prayers to eliminate anti-Christian sentiments. Buxtorf took a special interest in the textual history of the Hebrew Bible: he argued that the vowel points and accents had entered its text at a very early stage. This study shows that he developed his views in dialogue – and debate – with both Jewish and Christian scholars. Buxtorf often expressed contempt for both the Jews whose texts he read and those whom he occasionally met in Basel, Frankfurt, or elsewhere, yet his passion for Jewish literature of every kind never faltered.

Studies and Texts 239

Judaism in the Medieval and Early Modern World 5

Edited by
PIET VAN BOXEL
University of Oxford

JOANNA WEINBERG
University of Oxford

Johann Buxtorf, Impresario of Hebrew and Jewish Books

ANTHONY GRAFTON and JOANNA WEINBERG

Toronto
PIMS
PONTIFICAL INSTITUTE OF MEDIAEVAL STUDIES

Acknowledgment

The publication of this volume was made possible by funds generously provided by the Basilian Fathers of the University of St Michael's College.

Library and Archives Canada Cataloguing in Publication

Title: Johann Buxtorf, impresario of Hebrew and Jewish books / Anthony Grafton and Joanna Weinberg.
Names: Grafton, Anthony, author | Weinberg, Joanna, 1949– author | Pontifical Institute of Mediaeval Studies, publisher.
Series: Studies and texts (Pontifical Institute of Mediaeval Studies) ; 239. | Judaism in the medieval and early modern world ; 5.
Description: Series statement: Studies and texts ; 239 | Judaism in the medieval and early modern world ; 5 | Includes bibliographical references and index.
Identifiers: Canadiana (print) 20240520084 | Canadiana (ebook) 20240520122 | ISBN 9780888442390 (hardcover) | ISBN 9781771104401 (PDF)
Subjects: LCSH: Buxtorf, Johann, 1564–1629. | LCSH: Christian Hebraists – Switzerland – Basel – History – 17th century. | LCSH: Editors – Switzerland – Basel – History – 17th century. | LCSH: Printing, Hebrew – Switzerland – Basel – History – 17th century. | LCSH: Censorship – Switzerland – Basel – History – 17th century. | LCSH: Hebrew language – Study and teaching (Higher) – History – 17th century.
Classification: LCC PJ4534.B89 G73 2025 | DDC 492.4092–dc23

Pontifical Institute of Mediaeval Studies
59 Queen's Park Crescent East
Toronto, Ontario M5S 2C4
Canada
pims.ca

PRINTED IN CANADA

Contents

List of Figures

Abbreviations

BAV	Biblioteca Apostolica Vaticana
BSB	Bayerische Staatsbibliothek
BT	Babylonian Talmud
Leiden, UB	Leiden, Universiteitsbibliotheek
Basel, UB	Basel, Universitätsbibliothek
Zurich, ZB	Zurich, Zentralbibliothek

Acknowledgments

We thank the Katz Center for Advanced Judaic Studies in Philadelphia and the Max Planck Institute for History of Science in Berlin for hospitality, research support, and the opportunity to present the earliest versions of our work on Buxtorf to erudite and critical audiences. Special thanks to David Ruderman, then the Director of the Katz Center, and Lorraine Daston, then one of the Directors of the Max Planck Institute, for their generous interest in our project and helpful criticism of its earliest versions. An earlier version of Chapter 1 appeared in a volume sponsored by the Max Planck Institute for History of Science: "Johann Buxtorf Makes a Notebook," in *Canonical Texts and Scholarly Practices: A Global Comparative Approach*, ed. Anthony Grafton and Glenn W. Most (Cambridge, 2016), 275–298. It is reproduced here with changes and additions. Thanks to Princeton University as well for research support. We are also grateful to the friends who have helped us with advice on technical points: Paul Babinski, Edward Fram, Arthur Kiron, Yakov Mayer, Judith Olszowy-Schlanger, Lucia Raspe, Adam Shear, Joshua Teplitsky, and Eric White. Regine Buxtorf and Christoph Buxtorf showed us great kindness and helped us understand the history of their family. Fred Unwalla steered our manuscript into port, and the readers to whom he sent it offered helpful criticism. Alice Falk edited our text with great care and precision, John Schechtman-Marko polished it further, and Megan Jones at the Pontifical Institute of Mediaeval Studies expertly and patiently coordinated proof corrections and much more.

Many libraries have welcomed us and provided us with the materials we needed. Heartfelt thanks to the skillful and energetic librarians at the Katz Center and the Max Planck Institute for History of Science, and to the staffs of the Bodleian Libraries, the British Library, the Herzog August Bibliothek in Wolfenbüttel, the Princeton University Library – especially its Special Collections – and the Warburg Institute. The librarians of the Zentralbibliothek in Zurich and the Öffentliche Bibliothek of the University of Basel have welcomed us, year after year, with endless patience and generosity. We owe special thanks to Maria Benedicta Erny-Lopez Sanchez, Isabel Akagaç, Urs Leu, and Rainer Walter, but all of their colleagues have joined in helping us in every imaginable way over the years. Their kindness enabled us to examine vast amounts of source material in the best possible conditions, during short as well as long stays. To work in their collections was to enter the Republic of Letters in full operation, and we will be grateful forever for this experience. We owe a special debt, finally, to Louise Grafton, who encouraged us in our endeavors until her death in 2019, and to Piet van Boxel, who has advised and inspired us, with unflagging enthusiasm, since our work began.

INTRODUCTION

A Vocation for Hebrew

Early in the 1590s, Johann Buxtorf interrupted his lectures on the Hebrew text of Proverbs with a digression about modern Jews, inspired by the word בעל (lord, master):

> The rabbis use the term בעל הפסוק *Baal happasuk* for someone who has absolutely mastered the reading of the biblical text, as though he were the lord of the individual verses. For פסוק is a biblical verse, and by synecdoche the whole biblical text. Such a one was Elijah Levita, a Jew, who had almost the entire Bible at his fingers' tips. He also possessed such a strong locational memory of his book (which he had used for many years) that when Paul Fagius at Isny or Oswald Schreckenfuchs at Venice, both expert in Hebrew, cited some biblical verse, he could display the [relevant] page at once when he first opened his book with his thumbs. Similarly, last year I saw a Jew, born blind, who by constant listening in Jewish schools and synagogues had gained mastery of the Law of Moses (which is almost the only text customarily read and chanted in those places, along with some short readings from the Prophets, which they call Haphtaroth). When I offered him various verses at random from the Five Books of Moses, he rapidly named the Parascha that contained each text. A Parascha is a section, designated in the Hebrew text by the capital letter *Pe* written thrice.[1]

1. Johann Buxtorf, *Praelectiones Grammaticae in Proverbia Salomonis*, Basel, UB, MS Fr Gr III 22, a, fol. 22v: "Apud Rabbinos בעל הפסוק *Baal happasuk* dicitur, qui in textus biblici lectione versatissimus et exercitatissimus est, quasi singulorum versuum dominus. Nam פסוק Versus biblicus, et synecdochice Textus integer biblicus. Talis fuit Elias Levita, Judaeus, qui non tantum integra paene Biblia in promptu habuit, sed et in suo Codice multis annis trito, memoriam localem tam firmam, ut Paulo Fagio Isnae, et Osualdo Schreckenfuchsio Venetiis (viris hebraice doctissimis) versum aliquem biblicum adducentibus, ipse prima sui codicis per pollices apertione, illico paginam etiam ostenderit. Sic ego superiore anno vidi Judaeum caecum natum, qui assidua auscultatione in Scholis et synagogis judaicis tantum

Buxtorf's digression is typical in many ways. He admired the sixteenth-century Christian scholars who had preceded him in the study of Hebrew and enjoyed telling stories about them. He was fascinated by the Jewish lexicographer and grammarian Elijah Levita, whose work instructed and provoked him throughout his career. He thought his students should know how Jews divided the Bible for reading in synagogues. Most striking of all – and most characteristic of him – is the sovereign way in which he already ranged the centuries, moving easily from the Hebrew Bible to the reading practices of contemporary Jews, and did his best to induce his pupils to follow him. In 1593, he wrote to Jacob Zwinger, who was then in Padua. He asked his friend to obtain some copies of a small-format edition of the Torah, with the Haftarot – presumably to be used by the same students to whom he explained the Jewish form of Torah study.[2]

Buxtorf's range and skill are all the more impressive in light of his relative youth and inexperience. Born in 1564 to a minister in Westphalia, he had studied Hebrew with Johann Piscator at Herborn, rapidly becoming as expert as his master. His *Album amicorum* records his friendships with fellow students as well as with Piscator.[3] In 1588 he left Herborn, traveling to some of the centers of the Reformed world. The album records his meetings with the good and the great: the theologians Franciscus Junius in Heidelberg and Theodore Beza in Geneva, the Hebraists Heinrich Wolf in Zurich and Corneille Bertram in Lausanne, the astronomer and clock designer Conrad Dasypodius in Strasbourg.[4] At Basel, the renowned scholar Johann Jacob Grynaeus, whom he met in April 1588, offered him warm support and hospitality. He settled there, at first as a tutor in the Curio family. From 1588 onward he taught Hebrew in the university, and in 1590, after taking his MA, he was chosen professor of Hebrew. This was only the beginning of Buxtorf's long and largely successful career as teacher, scholar, and member of the local elite. In 1592, he married Margareta Curio, who bore him eleven children: his

sibi usum Legis Mosaicae (quae fere sola cum brevibus quibusdam lectionibus propheticis, quas *Haphtaroth* vocant, in illis legi et cantillari solet), comparaverat, ut mihi varios versus ex quinque libris Mosis ipsi saltuatim proponenti prompte indicarit, qua Parascha quisque scriptus haberetur. Parascha est sectio, quae in textu Hebraeo per tria majuscula Pe designatur."

2. Buxtorf to Zwinger, 4 August 1593, Basel, UB, Fr Gr II 9, 89: "Sunt quinque libri Mosis cum lectionibus propheticis minima forma excusi hebraice absque punctis, quale exemplar proculdubio aliquando vidisti per Pascalem mihi missum: avide ego adhuc quatuor vel sex exemplaria cuperem. Pentateuchum Iudaei appellant Chimmusch [*sic*] et ei adjunctas illas propheticas lectiones vocitant Haphtaroth, quomodo et tu eos petere poteris. Sunt parvi libelluli, parum ut spero molestiae tibi exhibituri."

3. Buxtorf, *Album amicorum*, Basel, UB, MS AN VI 26z (unpaginated), discussed by Stephen Burnett, *From Christian Hebraism to Jewish Studies: Johannes Buxtorf (1564–1629) and Hebrew Learning in the Seventeenth Century* (Leiden, 1996), 13–15.

4. Dasypodius entered the highly appropriate motto Τὰ μαθήματα πρὸς πάντα χρήσιμα.

daughters married members of the Grynaeus and Zwinger families and a printer with whom he worked extensively, Ludwig König. By the time he died in 1629, he had woven himself tightly into the fabric of Basel's academic patriciate. His son succeeded him as professor of Hebrew, and his family became one of Basel's most durable academic dynasties.[5]

Hebrew: The Meanings of a Language

Buxtorf developed themes from existing scholarship in his early lectures. He took care, for example, to show his students that Christian study of Hebrew had a solid genealogy. Support for Christian study of Hebrew came from Origen and Jerome in Christian antiquity and the Council of Vienne in 1311–12.[6] He praised past polyglots from the apostles to Cleopatra, with special emphasis on a Reformed master linguist, Theodore Bibliander.[7] He made clear that Johannes Reuchlin, who proclaimed that he had studied Hebrew for its own sake, not in search of profit, and spent his money freely on lessons in the language, was an exemplary figure.[8] And he attacked the Council of Trent for choosing the Vulgate as the authentic text of the Old Testament, citing errors from the Psalms.[9] He argued explicitly, moreover, that God Himself had revived Hebrew scholarship in the Christian world, to support the Protestant Reformation: "at the same time he aroused outstanding experts to restore the Hebrew language, so that by its aid the word of God would be blessedly illuminated and widely propagated."[10]

Buxtorf stuck with convention when he argued, in a formal *Encomion Hebraeae Linguae*, that Hebrew was the oldest language, given by God to Adam in Paradise.[11] It was also the most artfully formed: the name Adam showed that God had formed

5. Johann Buxtorf the Younger is extensively discussed in Jan Loop, *Johann Heinrich Hottinger: Arabic and Islamic Studies in the Seventeenth Century* (Oxford, 2013). For the fortunes of the Buxtorf family, see Rudolf Smend, *Vier Epitaphe: Die Basler Hebraistenfamilie Buxtorf* (Berlin, 2010).

6. Buxtorf, *Encomion Hebraeae Linguae*, Basel, UB, Ki Ar 189, pt. 4, fols. 1r–8r, at fol. 7r.

7. Buxtorf, *De quibusdam qui multas linguas calluerunt*, Basel, UB, Ki Ar 189, pt. 4, fol. 9r.

8. A lecture of November 1591, Basel, UB, Ki Ar 189, pt. 4, fol. 21r.

9. Buxtorf, *De necessitate linguae hebraeae*, 28 October 1595, Basel, UB, Ki Ar 189, pt. 4, ibid., fols. 23r– 32r, at fol. 26r.

10. Buxtorf, "Alia praefatiuncula, quibus necessitas linguae Hebraeae ex Novo Testamento potissimum evincitur. An. 1596. mense Novemb. recitata," Basel, UB, Ki Ar 189, pt. 4, fols. 32r–40r, at fol. 40r: "excitavit etiam simul praestantissimos artifices, qui linguam hebraeam restaurarent ut eius adminiculo felicissime verbum dei illustraretur et latissime propagaretur."

11. Buxtorf, *Encomion Hebraeae Linguae*, fol. 1v.

man from the earth; Havah, that Eve was the mother of all living humans.[12] Most important, perhaps, was another feature of the language: it was the source of all others, including Greek. In fact, Buxtorf argued – and here his claims became less conventional – that Christian scholars who did not know Hebrew were not competent to interpret their own Greek Scriptures:

> Not even the books of the New Testament can be properly understood without knowledge of this language. Evidence of this is provided not only by the many Hebrew words that are scattered everywhere, but also by an infinite number of idioms and manners of speaking drawn from the usage and customs of the Hebrews, which have often pushed those unskilled in the Hebrew language into divergent and contrary opinions. I could prove this with many examples, if your agreement on this point did not provide me with abundant confirmation.[13]

Here, Buxtorf developed arguments put forward forty years earlier by the Catholic scholar Angelo Canini, who in the mid-sixteenth century had taught Hebrew, Aramaic, and Syriac at Paris, when it was the center of Oriental studies in the Christian world.[14]

In a 1596 lecture on the importance of Hebrew for the study of the New Testament, Buxtorf followed Canini in discussing the Hebrew and Aramaic terms explicitly used in the Greek text.[15] He even noted that at least once, in Jesus's last words, "certain flaws arising from ignorance of the Hebrew language slipped into the text, and were transmitted to us through the carelessness of scribes."[16] In later years, Buxtorf might not have acknowledged that the text of Scripture could contain errors. Like Canini, he took the Talmud not only as a record of the debates

12. Ibid., fol. 3r.

13. Ibid., fol. 7r: "Atque haec adeo vera sunt, ut ne Novi quidem Testamenti libri absque cognitione hujus linguae recte intelligi possint. Cui rei testimonium praebent, non solum multa quae passim hebraica aspersa sunt vocabula, sed etiam infiniti idiotismi ac loquendi modi ex hebraeorum usu & consuetudine desumpti, qui hebraicae linguae imperitos in diversas saepe & contrarias rapuerunt sententias; quod quidem multis a me exemplis demonstrari posset, nisi vester consensus me hac in parte abunde confirmaret."

14. See Joanna Weinberg, "A Sixteenth-Century Hebraic Approach to the New Testament," in *History of Scholarship: A Selection of Papers from the Seminar on the History of Scholarship Held Annually at the Warburg Institute*, ed. Christopher Ligota and Jean-Louis Quantin (Oxford, 2006), 231–50, and Anthony Grafton and Joanna Weinberg, with Alastair Hamilton, *"I have always loved the Holy Tongue": Isaac Casaubon, the Jews and a Forgotten Chapter in Renaissance Scholarship* (Cambridge, MA, 2011), 283–87.

15. Buxtorf, "Alia praefatiuncula," fols. 32v–33v.

16. Ibid., fol. 33r: "Porro, nemo etiam hebraice peritus inficias ibit, in illis Christi verbis, Eli Eli lama sabachtani, menda quaedam ex ignoratione linguae hebraeae in scriptione obrepsisse, et ad nos librariorum incuria delata esse."

of learned Jews but also as evidence for "the vulgar and proverbial speech of the Jews."[17] He developed an elaborate and systematic analysis of what he saw as the echoes of Hebrew in New Testament Greek. Buxtorf characterized the "diction" and "character" of New Testament Greek as Hebraic.

> Only someone who knows the Hebrews' ways of speaking can properly understand the first words of the New Testament. The delicate ears of the Greeks would find this idiom unusual: "The book of the generation of Jesus Christ, the son of David, the son of Abraham." The character of the Latin language would not accept this manner of speech either unless it was drawn from elsewhere. "The book of the generation[s]" is a purely Hebrew way of speaking, which can be found in Genesis 5:1, for a list, enumeration, recitation, of a family, genealogy, ancestors. Βίβλος in this manner of speaking corresponds to the Hebrew ספר, which means not only simply book but a list, an enumeration, a catalog, a writing, a letter, and finally a book. The Hebrews use this word for that because useful doctrines or events are recounted and explicated in it, from the verb ספר, he recounted, he listed, etc.[18]

Buxtorf offered numerous examples, some very simple. When the Gospel authors wrote "in those days," for example, they echoed "what we read a hundred thousand

17. Ibid., 37v: "Sed et haec ex Iudaeorum vulgato et proverbiali sermone a Christo petebantur, quando Pharisaeus dicebat: Qua mensura mensi fueritis, remetientur vobis: quod totidem verbis extat in Tractatu Talmudico Sanhedrin במדה שאדם מודד מודדין לו [BT, Sanhedrin 100a], id est, ea mensura qua homo metitur, contra metientur ei: Quod etiamnum hodie alio tritissimo efferunt proverbio, מדה כנגד מדה Mensura pro mensura, vel Par pari. Huc refertur et illud a servatore nostro in eosdem Pharisaeos jactatum, qui minima quaeque et externa tantum observabant, omittentes quae in Lege erant maxima: nempe Ejice trabem ex oculo tuo. quod itidem extat totidem verbis in Talmud טול קורה מבין עיניך, Tolle trabem ex oculo tuo [This should read *ex oculis tuis*; Buxtorf probably had Matthew 7:5 in mind when he made it singular] [BT, Arakhin 16b]." Some of these parallels are also indicated in his copybook (discussed in Chapter 1, below), Basel, UB, A XII 20, 177.

18. Ibid., fols. 36r–v: "Adeo vero hoc verum et immotum est, quod initio posui: Neminem scilicet, Novum Testamentum recte et solide, ut theologum verum decet, intelligere et de proprietate sermonis judicare posse, nisi etiam linguae hebraeae mediocrem cognitionem habeat, adeo inquam hoc verum est, ut ne prima quidem N.T. verba recte quis intelligat, nisi qui hebraeorum loquendi morem noverit. Graecorum enim delicatis auribus hoc insolens dicendi genus est, βίβλος γενέσεως Ἰησοῦ Χριστοῦ υἱοῦ Δαυὶδ υἱοῦ Ἀβραάμ. Non minus etiam Latini sermonis proprietas talem loquendi modum admittit, nisi aliunde petitum. Est ergo Liber generationis purus putus hebraeorum loquendi modus, qui extat Genes. 5. 1. pro, Recensio, enumeratio, enarratio, familiae, prosapiae, majorum. βίβλος enim in hoc loquendi modo respondet hebraico ספר quod non tantum significat simpliciter librum, sed enarrationem, dinumerationem, catalogum, scripturam, Literas, librum denique qui sic hebraeis dictus, quod in eo utiles doctrinae sive res gestae enarrentur et explicentur, a verbo ספר, enarravit, enumeravit etc."

times in the Hebrew history, בימים ההם, 'in those days.'"[19] These departures from Hellenic purity, he argued, made the text more, rather than less, powerful. When the Gospel authors endlessly repeated the verb λέγων (speaking), in its various forms, they followed the Hebrew way of recounting events:

> the Hebrews usually bring in the characters themselves speaking, so that the event in question seems not so much to be an account of something past, as one that is taking place and being carried out. This certainly has far more rhetorical power [*energia*] than it would if sayings or actions were recounted in the way of the Greeks and Latins, in an oblique and naked fashion.[20]

Carlo Ginzburg has pointed out that early modern historians sought to give their accounts *enargeia* (vividness) by replacing descriptions of events and institutions with quotations from their primary sources.[21] Buxtorf – who seems to have had *enargeia* in mind – saw a different ancient model, the narratives of the Hebrew Bible – and the Greek New Testament for which they in turn had served as models – as charged with an *enargeia* that classical narratives lacked.

Buxtorf, moreover, defined the materials relevant to the study of Hebrew – as he would throughout his life – in an expansive way. Profuse citation of proverbs, for example, he took as characteristic not only of Hebrew but of a wide array of languages, once again including New Testament Greek: "The speech of the ancient Hebrews abounded in this sort of proverbial or parabolic or sententious speaking, and so indeed did that of the Oriental peoples. This is very clear from the books of both Testaments, as also from the books of the Talmudists and others."[22] Jewish writers continued to use this style, down to his own day, as Buxtorf showed by

19. Ibid., fol. 36v: "Praeterea quod in Evangelica historia subinde usurpari solet, ἐν δὲ ταῖς ἡμέραις ἐκείναις, in diebus illis, quid aliud est, quam quod centies millies in hebraica legitur historia, בימים ההם, in diebus illis, hujus vel illius."

20. Ibid.: "Vnde vero etiam est ista in N.To. petita consuetudo, quod toties repetitur λέγων, λέγοντες etc. nisi quod hebraei in suarum historiarum explicatione solent personas ipsas loquentes introducere, ut res ipsa non tam praeterita narrari, quam prorsus geri et in omnium oculis auribusve administrari videatur: quae res certe longe plus energiae habet, quam si Graecorum Latinorumque more oblique tantum et nude aliena dicta vel facta referantur." Buxtorf seems to be discussing *enargeia*, which denotes an especially vivid representation of a happening or event, rather than *energeia*, which refers only to the force of the expression in question.

21. Carlo Ginzburg, *Threads and Traces: True, False, Fictive*, trans. Anne Tedeschi and John Tedeschi (Berkeley, 2012), chap. 1.

22. Buxtorf, commentary on Proverbs, Basel, UB, Fr Gr III 22, a, fol. 3v: "Hoc vero dictionis genere sive proverbiali aut parabolico vel denique sententioso priscorum Hebraeorum, ac omnino orientalium populorum, sermonem abundasse, ex utriusque Testamenti sacri tum etiam Talmudicorum ac aliorum libris notissimum est."

hurling two distant examples very different from one another in time and style into a single basket:

> Thus we read in Josephus the son of Gorion [the tenth-century author of the *Yosippon*], "You have no part or inheritance in the land of Judah"; similarly, "What part or heredity do we have any more in the Law of the Lord?" Thus, in Elijah Levita in the preface to book 2 of the *Masoreth*, "Beyond the lace of the shoe you have neither a part nor an inheritance." These are the words of the painter to a cobbler too rashly offering a judgment of a whole picture, as the saying goes, "Let the cobbler stick to his last."[23]

Buxtorf's program for the study of Hebrew remained expansive throughout his life.

By the time Buxtorf died in 1629, he had become a locally prominent and internationally renowned professor. He played his part in the lavish rituals of university life and published textbooks on Hebrew grammar, letter writing, and Yiddish; issued further publications on a vast range of Jewish topics; and served as editor, corrector, and censor for dozens of Jewish books printed in Basel. He developed a style of scholarship that combined a staggering command of Jewish sources with a strikingly negative view of Jews and their religion and a strongly confessional inspiration. His erudition and his prejudices collided in a variety of ways. He claimed to know many Jews – and notably many rabbis – personally, and he wrote a detailed account of Ashkenazic Jewish life, which he presented as the results of eyewitness observation and conversation. Yet he excerpted most of the book's contents from a widely varied range of texts by Jews and Jewish converts. He worked closely with Jewish compositors and correctors in the Basel printing houses of Conrad Waldkirch and his successor Ludwig König, but when his relationship with them brought him into difficulties, he denied that he had been on intimate and friendly terms with them. Joseph Scaliger greatly admired Buxtorf's approach to Hebrew – especially his insistence on reading, and teaching others to read, a broad corpus of Jewish texts rather than sticking with the Hebrew Bible. Yet as he told his students, he found the German scholar's close relationships with Jews hard to explain: "It is remarkable how much the Jews love

23. Buxtorf, "Alia praefatiuncula," Basel, UB, Ki Ar 189, pt. 4, fol. 38v: "Sic apud Iosephum Gorionidem legimus: Non est tibi pars neque haereditas in terra Iehuda: Item, Quae nobis amplius pars aut haereditas in Lege Domini? Sic apud Eliam Levitam in praefat. 2. libri Masoreth, Ultra calcei corrigiam non est tibi pars neque haereditas: quae verba illic pictoris sunt ad sutorem temere nimis de tota pictura judicantem, q.d. Ne sutor ultra crepidam." For Elijah's interest in this last (classical) proverb, see Anthony Grafton, "Humanism and the Mishnah: Paul Fagius Edits Avot," in *The Mishnaic Moment: Jewish Law among Jews and Christians in Early Modern Europe*, ed. Piet van Boxel, Kirsten MacFarlane, and Joanna Weinberg (Oxford, 2022), 47–67.

Buxtorf: he censures them quite sharply in that *Synagoga Judaica*."[24] Buxtorf's life and work present us with more than a few paradoxes – as they did to some of his best informed contemporaries.

Scholars who try to resolve these must also struggle with another difficulty: most of Buxtorf's projects have attracted relatively little attention. In 1996, Stephen Burnett surveyed the full range of Buxtorf's activities in a pioneering and erudite book. Our debt to that study, and to Burnett's work, also pioneering, on Hebrew printing in early modern Germany, will be evident on every page of this volume.[25] Buxtorf's polemical ethnography of contemporary Jewish life and his theses about the text of the Hebrew Bible remain prominent subjects in contemporary scholarship.[26] Joseph Prijs's great survey of Hebrew printing in Basel sheds much light on Buxtorf's career in the printing house.[27] For the most part, though, his practices in a whole range of domains – as a reader of Jewish texts, a student of Jewish customs, an editor and censor of Jewish books, and a scholar – await exploration.

In the chapters that follow, drawing on the riches of the Basel collections and informed by recent scholarship in related fields, we do our best to show how Buxtorf worked at his various trades, and why. We reconstruct his practices as a scholar in detail. We show that he adapted techniques from contemporary Greek and Latin humanism, altering them in multiple ways to fit the different conditions of scholarship in this very different realm. But we also attend to the ways in which he drew on Jewish learning, often to put forward theses that contemporary Jews probably did not share. Throughout, we follow the many ways in which Buxtorf used one of the most commonly used humanist techniques, commonplacing: the collection and arrangement of excerpts from texts in commonplace books. It has

24. Joseph Scaliger, *Scaligerana, Thuana, Perroniana, et Colomesiana*, ed. Pierre des Maizeaux (Amsterdam, 1740), 2: 248–49, at 249: "Mirum quomodo Buxtorfius ametur a Judaeis: in illa tamen Synagoga Judaica illos valde perstringit."

25. Burnett, *From Christian Hebraism to Jewish Studies*. Several of Burnett's other studies on Buxtorf and on printing, all of them highly illuminating, are cited below. We also owe a special debt to the scholars – including Burnett – who collaborated to catalog the Hebrew manuscripts of the Basel University Library: Joseph Prijs, *Die Handschriften der Universitätsbibliothek Basel: Die Hebräischen Handschriften: Katalog auf Grund der Beschreibungen von Joseph Prijs*, ed. Bernhard Prijs and David Prijs, with Stephen G. Burnett and Thomas Willi (Basel, 1994).

26. See, e.g., Yaacov Deutsch, *Judaism in Christian Eyes: Ethnographic Descriptions of Jews and Judaism in Early Modern Europe*, trans. Avi Aronsky (Oxford, 2012); Timothy Twining, "The Early Modern Debate over the Age of the Hebrew Vowel Points: Biblical Criticism and Hebrew Scholarship in the Confessional Republic of Letters," *Journal of the History of Ideas* 81 (2020): 337–58.

27. Joseph Prijs, *Die Basler hebräischen Drucke, 1492–1866*, ed. Bernhard Prijs (Olten, 1964).

become clear that Buxtorf's complex and demanding scholarship in many areas rested on this simple-seeming practice.

We cannot hope to answer all the hard questions that Buxtorf's career and achievements pose. Knowing where he learned the methods that underpinned his scholarship does not help us understand his attitudes. Still, we hope that we have placed Buxtorf the scholar more firmly in his historical setting and illuminated the distinctive, and sometimes dismaying, qualities of his work. Parts of the territory remain to be explored – for example, the vast labyrinth of his work as a Hebrew and Aramaic lexicographer and textual critic. The known sources do not always allow us to separate his practices and conclusions from those of his son, who by his own account massively revised the lexicon that his father had begun before he published it.[28] We would be delighted if our studies attract others to map these rich and unexplored domains.

28. See Chapter 1, below.

CHAPTER ONE

Johann Buxtorf Makes a Notebook

Between 1559 and 1613, Christian scholars assembled an informal canon of books about Jews and Judaism – books ancient and modern, religious and secular, in Hebrew and Aramaic, and in Latin and German.[1] From many points of view, it was a strange list, its contents often dictated by the chances of transmission and the choices of printers. Though it included Maimonides and the medieval commentators on the Old Testament, it also emphasized such diverse areas as Kabbalah and chronology, without making clear how, or if, these works and approaches could be reconciled. Still, these were the books that scholars thoroughly engrossed in Jewish texts and traditions read, cited, and debated. They attracted the attention of others whose main interests lay in the literature and history of ancient Greece and Rome, and stimulated debate about everything from political thought to natural philosophy.[2] And they provided the instruments with which generations of Christians, through the seventeenth century, mastered Jewish learning to a level that compels admiration.[3]

No one did more to lay out the contents of this canon, or to make them accessible to Christians, than Johann Buxtorf (1564–1629). One of the textual

1. The dates are those of the two standard rabbinical bibliographies: Gilbert Génébrard, *Eisagoge ad legenda Rabbinorum commentaria* (Paris, 1559), and Johann Buxtorf, *Bibliotheca Rabbinica*, in his *De abbreviaturis hebraicis liber novus & copiosus* (Basel, 1613), 257–335.

2. On this loose canon, see in general Anthony Grafton and Joanna Weinberg, with Alastair Hamilton, *"I have always loved the Holy Tongue": Isaac Casaubon, the Jews, and a Forgotten Chapter in Renaissance Scholarship* (Cambridge, 2011), and *Jewish Books and Their Readers: Aspects of Jewish and Christian Intellectual Life in Early Modern Europe*, ed. Scott Mandelbrote and Joanna Weinberg (Leiden, 2016); for its impact, see Eric Nelson, *The Hebrew Republic: Jewish Sources and the Transformation of European Political Thought* (Cambridge, MA, 2010), and Avner Ben-Zaken, *Cross-Cultural Scientific Exchanges in the Eastern Mediterranean, 1560–1660* (Baltimore, 2010).

3. Gerald Toomer analyzes one such figure's work in depth in *John Selden: A Life in Scholarship* (Oxford, 2009).

tools that he forged to carry out this task was a copybook.[4] It is astonishingly varied in content. Buxtorf filled its pages with – among many other things – a vast number of excerpts from Jewish texts, some of them heaped up in no apparent order to serve the needs of anti-Jewish polemic, others organized under topical headings to support the writing of letters in Hebrew; excerpts from the Talmud, the vast compendium of rules, rabbinical opinions, and debates that forms the basis of Jewish law, and from its medieval Jewish commentators, the originals written out in Buxtorf's fluent Hebrew hand, and some of them translated into Latin as well; Latin texts by Christian scholars, some passages summarized and others copied out in full; brief inquiries into the state of the Hebrew text of the Bible and extended discussion of Hebrew prayer books, with collations of multiple editions and manuscripts; a Turkish word list; extracts on theological questions such as penitence and faith; and the full text of a story from the *Mayse bukh*, a collection of popular tales of Jewish sages in Yiddish.[5] The copybook seems to have had no internal logic, as Buxtorf himself tacitly admitted when he drew up the list of its contents: he made no effort to trace connections between the different sections, or even to list them in sequence.[6] On its unruly pages, however, we can watch Buxtorf mastering the bibliographical and textual material that he needed to compose a number of his books.

A rich body of scholarship on commonplace books and other forms of notebooks in early modern Europe sets the stage for this inquiry. Notebook making began, for many educated Europeans, at school. Erasmus and many others taught them to read pen in hand, marking passages of interest in their books and copying out those passages in topically organized notebooks for reuse in compositions of their own.[7] Many laid down material year after year through their lives, as trees

4. Buxtorf, copybook, Basel, UB, MS A XII 20.

5. For a brief description of the Buxtorf manuscripts, see Joseph Prijs, *Die Handschriften der Universitätsbibliothek Basel: Die Hebräischen Handschriften: Katalog auf Grund der Beschreibungen von Joseph Prijs*, ed. Bernhard Prijs and David Prijs with Stephen G. Burnett and Thomas Willi (Basel, 1994), 71–88 (by Stephen Burnett).

6. Buxtorf, copybook, 328. For a survey of the manuscript's contents, see Prijs, *Die Hebräischen Handschriften*, 89–98 (by Thomas Willi).

7. The best survey is Ann Blair, *Too Much to Know: Managing Scholarly Information before the Modern Age* (New Haven, 2010). For the pedagogical commonplace book and its development, see Ann Moss, *Printed Commonplace-Books and the Structuring of Renaissance Thought* (Oxford, 1996); Francis Goyet, *Le sublime du "lieu commun": L'invention rhétorique dans l'Antiquité et à la Renaissance* (Paris: Geneva, 1996); Kevin Sharpe, *Reading Revolutions: The Politics of Reading in Early Modern England* (New Haven, 2000); William Sherman, *Used Books: Marking Readers in Renaissance England* (Philadelphia, 2008); Angus Vine, *Miscellaneous Order: Manuscript Culture and the Early Modern Organization of Knowledge* (Oxford, 2019); and Matthew Eddy, *Media and the Mind: Art, Science, and Notebooks as Paper Machines, 1700–1830* (Chicago, 2023).

add rings and turtle shells grow larger.[8] In the late sixteenth century and after, as commentaries spawned commentaries, makers of notebooks began to concentrate as much on compiling bibliographies as on archiving valuable passages.[9] Critics of learning regularly worried that this culture of compilation promoted sterile pedantry.[10]

Yet notebook making need not deaden thought or bury it under heaps of erudition. Compilation helped transform the landscape of knowledge. In the seventeenth and eighteenth centuries, as historians of science have shown, learned men created compilations of many kinds – journals, waste books, critical assays of common-book collections, and the like.[11] Some of these still served as receptacles for textual extracts. But others brimmed with long series of precise, dated observations of everything from the stars to the weather, as well as notes on interviews with "ingenious" informants. The members of the Royal Society and their counterparts on the Continent used notebooks to perform a brilliant act of scholarly judo: they transformed the old, worn tools of erudition into the new, sharp ones of empirical science, founded on observation. As they compiled, they learned from experience that the mastery of nature depended on collaborative projects that would last for generations.[12]

Scholars' notebooks were also as varied as their makers, and as the scholars' careers and interests developed, so, often, did their ways of compiling information.

8. For a splendid case in point, see Sherman, *Used Books,* chap. 7.

9. The most original effort to keep tabs on all this information is reconstructed in Noel Malcolm, "Thomas Harrison and His 'Ark of studies': An Episode in the History of the Organisation of Knowledge," *The Seventeenth Century* 19 (2004): 196–232.

10. See, e.g., Martin Gierl, *Pietismus und Aufklärung: Theologische Polemik und die Kommunikationsreform der Wissenschaft am Ende des 17. Jahrhunderts* (Göttingen, 1997); Blair, *Too Much to Know;* and Martin Mulsow, *Knowledge Lost: A New View of Early Modern Intellectual History,* trans. Hans Christian Erik Midelfort (Princeton, 2022).

11. For Bacon's "Transportata ex comentario vetere" (1608), including a review of his notebooks, see *The Letters and the Life of Francis Bacon,* ed. James Spedding. (London, 1861–74), 4: 39–95, esp. 59–68, and the discussion by Michael Kiernan in Francis Bacon, *The Advancement of Learning,* ed. Michael Kiernan (Oxford, 2000), xxxiv–xxxvi.

12. For the turn from textual to observational notebooks, see Richard Yeo, "Between Memory and Paperbooks: Baconianism and Natural History in Seventeenth-Century England," *History of Science* 45 (2007): 1–46; Yeo, *Notebooks and English Virtuosi: Memory, Information, and Early Modern Science* (Chicago, 2014); Elizabeth Yale, "With Slips and Scraps: How Early Modern Naturalists Invented the Archive," *Book History* 12 (2009): 1–36; Lorraine Daston, "Perchè i fatti sono brevi?" *Quaderni storici,* n.s. 108 (2001): 745–70; and the studies collected in *Histories of Scientific Observation,* ed. Lorraine Daston and Elizabeth Lunbeck (Chicago, 2011); cf. Anthony Grafton, *Inky Fingers: The Making of Books in Early Modern* Europe (Cambridge, MA, 2020). For conversations with informants, see Lawrence Principe, *The Aspiring Adept: Robert Boyle and His Alchemical Quest* (Princeton, 1998), 91–137, 296–300.

Many scholars deposited the results of their reading not under the systematic headings found in formal commonplace books but, as they worked through one text or problem after another, in notebooks called "adversaria" – much as the Roman scholar and critic Aulus Gellius (second century CE), whose compendium, the *Attic Nights*, was very popular in Renaissance Europe, explained that he had done, centuries before.[13] Others recorded everything from travel notes to the local traditions of their cities.[14] The texts stored in commonplace books were often decontextualized and arbitrarily reused. But the texts copied in adversaria often became the object of intense critical scrutiny. Scholars like Angelo Poliziano in the fifteenth century and Isaac Casaubon a hundred years later not only copied texts in whole or part, but also corrected them, juxtaposed them to other texts, and interpreted them in novel ways – as when Poliziano connected fragments of lost Latin adaptations of Greek texts with their originals and Casaubon drew on the help of a Jewish tutor to teach him some of the basic terminology of the Talmud as he made his first notes on the text.[15] Recently, historians of science have identified notebooks as an epistemic genre – textual devices, paper tools that transformed the materials stored in them.[16] In fact, early modern notebooks belonged to a number of distinct epistemic genres, which served different ends and rang different changes on the extracts that they contained.

Between the rhetorical commonplace book and adversaria, a third type appeared that shared features with the others but differed from them both. In mid-sixteenth-century Magdeburg, Matthias Flacius Illyricus and Caspar von Nidbruck assembled a team to produce the first great Protestant history of the Christian church. They collected, copied, and annotated sources – some of them, from manuscripts that could not be borrowed from their libraries, obtained surreptitiously with the aid of the *culter Flacianus* (Flacian razor). Then young

13. Aulus Gellius, *Noctes Atticae* praefatio 2; for his imitators, see *The Worlds of Aulus Gellius*, ed. Leofranc Holford-Strevens and Amiel Vardi (Oxford, 2004).

14. See, e.g., the early sixteenth-century notebooks of the Lyon antiquary Claude Bellièvre, partially published as Bellièvre, *Souvenirs de voyages en Italie et en Orient, notes historiques, pièces de vers*, ed. Charles Perrat (Geneva, 1956).

15. See in general Jean-Marc Chatelain, "Les recueils d'*Adversaria* aux xvie et xviie siècles: Des pratiques de la lecture savante au style de l'érudition," in *Le livre et l'historien: Études offertes en l'honneur du Professor Henri-Jean Martin*, ed. Frederic Barbier et al. (Geneva, 1997), 169–86. For Poliziano, see Ida Maïer, *Les manuscrits d'Ange Politien* (Geneva, 1965); for Casaubon, Hélène Parenty, *Isaac Casaubon, helléniste: Des studia humanitatis à la philologie* (Geneva, 2009), 402–20, and Grafton and Weinberg, *"I have always loved the Holy Tongue."* Louis Delaruelle, "Un recueil d'adversaria autographes de Girolamo Aleandro," *Mélanges d'archéologie et d'histoire publiés par l'École française de Rome* 20 (1900): 3–21, remains a good case study.

16. See the works cited in n7, above.

research assistants made excerpts from the collected materials, century by century, filling commonplace books, topic by topic. More experienced scholars rewrote these notes into a full-scale church history. Flacius and his colleagues – who fairly soon ejected him from the project – distributed multiple sets of rules to help their assistants with these complicated tasks. Since antiquity, histories of the church had incorporated documents.[17] By Flacius's time, the canon of relevant sources had exploded in size, and collaboration in processing them was vital.[18] But – as Flacius's opponents soon noted – the categories in the excerptors' commonplace books determined their choice and interpretation of the historical evidence in advance. Extracts were selected and shaped to show that while the larger church had become corrupt over time, a small true church whose believers were persecuted as heretics had never ceased to exist.[19] Other readers sifted through, compared, and contemplated the axioms of Tacitean political counsel and the details of biblical geography. In these cases, "both modern and ancient accounts were processed in the same fashion, by redaction into brief, fragmentary notes."[20] Notebooks like these served neither rhetorical composition nor philological

17. See the classic studies of Eduard Schwartz, "Über Kirchengeschichte," *Nachrichten der Königlichen Gesellschaft der Wissenschaften zu Göttingen, Geschäftliche Mitteilungen* (1908), 106–22, repr. in Schwartz, *Gesammelte Schriften* (Berlin, 1938–63), 1: 110–30, and Arnaldo Momigliano, "Pagan and Christian Historiography in the Fourth Century A.D.," in *The Conflict Between Paganism and Christianity in the Fourth Century*, ed. Arnaldo Momigliano (Oxford, 1936), 79–99, repr. in Momigliano, *Terzo contributo alla storia degli studi classici e del mondo antico* (Rome, 1966), 1: 87–109. David DeVore has modified this account in "Character and Convention in the Letters of Eusebius's *Ecclesiastical History*," *Journal of Late Antiquity* 7 (2014): 223–52.

18. Martina Hartmann, *Humanismus als Kirchenkritik: Matthias Flacius Illyricus als Erforscher des Mittelalters* (Stuttgart, 2001); *Catalogus und Centurien: Interdisziplinäre Studien zu Matthias Flacius und den Magdeburger Centurien*, ed. Arno Menzel-Reuthers and Marina Hartmann (Tübingen, 2008).

19. Heinz Scheible, *Der Entstehung der Magdeburger Zenturien: Ein Beitrag zur Geschichte der historiographischen Methode* (Gütersloh, 1960); Hartmann, *Humanismus als Kirchenkritik*; Gregory Lyon, "Flacius, Baudouin, and the Plan for the Magdeburg Centuries," *Journal of the History of Ideas* 64 (2003): 253–72; Harald Bollbuck, *Wahrheitszeugnis, Gottes Auftrag und Zeitkritik: Die Kirchengeschichte der Magdeburger Zenturien und ihre Arbeitstechniken* (Wiesbaden, 2014); Bollbuck, "Searching for the True Religion: The Church History of the *Magdeburg Centuries* between Critical Methods and Confessional Polemics," *Renaissance Studies* 35 (2021): 100–117; Michael Hollerich, *Making Christian History: Eusebius of Caesarea and His Readers* (Oakland, 2021).

20. Nicholas Popper, *Walter Ralegh's "History of the World" and the Historical Culture of the Late Renaissance* (Chicago, 2012), 134; see also Lisa Jardine and Anthony Grafton, "'Studied for Action': How Gabriel *Harvey* Read His Livy," *Past & Present* 129 (1990): 30–78; Ann Blair, *The Theater of Nature: Jean Bodin and Renaissance Science* (Princeton, 1997); and Sharpe, *Reading Revolutions*.

analysis directly. Aggressively shaped for use, they were big tools of textual culture.[21]

Buxtorf's copybook is oblong in shape, roughly 20 centimeters (8 inches) high and 15 (6 inches) across, and designed and made as a receptacle: while many pages are full, some are almost blank, and give the impression that he intended to return to them. In content, it spans all three of the types just sketched. It includes passages in Hebrew collected under headings, for use in writing letters (see Fig. 1.1), and technical discussions of problems of biblical interpretation.

But it starts as an assembly of passages chosen for a single polemical end. At first glance what impresses is Buxtorf's astonishing capacity to scroll through the major and minor works of Jewish tradition, ancient and modern: Talmud and legal codes, Bible commentaries and literary compilations, kabbalistic and pietistic works. No text, however difficult or obscure, baffled him. He accessed the full, ever-changing canon of Jewish literature. Yet he seems to have made no distinction between major or fundamental texts of the tradition and lesser works. He consulted a contemporary homiletic commentary on the Pentateuch, which also constituted a super-commentary on the great medieval commentator Rashi (1040–1105) written by the sixteenth-century Polish rabbi Moses ben Abraham Mat, in conjunction with the Talmud and submitted it to the same kind of scrutiny.[22] We can almost visualize Buxtorf snatching books off his table haphazardly, scouring the texts until he happens upon the anticipated phrase or sentiment – for in the majority of cases, books were used as grist for his prefabricated theological mill.

In the middle of this weighty reader, one page, containing a mere three lines of abbreviated text, reveals a main category with which Buxtorf was approaching these books (Fig. 1.2). Under the heading Edom, the short notes refer to three distinguished medieval commentators:

1. the most eminent city of Edom is Rome, R. Sal. [i.e., Solomon ben Isaac, Rashi]
2. Kimchi [David Kimhi, 1160–1235] on the beginning of Obadiah and Isaiah 34:1 and 35:1
3. Bachai [i.e., Bahya ben Asher, d. 1340] in *Kad ha-kemah* [The Pitcher of flour], chapter *geulah* [redemption].[23]

Here Buxtorf singled out three exemplary authorities who occasionally – as the topic suggested – applied the rabbinic paradigm that equated the biblical

21. Cf. *Little Tools of Knowledge: Historical Essays on Academic and Bureaucratic Practices*, ed. Peter Becker and William Clark (Ann Arbor, 2001).

22. Buxtorf, copybook, 311. The discussion of Moses ben Abraham Mat, *Ho'il Mosheh* (Prague, 1611), on Genesis 49, comes directly after a long citation of BT, Sanhedrin 5a.

23. Buxtorf, copybook, 180.

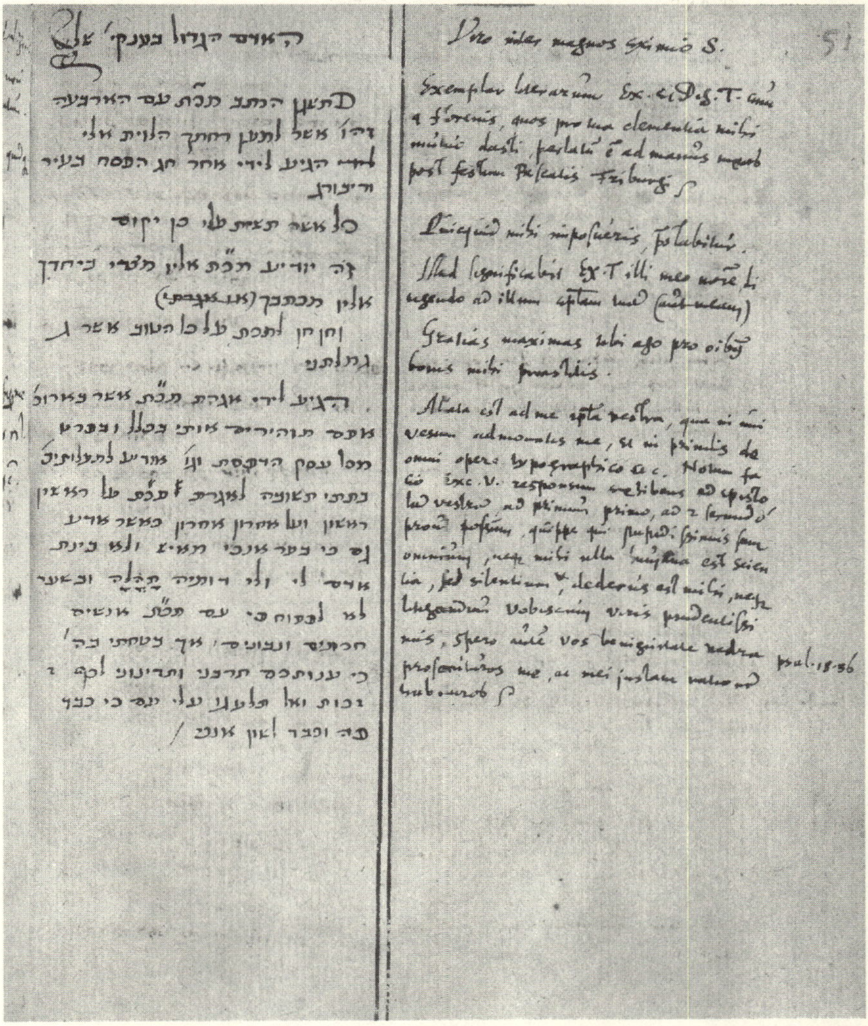

Figure 1.1. Buxtorf arranges excerpts from Hebrew letters for use by Christians in writing Hebrew letters of their own. Basel, Universitätsbibliothek, A XII 20.

Edom, an ancient kingdom in trans-Jordan that came to be identified with Esau and with Christian Rome, to the fourth and final kingdom in Daniel's prophecy (Dan. 2:31–43). Protestants normally interpreted the empires in Daniel as political communities – though they disagreed at times about which contemporary empire was the fourth. The Jewish scholars' typological application of Edom to Christianity irked Buxtorf, with its negative associations as the oppressor of Israel, which would

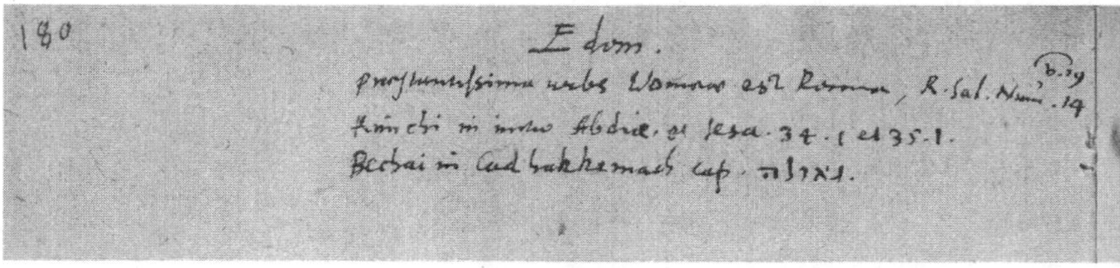

Figure 1.2. Three references to Edom in Buxtorf's copybook. Basel, Universitätsbibliothek, A XII 20.

receive its final comeuppance at the end of days. Ever on the outlook for the word *Edom*, he scanned multiple books and commentaries in an effort to root out the term. This reading practice is manifested on many of the pages of the copybook.

The very first set of notes, on the front pastedown of the copybook, sets the tone for the next fifty, as extract after extract from every text that Buxtorf chooses to read appears, on first glance, to reveal the endemic anti-Christianity and frivolous mendaciousness of the Jews (Fig. 1.3). At the bottom of the page is a citation from the Babylonian Talmud.[24] The centerpiece of rabbinic tradition, this was the bête noire of generations of Christian scholars, most of whom – with notable exceptions – had never set eyes on the text.[25] The extract Buxtorf selected from tractate *Baba Batra* (54b), part of a legal discussion, states: "The possessions of a Cuthean [i.e., a gentile] are like desert property, which becomes the possession of the one who takes it." The context from which he pulled this one-liner reads: "Rav Judah said in the name of Samuel: The property of a heathen is like desert land; whoever first occupies it acquires ownership. The reason is that as soon as the heathen receives the money, he ceases to be the owner, whereas the Jew does not become the owner until he obtains the deed of sale." Buxtorf must have realized that he had taken this passage totally out of context, for he added a note indicating that the Talmudic passage should be compared to the relevant law in the Book of Acquisition in Maimonides' *Mishneh Torah*.[26] In that definitive codification of rabbinic law, Maimonides (1137/8–1204) had given a succinct summation of the complex topic. In his treatment, the particular case discussed discursively in the Talmudic text becomes crystal clear:

> Hence if an Israelite buys a field from a heathen and pays the money, but before taking formal possession another Israelite comes and takes possession of the field, in the manner in which one takes possession of the property of a proselyte, then the second acquires title, and he pays the price to the first. For

24. Ibid., front endpaper. Buxtorf identifies the passage as "Bathra fol. 53. 2."

25. The exceptions included Angelo Canini, Konrad Pellikan, and Joseph Scaliger.

26. Buxtorf, copybook, front endpaper; "Confer Majemon Volumine ultimo Tractatum קנין fol. 96."

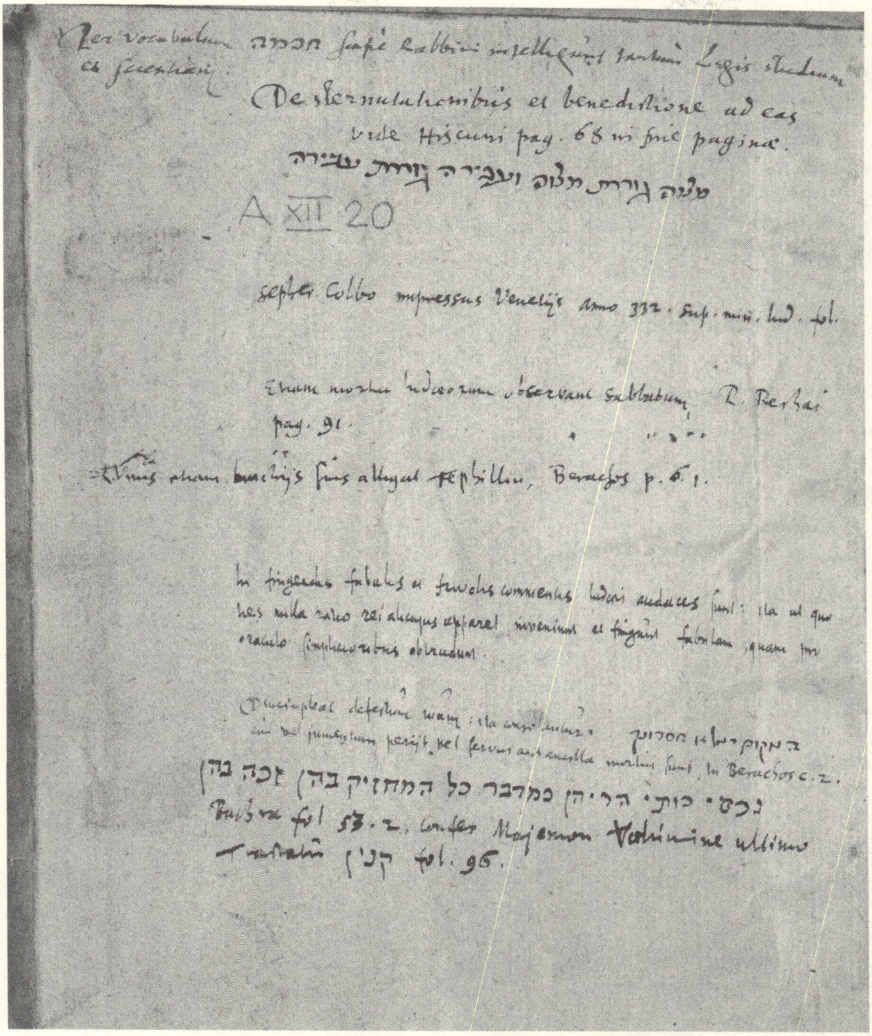

Figure 1.3. The first page of Buxtorf's copybook.
Basel, Universitätsbibliothek, A XII 20.

the heathen relinquishes his title once he takes the money, while the Israelite acquires title only when he receives the deed; hence this property is like property in the desert, which becomes his who first takes possession of it.[27]

In other words, the Talmudic passage does not define a heathen's property as ownerless or up for grabs. Rather, Jewish law deals with the intricacies involved

27. Moses Maimonides, *Mishneh Torah*, hilkhot zekhiyah 1,14.

in the transfer of property, and states at which stage in the transfer each party has legally taken possession. Why did Buxtorf not take down the rest of the text? Apparently first impressions count; and from a very superficial reading the text appeared to be upholding the anti-gentile attitudes and practices that he expected to encounter wherever he looked in Jewish life.

Single lines or short references to the Talmud, usually taken out of context, are scattered like bullet points throughout the copybook.[28] Two larger sections of Talmudic material unambiguously illustrate the nature of the enterprise in which Buxtorf was engaged. Disregarding the hundreds of pages of legal argument and creative biblical exegesis on a myriad of subjects, Buxtorf made a beeline for any text that either mentioned the authority of the rabbis and their books – a central theme of the first chapter of the *Jüden Schul* – or contained Jewish vituperation of Jesus or of Christians. As a staunch Protestant, who believed passionately in the authority of Scripture, Buxtorf regarded the Jewish rabbis, like the Catholic theologians who succeeded them, as foolish pedants who had burdened the consciences of men with superstitious rules and observances. He expended enormous effort in excerpting and translating these exemplary passages, which apparently required transcription and translation more than once.

Worthy of multiple copying, for example, was a notable passage in which one sage proffered an interpretation of a biblical verse, drawing the lesson that

28. Buxtorf knew that Jewish texts took many forms, some of which were normally read in distinctive ways, and he did his best to control all of them, even those that almost no Christian scholars had previously mastered. He worked his way through *Halikhot olam*, a guide to the techniques of Mishnah and Talmud written by the fifteenth-century rabbi Yeshuah Halevi, with great care. Buxtorf's heavily annotated copy of the text (Mantua, 1593) is Basel, UB, A V 32 (3). He noted technical terms and translated short passages that he took as essential for following a Talmudic debate: "Author hic dicit hoc, et author ille dicit aliud, et tamen non discrepant" (This author says this, and that author says that, and yet they do not disagree), and described the work as very helpful in the *Bibliotheca Rabbinica* (285): "Liber est aditum in Talmud facilem praebens, explicans rationes loquendi Doctorum Talmudicorum, modum objectionum & responsionum, & similia plurima, quae specialem habent usum apud Talmudicos." Buxtorf also annotated his copy of another guide to Talmudic argument, Joseph ibn Verga, *She'erit Joseph* (Mantua, 1593), Basel, UB, FA VI 18 (1). But unlike the Christian scholars of later generations – for example, Constantine L'Empereur (1591–1648), who translated Halevi's work, as well as whole Talmudic tractates, into Latin (Constantine L'Empereur, *Clavis Talmudica* [Leiden, 1634]) – he confined his actual reading and translation of Talmud to aggadic passages. For the development of systematic Talmudic study by Christians in the seventeenth century, see Carsten Wilke, "Splendeurs et infortunes du talmudisme académique en Allemagne," in *Les Textes judéophobes et judéophiles dans l'Europe chrétienne à l'époque moderne, XVIe–XVIIIe siècles*, ed. Daniel Tollet (Paris, 2000), 97–134, and *The Mishnaic Moment: Jewish Law among Jews and Christians in Early Modern Europe*, ed. Piet van Boxel, Kirsten Macfarlane, and Joanna Weinberg (Oxford, 2022).

"he who scoffs at the words of the sages will be punished by boiling excrement."[29] This text embodied for Buxtorf everything he detested about rabbinic Judaism: the presumed authority of the sages and what he deemed to be Jewish sacralization of rabbinic tradition. Its wording led him directly to another Talmudic passage, located in a different order and tractate, which contained the same offensive phrase (Babylonian Talmud Gittin 57a).[30] From extracting gobbets about rabbinic authority, Buxtorf passed to copying out a long passage in Aramaic of a very different kind from this text. It was a story, long infamous, about Onkelos son of Kalonikos, the son of Titus's sister. Seeking confirmation for his decision to convert to Judaism, he employed magical arts to conjure up various notable biblical personalities from the grave.[31] Among the chosen was "'Jesus the Nazarene'" – though Buxtorf left a blank space in the text and added the name of Jesus above the line. "This is not in the Talmudic text," he explained in a marginal note.[32] The critical words "Jesus the Nazarene" had undergone expurgation in the Basel edition of the Talmud (1578–81), which, though financed by a Jew, had been edited and censored by Christian scholars. Buxtorf made clear that he had examined more than one text of the passage in order to establish that it had originally named Jesus: "In the Venetian [edition of] the Talmud the word 'Jesus' fills this space; in another passage he is called 'Jesus the Nazarene.' Thus they cannot deny that this passage refers to our Jesus the Nazarene."[33] Buxtorf, in other words, had collated the Basel Talmud with the earlier Venetian text printed by Bomberg and corrected it.[34] As we will see, this scrupulous, even obsessive care for details was typical of him.

The story continues: "When asked: 'Who is important in that world?' he said, 'Israel.' 'What is your punishment?' 'Boiling excrement. For the master has

29. BT, Erubin 21b.

30. Buxtorf, copybook, 32–33, on rabbinic authority, leads to 34, with the extract from BT, Gittin 57a and a Latin translation on 35. As one of the referees for this volume rightly pointed out, the excessive authority of the rabbis had figured in Christian polemics against the Jews long before Buxtorf. The Genoese Carthusian Porchetus Salvagus (d. ca. 1315), in *Victoria adversus impios Hebraeos* (Paris, 1520), fol. XCv, quoted Rashi's comment on Deuteronomy 17:11, where he stated that if a rabbi says what seems to be right is left, or what seems to be left is right, then one must believe him. Luther, in turn, quoted this remark and harped on it in *Vom Schem Hamphoras* (1543), ed. F. Cohrs and O. Brenner, in *D. Martin Luthers Werke: Kritische Gesamtausgabe*, vol. 53 (Weimar, 1920), 573–648, at 589. Buxtorf, however, examines the sources in far more detail.

31. The story is a subtext of the narrative about Titus's divinely ordained death by gnat as punishment for destroying the Jerusalem Temple.

32. Buxtorf, copybook, 34: "Non est in textu Talmudico."

33. Ibid.: "Hoc spacium in Talmud Venet. complet vox ישו, quod alio loco dicitur ישו הנוצרי ita ut negare non possint de Iesu Nazareno nostro haec dici."

34. The Babylonian Talmud tractate Gittin printed in Venice (Bomberg, 1521), 57a, does indeed read ישו (Jesus).

said: "Whoever scoffs at the words of the sages is punished by boiling excrement." Go and see the difference between the sinners of Israel and the prophets of the nations of the world.'" In this passage, the rabbis appear to describe Jesus's fate in hell in harrowing terms, boiling him in excrement, the fate reserved for heretics. This became a crucial text for Buxtorf. He also transcribed Rashi's comment on it, which sent the reader back to tractate Erubin and its exhortation to treat the words of the scribes with more deference than the words of Scripture. The full Latin translation of the story that Buxtorf added on the following page undoubtedly reveals his purpose: exposing Jewish contempt for all things Christian. It also gave him the chance to insert the words "forgive me, O Christ," in parentheses after the point in the text where Onkelos used magic to raise Jesus from the dead.[35] And it was this text that he would invoke as he drew the polemical first chapter of the *Jüden Schul* to its triumphant end.[36]

A similar goal must have driven him to fill ten pages of the copybook with extracts, mainly in the original but occasionally translated into Latin, from a number of Talmudic tractates.[37] The criterion for entry into this collection of extracts appears to be that the text should contain negative statements about gentiles, explicit reference to Jesus, or apparent allusions to New Testament narratives. Buxtorf's collection of passages preserves an uncensored version of the glossed Talmud, which modern scholarship might find useful.[38]

Once again, the reader is struck by Buxtorf's ability to chart the Talmud's formidable waters, and to do so, despite his purpose, in Jewish fashion (see Fig. 1.4). For he read not only the text but also the standard glosses printed on either side of it. He knew that the Jews regarded Rashi, whose commentary was always printed on the inside margin of the page, as the essential companion for Talmud study. But he must also have realized that the opinions of the twelfth- and thirteenth-century glossators, the Tosafists, printed on the outer margin, also carried weight. His selection included the passage from tractate Berakhot (28b)[39] where Samuel the Small, one of the Tannaim, is said to have composed the blessing against heretics for the statutory Amidah prayer, recited three times daily – a blessing that has troubled so many commentators and offended so many gentiles.[40] The Talmudic text is terse, and as usual we are left in the dark

35. Buxtorf, copybook, 35: "Denique per magiam quoque produxit Iesum Nazerenum (ignosce o Christe mihi)."

36. Buxtorf, *Synagoga judaica: Das ist, Jüden Schul* (Basel, 1603), 90–91.

37. Buxtorf, copybook, 231–41.

38. Peter Schäfer, *Jesus in the Talmud* (Princeton, 2007).

39. Buxtorf, copybook, 238.

40. On the blessing against the heretics, see Ruth Langer, *Cursing the Christians: A History of the Birkat HaMinim* (Oxford, 2011).

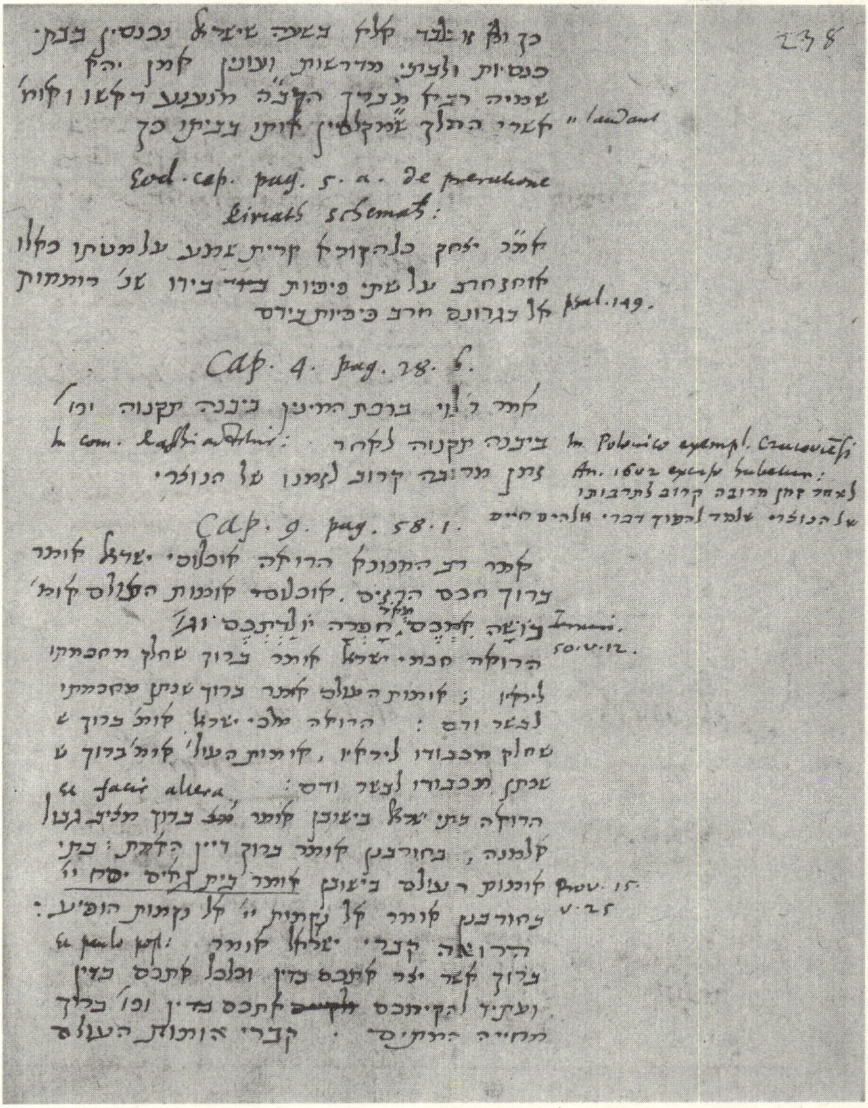

Figure 1.4. Buxtorf reads the Talmud with its commentaries. Basel, Universitätsbibliothek, A XII 20.

as to the identity of the damned heretics. Buxtorf wrote out a short extract from the beginning of the discussion: "Rabbi Levi said: The benediction about the *Minim* (heretics) was instituted in Javneh [Jamnia]" – the city in Palestine where, according to tradition, Rabbi Johanan ben Zakkai established a rabbinical school after the fall of Jerusalem to the Romans in 70 CE. He added Rashi's comment: "They arranged it in Javneh much later, around the time of the Christian." In the margin Buxtorf noted the variant text in the Cracow edition of 1602, which reads: "in Yavneh, a long time after, in the time of the depraved teaching of the Nazarene who taught them to overturn the words of the living God."[41] Buxtorf, as we will see, served as an official censor of Jewish books in Hebrew and Yiddish printed in Basel. Here we witness Buxtorf the censor at work, reading uncensored texts of the Talmud printed both before and after the production of the Basel edition.[42] That Buxtorf was able to consult all these enormous volumes and always put his finger on the relevant bit of text confirms Piet van Boxel's argument that neither the burning of copies of the Talmud in Italy in 1553 and after nor censorship was truly effective in preventing anyone who needed to from gaining access to the full text.[43] It also demonstrates Buxtorf's uncanny brilliance in deciphering the technical and rhetorical language of the Talmud, as well as his blatant disregard for the main thrust and parry of the Talmudic argument.

Buxtorf submitted the rich literature of Midrashim, sermons, moralistic compilations, and medieval commentaries to the same eclectic reading. Take his excerpt from the midrashic compilation Yalkut (Shimoni), which he had come across in the Cracow edition of 1605 (32 col.b): כל האומות שונאים את ישראל "All the nations hate Israel." The transcription of the Hebrew phrase is followed by a note in Latin: "that is why they direct their prayers against those who hate them."[44] But the context is provided by the Midrash, which presents various ways of interpreting the verse "Two nations are in your womb" (Genesis 25:23). They are certainly not all in the vein of the one from which Buxtorf is extracting his one-liner. The text states as follows: "Two (*shenei*) peoples hated (*senuei*) by the nations are in your womb: all nations hate Israel and all nations hate Esau. The one whom your Creator hates is in your womb as it is written, 'But Esau I hated' (Mal. 1:3)."[45] This passage, which is already found in the fifth-century Genesis Rabbah Midrash, expressed even more virulent hatred toward Esau – that is, Christian Rome – than the bit transcribed by Buxtorf would seem to indicate. But Buxtorf devoted a large part of his activity to scanning all printed Ashkenazic prayer books (as we will see, the copybook contains sheaves of notes

41. Buxtorf, copybook, 238: "In Polonico exempl. Cracoviensi An. 1602 excuso habetur:..."

42. Rashi's text is expurgated in the Basel Talmud, reading: "They instituted it in Yavneh a long time after" (BT, Berakhot [Basel, 1579], 28b).

43. See Piet van Boxel, "Hebrew Books and Censorship in Sixteenth-Century Italy," in *Jewish Books and Their Readers*, ed. Mandelbrote and Weinberg, 75–99.

44. Buxtorf, copybook, 2: "Hinc contra osores suos preces suas dirigunt."

45. *Yalqut* (Cracow, 1595), 1:32 col. b.

that Buxtorf took down from the entire Jewish liturgical tradition). Without bothering to fathom the Midrash, he makes a superficial connection with some Jewish prayers. The play on words, the variety of exegetical approaches to the verse in question, and even the violent anti-Roman sentiment in the Midrash are all overlooked.

The margins of Buxtorf's surviving books are not filled with the labyrinths of annotation that many of his contemporaries – for example, Isaac Casaubon – constructed in their favorite texts. Still, enough of his marginalia remain to show that his method of reading was as rigidly selective as his method of excerpting. In his copy of the second Rabbinic Bible, for example, he found little but grammatical points to note in the massive commentaries of Rashi and Ibn Ezra. Yet at Psalm 2:1, "Why do the nations roar?" Rashi wrote, "Our rabbis explain that this refers to the King Messiah. According to its real meaning – and it serves as a rejoinder to the heretics [*Minim*] – it should be interpreted as a reference to David."[46] Here Buxtorf inserted a marginal note in a distinctive, large script: "He calls Christians who explicate this Psalm as about Jesus Christ *Minim*, heretics."[47] The laser focus that enabled Buxtorf to distill his vast reading into the narrow range of materials in the copybook was already at work while he scanned the massive volumes that preserved Jewish learning. It is no accident that he collected other passages in which Jewish scholars called Christians heretics in the copybook.[48] His approach to Jewish texts was all of a piece.

The copybook, however, makes it possible to reconstruct the practices that Buxtorf brought to bear on a very wide range of texts, his copies of which usually do not survive. Conspicuous by his absence from the extracts is Maimonides, the only Jewish author who received the highest accolade from all Christian scholars without exception. Buxtorf seems to have ignored (purposefully or not) the purple anti-Christian passages that he could certainly have retrieved from Maimonides' works. Yet he did read the *Guide of the Perplexed* rather carefully.[49] Otherwise he could not have excerpted the annotations of the fourteenth-century commentators Ephodi and Narboni on part 1, chapter 66, of the *Guide*.

46. Volume 2 of Buxtorf's copy of this edition of the bible (Venice: Bomberg, 1525) is Basel, UB, A 1 17, which belonged first to Albert Sulzer and then to him.

47. "Minim, Haereticos. Christianos hic vocat qui de Iesu Christo hunc psalmum explicant." At some point Buxtorf also noted and recorded Kimhi's comment on the passage, not in the margin of his Bible but in his copybook, 320: "Psalmum 2 recentiores non de Messia exponunt, quod fecerunt veteres teste kimchi, sed dicunt Davidem loqui illic de se ipso. Vide kimchi."

48. Buxtorf, copybook, 3: "Quod Iudaei Christianos haereticos appellant liquet ex Kimchi in cap. 7 Ies. V. 14 et seqq." In the margin he also cites Rashi on Deuteronomy 32:21.

49. His copy of the *Guide*, published in Sabbioneta in 1553 and acquired (as Buxtorf noted, unusually) in 1615, is Basel, UB, FA IV 17 1.

Both of them, like ibn Ezra before them, explained the etymology of Sinai on the basis of the word סנה, "bush" – the locus of God's revelation to Moses.[50] Buxtorf must have been reading a manuscript copy of the Narboni commentary, which later became embedded in the margin of the standard printed page alongside other commentaries. His Hebrew transcription from Ephodi was followed by a Latin translation of a similar account given by Moses Narboni, the commentator famed for exposing the secrets of the *Guide*. According to Narboni, a member of the aristocratic Hasdai family had returned to Barcelona from the Holy Land with an amazing story about the numinous nature of the stones found on Mount Sinai, which were all engraved with a form of the bush. However many times one of these stones might be broken, an image of the bush would reappear on the fragments. This explanation, Narboni asserts, assisted the reader to understand Maimonides' (rather obscure) chapter on the way God's will was impressed on the natural world.

Here we see Buxtorf reading difficult medieval philosophical commentary – but to what purpose? Buxtorf was a philologist who avoided, or rather spurned, the more fantastic realms of Renaissance etymology. He just wanted to get to the root – the grammatical one. Yet he was apparently intrigued by the etymologies that these medieval philosophers suggested. Traces of his reading remerge in his son's reprinting of Buxtorf's *Lexicon* of Hebrew and Aramaic roots in 1631. Under the entry for סנה, he informs the reader that according to the commentators on Guide 1.66, the name of Mount Sinai derives from the bush, images of which were inscribed on the sherds, as stated in the eyewitness report of Ephodi, one of the commentators. Here the younger Buxtorf referred to the wrong commentator, confusing Ephodi, who did not claim to have actually seen the stones, with Narboni. Apparently when he compiled the *Lexicon* he did not scrutinize his father's extracts with sufficient rigor.[51]

Buxtorf integrated some of these notes into his *Jüden Schul*. In chapter 1, Buxtorf expatiated on the Jews' rejection of the true word of God by their deference to the rabbinic interpretation of the written law. The seventeenth-century English translator did a good job of conveying Buxtorf's sarcasm:

> This is the very means and plot, wherby the Divel first of all seduced the Jews to forsake the Word of God, and after his magisteriall manner compelled them to imbrace the Traditions of men, and that with such a tenacious grasp, that

50. Buxtorf, copybook, 204. He first gives extracts from Kimhi's *Shorashim*, then transcribes Ephodi in Hebrew, and then translates Narboni's comment into Latin.

51. Johannes Buxtorf, *Lexicon Hebraicum et Chaldaicum* (Basel, 1631), 514, s.v. סנה: "Sinai *m*. Rubus … Hinc סיני Sinai, montis nomen, a Rubi copia … quod lapides inventi in eo figuratum in se habuerint rubum, ut scribunt commentatores in librum More nebhuchim, Par. 1. cap. 66: adeo ut etiam in fragmentis lapidum istorum, figurae rubi apparuerint, quod se Ephodeus, alter istorum commentatorum, vidisse scribit."

> neither *Esay, Christ,* nor any other to this very day could by any means unclasp their arms. Go to then, my prudent and skilfull Rabbine, where shall we find a true Exposition of the written Word? In *Weckers* Books of Secrets? no verily: In *Reuchlines* Caballistical Art? no such matter: or lastly, in *Marcolfus*? much less; but read the *holy Talmud* and there you may find it.[52]

Buxtorf, as we have shown, was an omnivorous reader of Jewish texts. Underlying this passage was one phrase from a book that he had transcribed in his copybook: "In Sefer hachaiim (printed in Cracow in quarto) the Talmud is called a holy book ספר הקדוש on page 5b."[53] Buxtorf's omnivorous reading habits had led him to a book that the later Hebraist Giovanni Bernardo de Rossi pronounced as "most treasured by Jews."[54] The work was written in 1578 and published in Cracow in 1593, some years after the death of the author, Hayyim ben Bezalel, the brother of the more prominent Maharal of Prague. Known for his opposition to codification and, in particular, the work of Moses Isserles, Hayyim was also, like his brother the Maharal, a pedagogue.[55] He wrote a Hebrew grammar to ensure that Jews did not fall behind Christians in their mastery of their holy tongue. People accused Jews, he stated, of not knowing Hebrew grammar because they devoted time to educating their children in "this holy book [i.e., the Talmud]."[56] Though he polemicized against Christianity, he could be similarly antagonistic to Jewish philosophers whose reliance on human intellect he castigated.

But Hayyim did not take these attitudes to extremes, as his account of the Reuchlin affair in the *Sefer ha-hayyim,* which resembles that of Josel of Rosheim, indicates:

> I heard from elders that in the past certain wicked members of our people attacked the "precious vessel" verbally and physically and devised a plan

52. Johannes Buxtorf, *The Jewish Synagogue or an historical narration of the State of the Jewes at this day dispersed over the face of the whole Earth, in which their religion, education, manners, sects, death and burial are fully delivered, and that out of their own writers,* trans. 'A.B.' (London, 1663), 27.

53. Buxtorf, copybook, 5: "In sepher hachaim (impressus Cracoviae in 4°) Talmud vocatur ספר הקדוש Liber sanctus. pag. 5. b."

54. Giovanni Bernardo de Rossi, *Dizionario storico degli autori ebrei e delle loro opere.* (Parma, 1802), 1:67.

55. On Hayyim, see Haim Hillel Ben-Sasson, "The Reformation in Contemporary Jewish Eyes," in *Proceedings of the Israel Academy of Sciences and Humanities* 4 (1971): 239–326; Eric Zimmer, *Rabbi Hayyim b. R. Betsalel me-Friedberg, aḥi ha-Maharal mi-Prag* (in Hebrew) (Jerusalem, 1987); and Byron L. Sherwin, "In the Shadow of Greatness: Rabbi Hayyim Ben Betsalel of Friedberg," *Jewish Social Studies* 37 (1975): 35–60.

56. See Hayyim ben Bezalel, *Ets hayyim,* ed. Shraga Schneebalg (Jerusalem, 1973), 8. Hayyim singles out the medieval Ashkenazic pietists (*hasidei Ashkenaz*) who spent all their time on study of the Talmud.

> to assign it to the flames – this was because they despised the bewildering Aggadot because of their own limited understanding. They had hardly hatched their scheme when a Christian stood up before princes and the peoples in defense of *this holy book*. He said that the bewildering Aggadot are like bitter and poisonous herbs that can kill which are found in the shops of perfumers with all kinds of other spices, but that these too have great healing qualities. As King Solomon said: "The bruise of a wound cleans away evil." This means that every evil thing, such as wounds and bruises, can be cleansed and dispelled only by a similar ill. But these matters are known only to expert physicians, for they know how to use them in the appropriate manner. Similarly, all the rhetorical sayings of the sages, blessed be their memory, are enigmas.[57]

We do not know if Hayyim read Reuchlin's *Gutachten über das jüdische Schrifttum*, but his representation of Reuchlin's words is most intriguing for its inversion of a notable passage in Reuchlin's report:[58]

> We may conclude that it is permissible to read and study good and evil side by side: the evil so as to combat it with reasons and the good strewn among the evil like roses among thorns to be extracted.... Now there is no one who can or may in good conscience avow of the Talmud that it is altogether evil. For it contains many useful medicinal prescriptions and a knowledge of herbs and roots.... From all this we may well conclude that the Talmud is no barren tree.... Rather it contains much that is good and learned experts can derive much good from it in the spirit that St Jerome writes in his letter to Athleta [Laeta], "The wise man can find gold in dung."[59]

It is impossible to prove that Hayyim deliberately manipulated Reuchlin's words. But he seems to sing the Christian's praises not only for saving the Talmud from the pyre but also for grasping the nature of the "bewildering Aggadot," a central theme in the passage in the *Sefer ha-hayyim* from which Buxtorf extracted the phrase "holy Talmud."

Hayyim wrote the *Sefer ha-hayyim* as an act of expiation in response to the plague of 1578, which had killed his maid, his daughter, and the young man who was boarding in his home. He constructed the sections of the book according to the elements of the *avinu malkenu* prayer recited on the Day of Atonement. The

57. Hayyim ben Bezalel, *Sefer ha-hayyim* (Cracow, 1593), part 1, chap. 2, fol. 6r.

58. Johannes Reuchlin, *Recommendation Whether to Confiscate, Destroy, and Burn All Jewish Books*, trans. Peter Worstman (New York, 2000), 54. For the original, see *Johannes Reuchlins Sämtliche Werke*, Band 4: *Schriften zur Bücherstreit*, Teil 1: *Reuchlins Schriften*, ed. Widu Wolfgang Ehlers et al. (Stuttgart-Bad Cannstatt, 1999), 408, 29–30, and 410, 1–3.

59. Jerome, *Epistolae* 107.12.

passage about the Talmud occurs in chapter 2 of the first section, the "Book of Merits." Hayyim insists upon zealous observance of the *mitsvot*. The prohibitions and enactments (*gezerot* and *takkanot*) form an inherent part of what God commanded. Talmud is the quintessential truth that is uniquely served by Israel. Jews should therefore endeavor to uphold it through study in the synagogue.

Hayyim continues:[60]

> And those who scorn find an opportunity to place their scorn on '*this holy book*' on account of the esoteric aggadot that it contains. Apparently, it would appear that they are not only not beneficial but actually harmful. However, our rabbis of blessed memory already made an analogy in connection with this issue in the chapter "*bameh madlikin*."[61] They compared it to a person who ordered his servant to bring a *kor* of wheat to the upper chamber, which he duly did. He then told him to put a *kav* of *humtin* inside it, which he refused to do. He said to him: "Then it would have been better had you not brought the wheat."[62] ... For *humtin* is fine sand that does not cause plants to grow, and yet the fine sand is the only means by which the wheat is preserved. So, too, the produce of Torah is preserved only by means of the bewildering Aggadot, which ensures that their wisdom is concealed. Otherwise, the enemy would have already set his hand and tongue against the sages of blessed memory and turned their words into heresy – this was what was done with the written Torah – with the consequence that nothing would be left that could distinguish us from all the nations. But now, by means of the bewildering aggadot, *this holy book* is despised in their eyes and they will therefore not transpose its words into their words. Consequently, these words are left in our hands in their integrity for the dogs cannot touch them, and through them it is shown that we alone are the people of the Lord and his flock.[63]

Hayyim transforms a Talmudic parable into an elaborate defense of Aggadot – one surprising in its novelty, given the many attempts from medieval times onward to explain their true meaning to both Jews and non-Jews. Gentile appropriation of Scripture requires defensive action. Hayyim turns the old argument that the Oral Torah is God's secret, already articulated in Midrash Tanhuma, in a new direction.[64]

60. Hayyim ben Bezalel, *Sefer ha-hayyim*, fols. 5r–v.

61. BT, Shabb. 31a.

62. The Talmudic text is commenting on the verse "the fear of the Lord is his treasure" (Isaiah 33:6). It then draws a parallel with the idea that the preservation of the wheat requires the intermingling of the *humtin*, just as Torah and the commandments are useless unless accompanied by fear of the Lord.

63. Hayyim ben Bezalel, *Sefer ha-hayyim*, fols. 5r–v.

64. Salomon Buber, ed., *Midrash Tanhuma* (Vilna, 1885), Vayera, 6.

It is necessary to mystify the Aggadot in order to prevent aliens from reaching their inner truths. The unnatural or strange nature of the Aggadot ensures that the "malign," those who pour scorn, perceive only the outer, superficial layers of the Talmud. It was to this complex passage that Buxtorf reacted when he excerpted the phrase "holy Talmud" in his scathing attack on rabbinic tradition. He had succeeded in unearthing a contemporary work that contained one word, used in passing, that confirmed his view of the Jewish sacralization of a human work. He had also somehow absorbed and then reacted to Hayyim's powerful attempt to exclude non-Jews from entering the truly "sacred" domain of Jewish existence.

One author whose name appears on countless pages of the copybook and many times in the *Synagoga,* and who received an entry in Buxtorf's later bibliographical tool, the *Bibliotheca Rabbinica,* is the commentator and kabbalist Bahya ben Asher. The very conspicuous presence of this most popular writer in Buxtorf's copybook, in the form of swathes of excerpts both in Hebrew and in his own Latin translation, attests to a wider phenomenon. Bahya's writings, particularly his Bible commentary written in 1291 and his encyclopedia of fundamental Jewish concepts, the *Kad ha-kemah,* should be placed in the top ten of medieval Jewish writings. Diffused widely, Bahya's works, with their lucid exposition and uncomplicated style, attracted readers of all types, whether kabbalistically inclined or not. For though Bahya propounded the four hermeneutical modes – literal, midrashic, philosophical, and kabbalistic – with a distinct emphasis on the fourth level that veered on high, he often clothed the mystical in a homiletic form that was accessible to all, including Christian readers.[65] Already in the fifteenth century Bahya's fourfold method of exegesis was known to the philosopher and eager student of Jewish esoteric traditions Pico della Mirandola.[66] By the sixteenth century all kinds of Christians were perusing this very Jewish commentary. In Rouen we find evidence of Guillaume Postel's extensive study of the 1546 edition of the commentary. His annotations sometimes approved, but sometimes expressed contempt for, what he (like many others) regarded as the frivolous excesses of rabbinic creativity.[67] Bahya's works were revered, criticized, studied routinely, and printed over and over again throughout the sixteenth century, both in Italy and in Poland. Commentaries on them were written and sometimes printed – such as those of Naftali Treves and Manoah Hendel bar Shemariah – in order to

65. On Bahya's hermeneutics, see Gershom Scholem, *On the Kabbalah and Its Symbolism* (New York, 1965), and Ephraim Gottlieb, *The Kabbalah in the Writings of R. Baḥya ben Asher ibn Halawa* (in Hebrew) (Jerusalem, 1970).

66. Scholem, *On the Kabbalah,* 62 n1.

67. Valerie Neveu, "De Guillaume Postel à Richard Simon: Zohar et autres sources hébraïques dans les collections de la Bibliothèque municipale de Rouen," *Revue des études juives* 155 (1996): 75–105.

uncover the jewels that this esteemed author had enclosed in his works. One of Buxtorf's correspondents, Elijah Loanz – the rabbi of Frankfurt, later of Hanau and Worms – joined Bahya's entourage.[68] In the preface to his commentary he spoke of the many people who studied Bahya every week and of his own modest attempt to enlighten readers by building on the wisdom of experts in Kabbalah.[69] Similar statements can be found in the works of such early modern rabbis as Yuspa Hahn and Yom Tov Lipmann Heller.[70]

Buxtorf grasped the central role that Bahya's works played in the religious and intellectual life of both Jews and Christians.[71] His intense perusal of Bahya's works should therefore not elicit surprise. But how did he read Bahya? Not as his correspondent Elijah Loanz or even as Postel did. For Buxtorf had no inclination for mystical interpretations, particularly Jewish ones. Bahya, like so many other Jewish authors who lived in Spain, had a pronounced sense of exile. Every so often he expressed notions of Jewish exclusivity and hope for the downfall of oppressors.[72] Hence his books were often ruthlessly censored. It was sentiments like these that Buxtorf noted down, as often as not, with a cavalier disregard for context; and in so doing he transformed Bahya's consoling and uplifting discourses, addressed to his fellow Jews, into haphazard and vicious attacks on Christianity (see Fig. 1.5). When he transcribed an extract from Bahya's entry for "the Hanukkah candle," the festival that celebrates the rededication of the Temple in the time of Antiochus Epiphanes, Buxtorf noted a significant element in Bahya's treatment of the topic in the margin – that future redemption would parallel the first deliverance from Egypt.[73] According to this scheme, Israel's adversaries would, like the Egyptians of old, be subjected to divine wrath, as described in Isaiah 27:10: "For the fortified city is solitary, a habitation abandoned and forsaken, like the wilderness." But Buxtorf did not write out the remainder of the passage, which ends not on a vindictive note but rather with the promises inscribed in Daniel's prophecies: "And they that are wise shall shine as the brightness of the

68. For Buxtorf's use of Elijah's letters, see Johann Buxtorf, *Institutio epistolaris hebraica* (Basel, 1610), 28.

69. The autograph manuscript of *Ma'aglei tsedek* is Oxford, Bodleian Libraries, MS Opp. 216. It was completed in 1613 and contains references to contemporaries.

70. Yom Tov Lipmann Heller, *Megillat Eivah* (Breslau, 1836), 3: "And because many of the students were accustomed to read and study the work of our Rabbi Bahya, every sabbath I too became like them and my heart was aroused to explain kabbalistic matters."

71. Konrad Pellikan translated Bahya's commentary on Genesis and Exodus into Latin. The translation remains in manuscript in Zurich, ZB, Car C 63.

72. Bahya strongly denounced the fourth and last empire, Rome, and, by insinuation, Christian domination.

73. Buxtorf, copybook, 14. "Liberatio Judaeorum similis erit liberationi Aegyptiacae." The passage is taken from Bahya ben Asher, *Kad ha-kemah* (Venice, 1545), 47 col. 4–48 col. 1.

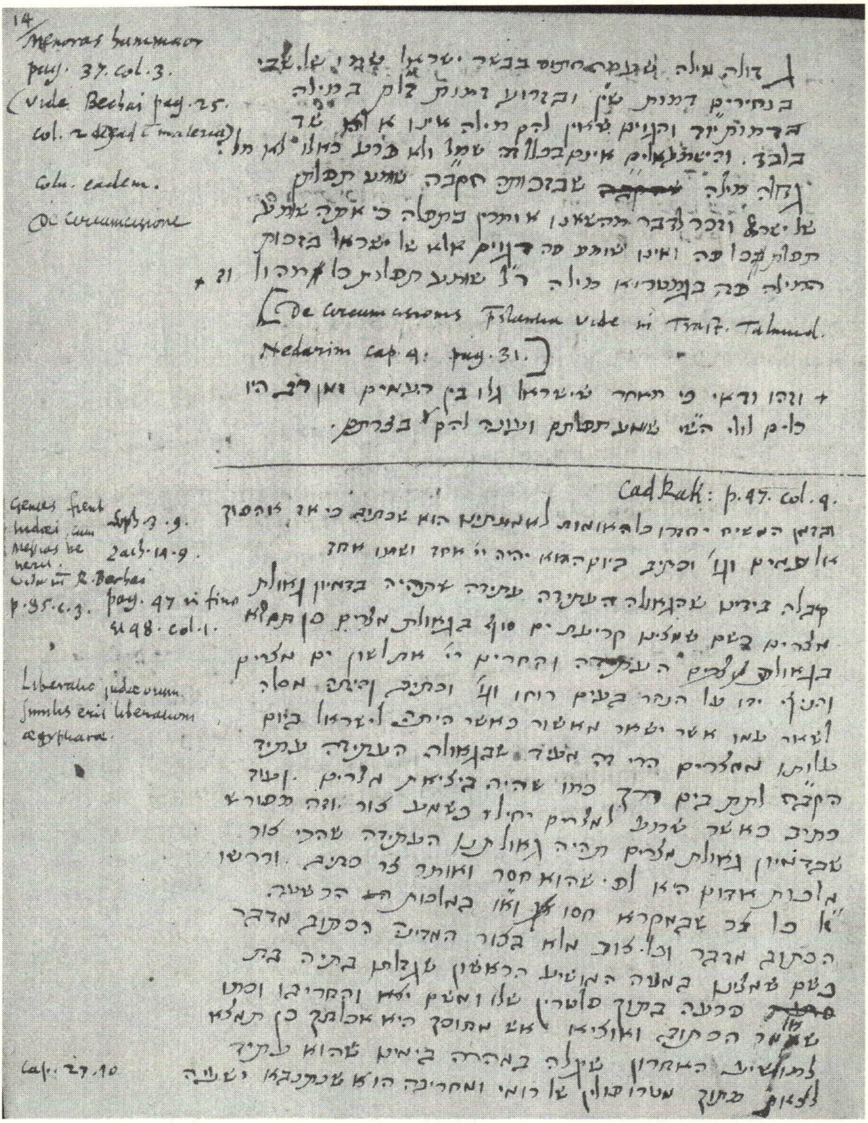

Figure 1.5. Buxtorf takes extracts from Isaac Aboab and Bahya ben Asher. Basel, Universitätsbibliothek, A XII 20.

firmament; and they that turn the many to righteousness as the stars for ever and ever" (Daniel 12:3).[74]

74. The entire passage is much more nuanced than indicated here, with an emphasis on the correspondence of the number eight to the eight-day festival of Hanukkah, and its

In a letter of 1606 to the Bremen theologian Matthias Martinius, Buxtorf offered an illustrative snippet. He had come across "a dreadful explanation" for the unusual form of the letter *ayin*, which was suspended above the line in the word "forest" in Psalm 80:14 ("The boar from the forest ravages it"). Bahya had suggested that the suspended letter symbolized "those who worship [Buxtorf writes 'believe in'] the one who was ḥung," i.e., Christ, who was represented in the Psalm by the boar who had trampled on the vineyard of Israel. This interpretation, Buxtorf noted, had serious implications for Christian printers. Those who retained the suspended letter in the Hebrew text "throw dung on their own faces."[75] With this rhetorical adaptation of Malachi 2:3, in Hebrew, Buxtorf revealed the purpose of his whole enterprise. Hebrew texts had to be purged of their Judaic aberrations. The offensive text from Bahya's discussion of redemption was just one example of the many that fill numerous pages of the copybook.

Buxtorf was equally arbitrary – if not equally critical – when he copied passages from Christians whose expert knowledge of Hebrew or Jewish practices he respected. Especially revealing is a note in which Buxtorf quoted statements of R. Yehiel b. Joseph from the 1240 disputation in Paris. Yehiel, he recorded, stated that Jews studied Talmudic Aggadot in other to gain eloquence, that the Talmud said nothing against Mary "who was of our bones and flesh," and that attacks on Jesus in the Talmud referred to individuals other than the Savior.[76] Buxtorf indicated only in passing that he took this material from Reuchlin, and he transformed the setting Reuchlin had crafted for these statements. Reuchlin explained, with a characteristically evocative choice of wording, that the Christian counterpart to Aggadot was "studia humanitatis." He explicitly compared the meaning of these materials for Jews to that of ancient poets, orators, philosophers, and historians for Christians, and noted that according to Yehiel, Jews were not obliged to believe in the truth of Aggadot, any more than Christians had to believe in the pagan fables they read in school.[77] Buxtorf reduced Reuchlin's effort to legitimate

---

messianic import. Buxtorf did not transcribe this part and started from Bahya's statement that the nations would return to "our faith." It should be noted that Buxtorf was also examining a work by another favorite author, Isaac Aboab's *Menorat ha-ma'or*, on the same page of his copybook.

75. Buxtorf to Martinius, 4 September 1606; published in *Bibliotheca historico-philologico-theologica* 4, no. 3 (1721): 602.

76. Buxtorf, copybook, 137: "Rabbi Iechiel filius Iosephi de libris Talmudicis: in disputatione Gallicana."

77. Johannes Reuchlin, *Defensio… contra calumniatores suos Colonienses* (Tübingen, 1514), sig. M iiv: "Enimvero sunt in eis sermones agada hoc est fabularum (sicut nos dicimus studia humanitatis) ad inducendum cor hominis ut intelligat eloquentiam, si voles credes, si non voles non credes, haec ille. Ita nos poetarum, oratorum, philosophorum et historicorum libros, etiam paganorum, propter eloquii venustatem legimus, et non comburimus."

Jewish tradition and the Christian study of it to one snide remark: "trivial fables for the sake of learning: thus we too study the fables of poets and orators to gain eloquence."[78] After citing Yehiel's argument about Jesus, moreover, Buxtorf added a characteristic note: "However this may be, nevertheless the Talmud speaks explicitly against our Jesus. I have noted these passages elsewhere."[79] In cases like this, Buxtorf's method of compilation resembles that of the humanist copybooks described by Ann Blair, which reduced rich stories and vivid quotations to pellets of information that were functionally equivalent in authority, whatever their origins.[80] The varied materials that go into Buxtorf's notebook, Jewish or Christian, originally critical or originally favorable, often come out making the same point.[81] That may perhaps explain why it is so hard to understand the principles that made Buxtorf copy extracts in some cases and translate them into Latin in others.

One formidable exception proves this rule. It also suggests that Buxtorf may have evaluated Christian scholars' authority in the realm of Jewish learning differently than do modern historians of Christian scholarship. He copied three passages from Joseph Scaliger, in full: a discussion of the silliness of rabbinical discussions of astronomy and one of the rules for postponement of holidays in the Jewish calendar, both from the first edition of the *De emendatione temporum*, published in 1583 and reprinted in 1593, and Scaliger's short treatise on Jewish tithes, from his *Opuscula*, published in 1610 and reprinted in 1612.[82] These passages Buxtorf transcribed almost word for word – a curious decision in the case of the passage on the calendar, since Scaliger had transcribed most of this, in his turn,

78. Buxtorf, copybook, 137: "Rabbi Iechiel filius Iosephi de libris Talmudicis: in disputatione Gallicana . . . [marginal note right: 'fabulae nugatoriae eloquentiae addiscendae gratia. ita nos quoque fabulas poetarum et oratorum propter eloquii venustatem legimus. Vide Reuchlini libellum contra Colonienses']."

79. Buxtorf, copybook, 137: "¶ Hoc ut ut sit, tamen alibi expresse de nostro Iesu Talmud loquitur, quae loca alibi a me notata sunt."

80. Ann Blair, "Humanist Methods in Natural Philosophy: The Commonplace Book," *Journal of the History of Ideas* 53 (1992): 541–51, at 545: "The commonplace book thus accumulates 'facts' from an indefinitely large and disparate range of sources and treats each fact, whether traditional or of recent origin, bookish or directly observed, as equivalent to every other."

81. The shaping ideological impetus visible in Buxtorf's copybook was generally not present in the reference works analyzed by Blair.

82. Buxtorf, copybook, 173–74 (copied with small variants and one error from Joseph Scaliger, *Opus novum de emendatione temporum* [Paris, 1583], 300: note that Casaubon wrote "Rabbinorum μυθολογία καὶ περιεργία" in the margin of his copy [London, British Library 582.l.9] at this point); 176 (from Scaliger, *De emendatione temporum*, 316); and 159–70 ("Josephi Scaligeri Diatriba De DECIMIS in lege Dei. ex Opusculis ejus excerpta," from Scaliger, *Opuscula* [Paris, 1610], [61]–70; [Frankfurt, 1612], 55–68. Selden's copy of the latter edition, signed J. Selden on the title page, is Oxford, Bodleian Libraries, 8º S 37 Art. Seld.).

from the *Kalendarium Hebraicum* of Buxtorf's predecessor in Basel, Sebastian Münster. Particularly interesting is the way that Buxtorf struggled with Scaliger's long discussion of tithes. Scaliger emphasized the importance of studying Jewish law as he had learned, from the jurist Jacques Cujas, to study Roman law: to scrutinize the Jewish "corpus juris" or Pandects (by which he meant the Talmud, and more particularly the Mishnah) and extract the surviving records of early Jewish law and the statements of authoritative figures.[83] Buxtorf annotated these passages with special interest, identifying halakhic texts (though not the Mishnah passages themselves) in which the principles Scaliger quoted could be found. In his *De abbreviaturis* of 1613, Buxtorf described the Talmud as a "full *corpus juris*" before giving his readers a neat lesson in how to use it, starting from a passage in the Mishnah and moving to the Gemara and then to the commentators.[84] It is quite possible that he derived his comparative vision of the Talmud – though not his competence in using it – from Scaliger. It seems certain, at any rate, that Buxtorf accorded Scaliger a different and higher level of authority, as an interpreter of Jewish texts and traditions, than he did either to the converts of the early sixteenth century or to Reuchlin.

Though Buxtorf transcribed Scaliger's treatise word for word, one small deviation calls for comment, and suggests that he felt provoked, as well as impressed, by Scaliger's expeditions into his domain. While visiting Avignon in the 1590s, Scaliger enjoyed lunch and conversation with a Jewish woman. She told him that "we are all mixed together, and if we had a sanctuary, we would not sacrifice, but we would pray, because we have no priests."[85] He incorporated this statement in his discussion of the tithes, where it attracted Selden's attention:

> the great *Ioseph Scaliger* saies, he askt some of them whether, if they might again build their Temple (as after the captiuite they did) their Laws of Sacrifices, first Fruits, & Tithes, would be then reuiud; and their answer

83. See Joanna Weinberg, "Crossroads in Hebraism: Johann Buxtorf Gives a Hebrew Lesson to Philippe Duplessis-Mornay," in *Jewish Books and Their Readers*, ed. Mandelbrote and Weinberg, 151–68; Anthony Grafton, "'Pandects of the Jews': A French, Swiss and Italian Prelude to John Selden," in *Jewish Books and Their Readers*, ed. Mandelbrote and Weinberg, 169–88.

84. Johann Buxtorf, *De abbreviaturis Hebraicis liber novus & copiosus: Cui accesserunt operis Talmudici brevis recensio: Item Bibliotheca rabbinica nova*, ed. Johann Buxtorf the Younger (Basel, 1640), sigs.):(, fols. 4v–5v. For a fuller account, see Chapter 3, below.

85. Joseph Scaliger, *Scaligerana, Thuana, Perroniana, Pithoeana, et Colomesiana*, ed. Pierre des Maizeaux (Amsterdam, 1740), 2: 409–10: "Dicebat mihi, nullos habemus Sacerdotes, nam multi se dicunt, sed non possunt probare, sumus confusi, & si haberemus Sanctuarium, non sacrificaremus, sed oraremus, quia non habemus Sacerdotes; non possunt extra templum sacrificare."

> was, that to build it again were to no purpose, because they had no lawfull Priesthood, there being not one of them that can proue himselfe a Leuit, though many pretend to be so, and some bear also the office of a kind of Priesthood amongst them.[86]

Buxtorf – as we have seen – emphasized his own personal contacts with Jews, and Scaliger and Casaubon both acknowledged his superiority in this domain. When Buxtorf transcribed this passage, he tacitly changed the tense of Scaliger's verb from imperfect to future: the Jews, in his version, "if asked, will answer" – as if he could not accept that even the great Scaliger had also learned from conversation as well as from reading.[87]

Buxtorf cast his net widely. He noted that Jacob Mocherseforim (literally, "the bookseller") of Meseritz lent him a manuscript copy of the *Mayse bukh* (Book of stories), a massive collection of legends and traditions in Yiddish.[88] One story, full of pathos but with a happy ending, caught Buxtorf's eye, and he transcribed it in full, in Latin characters. Rabbi Ibn Ezra, so the story goes, went to Rome, where he came across two Jewish boys wailing and bemoaning their lot. On asking the reason for their state of mourning, the rabbi learned that the lads were to be the victims of a papal mandate. As revenge for the death of the Christian God, every year on the day of the crucifixion of the Talui (the term used to refer to Jesus as "the hanged one"), two Jewish boys were to be transported to a high mountain outside Rome and hacked to pieces by the pope's cortege.[89] Without further ado the rabbi sought an audience with the pope. To cut a long story short, he had a discussion with Jesus, in a church, in the presence of twelve cardinals. Jesus admitted that he did not wish to see innocent blood spilled, and this resulted in the immediate release of the boys. Oxen were brought in to serve as surrogates, upholding the mandate. The story ends with the plea: "May God allow us to

86. John Selden, *The Historie of Tithes* (London, 1618), 21–22, summarizing Scaliger, *Opuscula* (1612), 65: "Sed neque certum genus Levitarum aut sacerdotum habent, quamvis multi Levitarum nomen praeferant. Nemo enim illorum tam confidens est, ut se Levitam verum aut Sacerdotem deierare audeat. Quamobrem a nobis interrogati, si illis data esset facultas instaurandi Sanctuarii in monte Moria, ut antea, an victimas immolaturi fuerint, respondebant frustra hoc fore, quia inquiebant אין היום כהונה בישראל οὐκ ἔστιν σήμερον ἱεράτευμα ἐν Ἰσραιλ Non est sacerdotium hodie in Israele."

87. At copybook, 169, Buxtorf turns *respondebant* into *respondebunt*.

88. Buxtorf, copybook, 248. On Jacob ben Abraham Pollack or Buchhändler, see Joseph Prijs, *Die Basler hebräischen Drucke, 1492–1866*, ed. Bernhard Prijs (Olten, 1964), 254–57, 272–73, 275–77, 517–20. On this extraordinary tale from the tradition of the *Mayse bukh*, see below and Erika Timm, "Abraham ibn Ezra und das *Maiśebuch*," in *Leket: Yidishe shtudyes haynt: Jiddistik heute: Yiddish Studies Today*, ed. Marion Aptroot et al. (2012), 1: 281–308.

89. This version of the story seems to conflate Christmas and Easter.

benefit from his [ibn Ezra's] merit for all times, and may the evil decrees and order be *mevattel* [annulled]. Amen."[90]

Why did the Jew Jacob ben Abraham Pollack, who preferred or vaunted his professional name Mocherseforim, lend his manuscript to Buxtorf? Why did the Christian inscribe six and a half pages of his copybook with this particular story, translating some of the typical Hebraic Yiddish phrases?[91] For it is clear by now that Buxtorf could not have been attracted to the tale for its praise of its rabbinic hero, whose successful sabotage of Christian power must have cheered Jewish readers. As his name betrays, Jacob was a bookseller. From 1598 to 1603 he also served as editor and corrector in Conrad Waldkirch's printing house in Basel, normally a city without Jews. Like other Jews, he provided Buxtorf with Hebrew books, and thus avoided being placed in the category of the "pessimus Judaeus," a term that Buxtorf reserved for Jews who refused to lend or sell him books.[92] The official first printed text of the *bukh* was printed under Jacob's aegis. But this precious volume, with its woodcut frame by Holbein on the title page, did not contain the story that Buxtorf had transcribed that very same year. Buxtorf's selection of this story from the *Mayse bukh* typifies his reading of Jewish literature, ever on the lookout for the anti-Christian sentiment. This reading was an "appropriation" in the most extreme sense: also one that led to censorship, for it is highly probable that in his capacity as censor, Buxtorf insisted that the story that he copied should be omitted from the printed text.

Buxtorf's encounter with the *Mayse bukh*, in other words, followed the same lines as most of his readings of Jewish texts. But the unusually precise way in which he identified his source reveals that not all his raw material came from texts – and calls attention to another striking feature of Buxtorf's collection of extracts. In addition to extracts from books, he recorded and dated meetings with Jews. On 1 July 1596, he asked "a certain Jew" about how Jews understood the phrase "behind one in the midst" (*ahar ahad [ahat] ba-tavekh*) in Isaiah 66:17 ("They that sanctify themselves, and purify themselves in the gardens behind one in the midst"), a verse that had puzzled many commentators. His unnamed interlocutor

90. Buxtorf, copybook, 248. For the versions of the story, see Timm, "Ibn Ezra und das *Maiśebuch*."

91. Buxtorf promoted Yiddish studies: see Johann Buxtorf, *Thesaurus Grammaticus linguae sanctae Hebraeae* (Basel, 1609), 648–71; Jerrold Frakes, *The Cultural Study of Yiddish in Early Modern Europe* (New York, 2007); Aya Elyada, "Protestant Scholars and Yiddish Studies in Early Modern Europe," *Past & Present*, no. 203 (May 2009): 69–98; and Elyada, *A Goy Who Speaks Yiddish: Christians and the Jewish Language in Early Modern Germany* (Stanford, 2012). On this passage in the copybook see Andreas Berger, *Stadt ohne Juden? Präsenz und Absenz in der Frühen Neuzeit* (Basel, 2024), 271–80.

92. Buxtorf to Waser, 1 May 1594; Zurich, ZB, S-150, 32: "Vnum hujusmodi [a book on Hebrew letter-writing] vidi, sed nec prece nec pretio impetrare a pessimo Judaea potui."

"answered a *tselem*, that is, an idol, that is, of the cross."[93] On 10 May 1600, Buxtorf described at length the rules that governed the proper way for a Jew to cut his nails. He noted that "I learned this from a Jew"[94] (see Fig. 1.6). Two and a half weeks later, a rabbi and corrector, Menachem ben Jacob, whom Buxtorf described as "outstandingly learned in the law," showed him that the expression *vehikahu* (and he shall smite him) used in Deuteronomy 19 was a palindrome. Buxtorf used this fact to explicate the passage in question and its legal import: "read backward, it means the same thing: it is permitted to strike the one who struck first."[95] It seems likely that Menachem also provided the text of a slaughtering license, signed by him, that appears in the copybook.[96]

For one discussion Buxtorf even provided a location in space. On 29 May 1600, he noted that if Christians attended a Jewish wedding and the Jews toasted them with *Lechayim Tovim* (to a good life), they were really wishing their guests *Kelalah* (a curse): "For by the art of Gematria [which assigns numerical values to Hebrew words and identifies those of equivalent value], those words are the same as *Kelalah*." "I learned this," he explained, "from a young Jew, 29 May 1600, in the district of Allschwil."[97] Most – if not all – of these conversations probably took place in Allschwil, a suburb of Basel, where a small Jewish community remained until the second decade of the seventeenth century. As we will see, they formed part of the preparation for Buxtorf's tendentious but informed survey of Ashkenazic Judaism.

Buxtorf was not the only late Renaissance humanist who worked with informants. He shared at least some of his practices with contemporary German

93. Buxtorf, copybook, 29: "Sequitur ibidem. Rectissimum autem est quod etsi nos quidem illorum (Iesuaolatorum) potestati subjecti simus, proprie tamen ad Edomem nostra captivitas refertur. Ipse enim 1 Julij 1596, cum quaererem a quodam Iudaeo quid per ista verba intelligerent: respondit unum צלם id est idolum, scilicet crucis."

94. Buxtorf, copybook, 287: "A Iudaeo didici 10 Maij anno 1600." Buxtorf used related material in chapter 10 of the *Jüden Schul*, 245.

95. Buxtorf, copybook, 314: "Vim vi repellere licere. R. Menachem praeclare in lege Judaica doctus, mihi ostendit ex eo quod scribitur Deut. 19. 6. וְהִכָּהוּ et percusserit eum, quae vox ordine retrogrado lecta, idem sonat, nempe qui primum percusserit, eum rursus percutere licet. 28 Maij anno 1600." Buxtorf's informant advised on the 1600 Waldkirch edition of Elijah Loanz's commentary on the Song of Songs. Prijs, *Die Basler hebräischen Drucke*, 271–72.

96. Buxtorf, copybook, 324. The younger Buxtorf entered this in chap. 36 of his expanded Latin version of the *Jüden Schul* (Johann Buxtorf, *Synagoga Judaica*, ed. Johann Buxtorf the Younge. [Basel, 1680], 612–13), as he noted in the manuscript.

97. Buxtorf, copybook, 282: "Iudaei propinanti poculum solent respondere לְחַיִּים טוֹבִים, id est ad vitam bonam, hoc est salutare sit vitae poculum istud. At cum Christianis ita respondent, tum intelligunt, Maledictum tibi sit poculum istud. Nam per artem Gematriam idem sunt verba ista quod קְלָלָה id est maledictio. A Iudaeo juveni hoc didici 29. Maij An. 1600. in pago Alswiler."

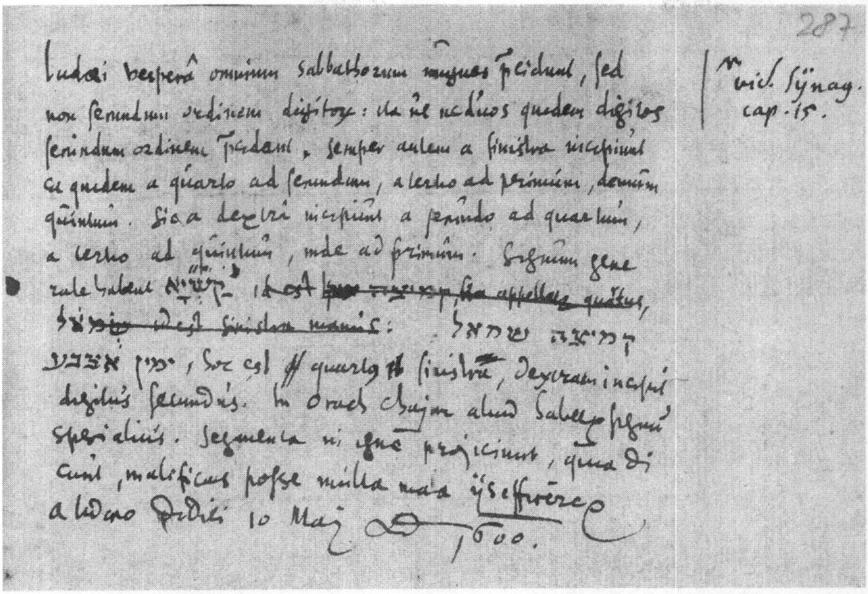

Figure 1.6. Buxtorf converses with a Jew. Basel, Universitätsbibliothek, A XII 20.

scholars whose love of Greek matched his "severus . . . amor" for Hebrew: Martin Crusius (1526–1607), a longtime professor of Greek in Tübingen, and Matthias Bernegger (1582–1640), a celebrated political writer and student of Tacitus at Strasbourg.[98] Though of different generations and interests, these men shared a passion for Greek: classical, Byzantine, and the contemporary version, as spoken and as written in popular vernacular texts. As in Buxtorf's case, so in Crusius's: ideology inspired interest. Starting soon after 1571, when a European fleet defeated the main fleet of the Ottoman Empire at Lepanto, Crusius began to study modern Greek and Greece systematically. He hoped that the Greek Orthodox Church might reunite with the Lutherans. Long theological discussions quickened and sustained his interest.[99]

98. Buxtorf's words are from Buxtorf to Casaubon, 24 December 1609; London, British Library, Burney MS 363, 129r. For Crusius, see in general Klaus-Henning Suchland, *Das Byzanzbild des Tübinger Philhellenen Martin Crusius (1526–1607)*, Dissertation Würzburg (Würzburg: n.p., 2001), and Richard Calis, "Martin Crusius and the Discovery of Ottoman Greece" (PhD diss., Princeton University, 2020; revised version forthcoming); for Bernegger, see Wilhelm Kühlmann, *Gelehrtenrepublik und Fürstenstaat: Entwicklung und Kritik des deutschen Späthumanismus in der Literatur des Barockzeitalters* (Tübingen, 1982).

99. Panagiotis Toufexis, *Das Alphabetum vulgaris linguae graecae des deutschen Humanisten Martin Crusius (1526–1607): Ein Beitrag zur Erforschung der gesprochenen griechischen Sprache im 16. Jh.* (Cologne, 2005), 61–93; for relations between the churches, see

When Crusius set out to learn modern Greek and to inquire about the Greek world under Ottoman rule, and when Bernegger went to work, they used living tutors. Crusius listed "the Greek men, who were with me (in Germany) at different times, from whose mouths I noted down everything that had to do with the modern corrupt Greek language."[100] He named his helpers and dated their stays with him: "Andreas Darmarios, a Laconian from Epidauros, came to Tübingen on Sunday, 30 August. He entered my study a little after 12:00 noon, and I greeted him most humanely."[101] With the help of Darmarios, an erudite scholar and scribe, and that of other friends and patrons he wove a larger network of informants, both Greek and German.[102] From their letters and his notes Crusius compiled his massive survey of the condition of the Greeks in his own time, the *Turcograecia* – a rich anthology of texts and commentary that, though different in form and style, provided for the Greek Orthodox world something of what Buxtorf provided for the Jews of Ashkenaz.[103] Bernegger, similarly, recorded in his copy of a work on the lexicon of later Greek by the Leiden Hellenist Joannes Meursius that "[i]n July and August of the year 1627, the reverend and most learned gentlemen Metrophanes Kritopulus of Beria, a Macedonian, stayed in my house. From the conversations we had at table I hastily recorded these notes."[104]

Both Buxtorf and Crusius, finally, collected and studied letters by contemporary Jews and Greeks. Buxtorf's copybook contains almost thirty pages of

---

Dorothea Wendebourg, *Reformation und Orthodoxie: Der ökumenische Briefwechsel zwischen der Leitung der Württembergischen Kirche und Patriarch Jeremias II. von Konstantinopel in den Jahren 1573–1581* (Göttingen, 1986), and, for the wider context, Richard Calis, "The Impossible Reformation: Protestant Europe and the Greek Orthodox Church," *Past & Present*, no. 259 (May 2023): 43–76.

100. Crusius, colophon, Tübingen, Universitätsbibliothek, MS Mb 37, fol. 83r; reproduced and discussed in Toufexis, *Das Alphabetum vulgaris linguae graecae*, 161–62. For a rich account of Crusius's ways of working with his Greek visitors, see Richard Calis, "Reconstructing the Ottoman Greek World: Early Modern Ethnography in the Household of Martin Crusius," *Renaissance Quarterly* 72 (2019): 148–93.

101. Crusius, *Diarium*, quoted by Toufexis, *Das Alphabetum vulgaris linguae graecae*, 166.

102. From the rich literature, see esp. Ulrich Moennig, "Matthias Berneggers Handexemplar des Glossarium graecobarbarum des Ioannes Meursius mit Korrekturen des Metrophanes Kritopoulos," in *Graeca recentiora in Germania: Deutsch-griechische Kulturbeziehungen vom 15. bis 19. Jahrhundert*, ed. Hans Eideneier (Wiesbaden, 1994), 161–98; Suchland, *Das Byzanzbild des Tübinger Philhellenen Martin Crusius*; Hans Eideneier, "Von der Handschrift zum Druck: Martinus Crusius und David Höschel als Sammler griechischer Venezianer Volksdrucke des 16. Jahrhunderts," in *Graeca recentiora in Germania*, ed. Eideneier, 93–111; and above all Toufexis, *Das Alphabetum vulgaris linguae graecae*, and Calis, "Martin Crusius."

103. Martin Crusius, *Turcograeciae libri octo* (Basel, 1584).

104. Bernegger, notes, in Wolfenbüttel, Herzog August Bibliothek, MS Gram. 73, loose sheets at start. See also Wolfenbüttel, Herzog August Bibliothek, MS 81.30 Aug. 2°, and Moennig, "Matthias Berneggers Handexemplar," esp. 169–70.

extracts from Hebrew letters, arranged in columns next to Latin translations. He described them, in the textbook on writing Hebrew letters that he published in 1610, as "the manuscript letters of very learned Rabbis from Italy and Germany, which were sent thirty years ago, at the time when the Talmud was published [in Basel (1578–80)], to a Jewish corrector. At last, through a strange fate, they reached me, from the shop of perfume vendors."[105] Many – though not all – of them were in fact written to or by the Jewish print corrector Israel Zifroni, who worked on the Talmud and other Basel projects from 1578 onward.[106] The heading of this section of Buxtorf's copybook reads: "Exordia of letters: models for greetings at the beginnings of letters."[107] As his terminology suggests, he was collecting set phrases and model sentences that Christians could cite or adapt when they wrote Hebrew letters. He stripped out names, including those of almost all senders and recipients, and concentrated on useful models for imitation, devoid of excess detail. Still, enough references to the Frankfurt book fair and complaints about thin ink, weak paper, and broken promises by publishers survive to show that most of the letters emerged from Zifroni's world of book producers – as Abraham Habermann saw more than forty years ago.[108] Crusius, for his part, assembled the letters of contemporary Greeks – ranging from scholars to an unnamed and unlearned woman – on a much grander scale in his *Turcograecia*. Yet his purposes were different: he regarded these texts not only as, at least in some cases, models of learned Greek but also as key sources for anyone who wished to understand the present state of the Greeks in the Ottoman world.[109] Buxtorf did not see his Hebrew letters in this richly historical light.

Buxtorf's attitudes and practices were not identical to those of Crusius and Bernegger. Both men named their Greek helpers, worked with them for sustained

105. Buxtorf, *Institutio epistolaris Hebraica*, ep. ded., sig.):(3r: "Hujus institutionis occasionem praebuerunt doctissimorum Rabbinorum ex Italia & Germania manuscriptae epistolae, quae tempore editionis Talmud ante trigir.ta annos Basileam ad Correctorem Judaicum missae, tandem singulari fato ex aromatariorum officinis ad manus meas pervenerunt." For the larger context and later development of Buxtorf's work on Hebrew epistolography, see Theodor Dunkelgrün, "The Humanist Discovery of Hebrew Epistolography," in *Jewish Books and Their Readers*, ed. Mandelbrote and Weinberg, 211–59.

106. Buxtorf, copybook, 77–50 (pages in Hebrew order). On 75 an extract is headed: "Ex Pauli Fagii epistola ad Osiandrum, cujus habeo integrum exemplar." Buxtorf printed selections from these texts and translations in *Institutio epistolaris*, 30–39.

107. Buxtorf, copybook, 77: "Exordia epistolarum, Formulae salutationum initio literarum."

108. Abraham M. Habermann, "The Printer Israel Zifroni and His Son Elishma and Their Books," in *Studies in the History of Hebrew Printers and Books* (in Hebrew) (Jerusalem, 1978), 215–92, at 272. Habermann transcribes the Hebrew texts from Buxtorf's copybook, 272–89.

109. Crusius, *Turcograeciae libri octo*.

periods, and liked and admired at least some of their visitors. Crusius described them as "humanus," "pius," and "ein feiner mann." In 1598 he recorded how he spoke with two of them "with Christian warmth," and gave them "a good lunch."[110] Buxtorf, by contrast, recorded abstract, one-sided exchanges: unnamed Jews admitted things that discredited their religion. He must have pressed them hard to give the right answers. This was ethnography, but of a special early modern sort – that practiced, as Carlo Ginzburg has suggested, most often by inquisitors.[111] Crusius occasionally complained that Greeks had invented the relatives, supposedly captured by pirates, for whom they needed ransom money, or had forged their credentials. But his enthusiasm for his visitors did not disappear, or even diminish over time.[112]

The partial resemblances between the bodies of material that Crusius and Bernegger assembled and those that Buxtorf compiled are suggestive in a larger way as well. Copybooks were by no means the only receptacles that scholars of the decades around 1600 used to compile information. Both Crusius and Bernegger made existing Greek dictionaries into conveniently alphabetized armatures on which they could string their own definitions of words in vulgar Greek. Crusius collected information partly through conversation, partly by persuading Greeks to gloss the difficult words in contemporary Greek works in print. He then compiled their glosses in the margins of his copy of the *Thesaurus cornucopiae,* a collection of excerpts from Greek grammatical texts, in alphabetical order, which Aldus Manutius had printed in 1496.[113] Similarly, Bernegger recorded the results of his conversations with Metrophanes Kritopulus, who lived with him during July and August 1627, about contemporary Greek in the margins of Joannes Meursius's "Graeco-Barbaric" lexicon, which dealt with the Byzantine form of the language.[114] In each case, the printed book metamorphosed into a manuscript of a sort.

110. E.g., Toufexis, *Das Alphabetum vulgaris linguae graecae,* 185; Martin Crusius, *Diarium,* ed. Wilhelm Göz et al. (Tübingen, 1927–61), 3: 273, 761–62.

111. Carlo Ginzburg, *Clues, Myths, and the Historical Method,* trans. John Tedeschi and Anne Tedeschi (Baltimore, 1992),156–64.

112. For Crusius's concerns about visitors, see Crusius, *Diarium,* 1: 369–70, 3: 117, 3: 123; for more enthusiastic descriptions in the same period, see 2: 254, 2: 255, 3: 273–74, 3: 761–62. By contrast, Crusius responded to reports that Jews were at large in the land – and especially to reports of the magical dealings of a Jewish alchemist and entrepreneur, Abramo Colorni – with a horror that the Greeks never provoked from him. Crusius, *Diarium,* 1: 407, 2: 24, 2: 28; cf. Daniel Jütte, *Das Zeitalter des Geheimnisses: Juden, Christen und die Ökonomie des Geheimen (1400–1800)* (Göttingen, 2011), 300–301.

113. Crusius's copy of *Thesaurus cornucopiae* is New Haven, Beinecke Library, Zi +5551 Copy 3: see Toufexis, *Das Alphabetum vulgaris linguae graecae,* 105–23.

114. Joannes Meursius, *Glossarium Graeco-Barbarum* (Leiden, 1614); Wolfenbüttel, Herzog August Bibliothek Gram. 7.3.

Buxtorf too made a dictionary into one of his major repositories. In his case, it was the *Tsemah David* (The Shoot of David), a vast dictionary of Hebrew, Latin, and Italian published in 1587 by the Jewish scholar and medical man David de Pomis.[115] Year after year, as is clear from shifts in handwriting and layers of revision, Buxtorf filled its margins with inky layers of references, further meanings, words and definitions from other texts, and short comments of his own (see Fig. 1.7). These notes served as the foundations for the great lexicon of Hebrew and Aramaic that his son would complete and publish, more than a decade after his death. To imagine Buxtorf at work, then, we cannot simply think of him reading Jewish texts and entering extracts in the copybook and, perhaps, others like it. He moved back and forth between his notebooks and the de Pomis and other printed books, which he had transformed into tools as distinctive and as serviceable as the copybook.

Here too, a comparison with Crusius is helpful. Scholars have only begun to map the immense apparatus of texts that filled what Crusius called his "Museum." It included his records of 6,588 sermons by sixty-seven preachers, which he translated into Greek as he listened to them, to practice his composition and to prevent himself from dozing, as well as dozens of his letters.[116] Crusius also recorded his dreams and his daily life in a very detailed diary.[117] And each part of this apparatus was connected with the rest. The diary, for example, served as a repository for draft letters and summaries of letters received – and thus as part of the preparatory material for the *Turcograecia*.

In some respects, the two men's practices differed widely. Buxtorf himself wrote a fluent Hebrew script, not at all like the square-charactered printing used by most Christians (including Isaac Casaubon). But he showed no serious interest in examining or recording the various material forms of Hebrew writing that he encountered. Crusius, by contrast, not only took notes on the calligraphic signatures and seals used by the Greeks whose letters he collected but printed them with the letters in his *Turcograecia*. Buxtorf discussed the material qualities of Jewish books in his rabbinic bibliography, and sometimes provided examples

115. Buxtorf's copy, with an affectionate inscription from Pascalis Gallus on the title page, is Basel, UB, FA IV 23; see also Buxtorf to Jacob Zwinger, 4 August 1593; Basel, UB, Fr Gr II 9, 89r.

116. The best (and only) guide is the magnificent catalogue of his manuscripts and collection of printed books by Thomas Wilhelmi, *Die griechischen Handschriften der Universitätsbibliothek Tübingen: Sonderband Martin Crusius: Handschriftenverzeichnis und Bibliographie* (Wiesbaden, 2002). For the sermons, see 25–172.

117. Crusius, *Diarium*; for Crusius's dream reports, see Hubert Cancik, "Crusius contra Frischlinum: Geschichte einer Feindschaft," in *Nicodemus Frischlin (1547–1590): Poetische und prosaische Praxis under den Bedingungen des konfessionellen Zeitalters*, ed. Sabine Holtz and Dieter Mertens (Stuttgart, 1999), 261–95.

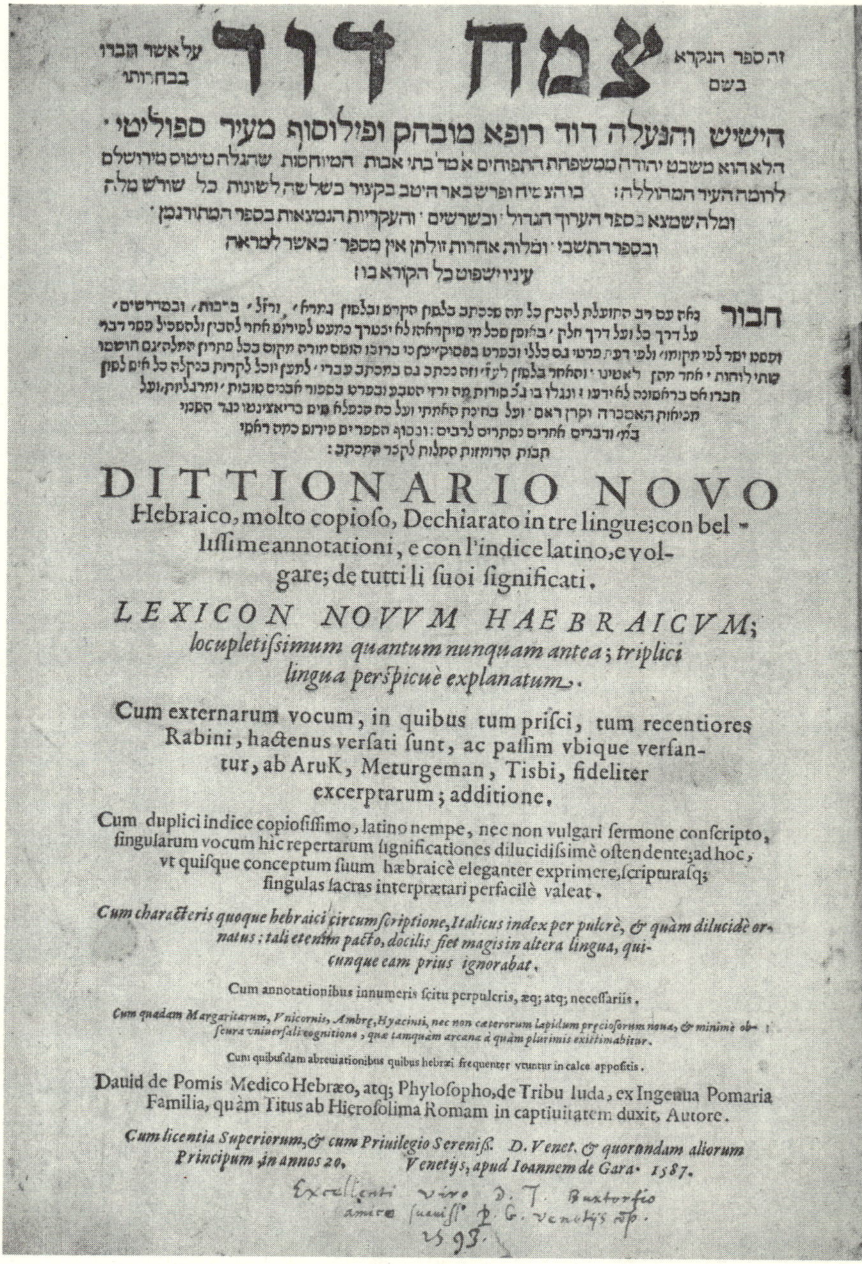

זה ספר הנקרא בשם צמח דוד על אשר חברו בבחרותו

הישיש והנעלה דוד רופא מובהק ופילוסוף מעיר ספוליטי ·

הלא הוא משבט יהודה ממשפחת התפוחים אשר בתי אבות המיוחסות שהגלה טיטוס מירושלם

לרומה העיר המהוללה: בו הצמח ופרש באר היטב בקצור בשלשה לשונות כל שורש מלה

ומלה שנמצא בספר הערוך הגדול · ובשרשים · והעקריות הנמצאות בספר המתורגמן ·

ובספר התשבי · ומלות אחרות זולתן אין מספר · כאשר למראה

עיניו ישפוט כל הקורא בו:

DITTIONARIO NOVO

Hebraico, molto copioso, Dechiarato in tre lingue; con bellissime annotationi, e con l'indice latino, e volgare; de tutti li suoi significati.

LEXICON NOVVM HAEBRAICVM; locupletissimum quantum nunquam antea; triplici lingua perspicuè explanatum.

Cum externarum vocum, in quibus tum prisci, tum recentiores Rabini, hactenus versati sunt, ac passim vbique versantur, ab AruK, Meturgeman, Tisbi, fideliter excerptarum; additione,

Cum duplici indice copiosissimo, latino nempe, nec non vulgari sermone conscripto, singularum vocum hic repertarum significationes dilucidissimè ostendente; ad hoc, vt quisque conceptum suum hæbraicè eleganter exprimere, scripturasq; singulas sacras interpraetari perfacilè valeat.

Cum characteris quoque hebraici circumscriptione, Italicus index per pulcrè, & quàm dilucidè ornatus: tali etenim pacto, docilis fiet magis in altera lingua, quicunque eam prius ignorabat.

Cum annotationibus innumeris scitu perpulcris, æq; atq; necessariis.

Cum quadam Margaritarum, Vnicornis, Ambre, Hyacinti, nec non cæterorum lapidum preciosorum noua, & minimè obscura uniuersali cognitione, quæ tamquam arcana à quàm plarimis existimabitur.

Cum quibusdam abreuiationibus quibus hebræi frequenter vtuntur in calce appositis.

Dauid de Pomis Medico Hebræo, atq; Phylosopho, de Tribu Iuda, ex Ingenua Pomaria Familia, quàm Titus ab Hierosolima Romam in captiuitatem duxit, Autore.

Cum licentia Superiorum, & cum Priuilegio Sereniss. D. Venet. & quorundam aliorum Principum, in annos 20. Venetijs, apud Ioannem de Gara. 1587.

Excellenti viro D. J. Buxtorfio amico suauiss. D. G. Venetijs dp. 1593.

Figure 1.7. Buxtorf's copy of David de Pomis's *Tsemah David*. Basel, Universitätsbibliothek, FA IV 23.

of Jewish writing or printing to instruct his readers on how to read the so-called Rashi script and, in particular, Yiddish characters.[118]

These cases show that the practices of textual scholars, well before 1600, could be as flexible in form and content as those of seventeenth-century natural philosophers. Much less of Buxtorf's apparatus is preserved than of Crusius's, and typically for the man, it contains no revelations about what he dreamed.[119] But enough remains to make clear that Buxtorf, like Crusius – and even more like Robert Boyle – lived his life cocooned in a mass of paper that he too might have described as "a Chaos, rude & indigested."[120] Buxtorf's few surviving manuscripts include groups of loose but related papers; letters kept for their bibliographical or technical content; tiny sheets, like the pages of a modern scratch pad, bearing notes on a single subject; and larger sheets folded into gatherings but not bound.[121] Most of his printed books are not so thickly hedged with manuscript notes as his de Pomis. But they show many signs of work – from the pointing hands with which he identified especially unpleasant items of Jewish belief, law, or language, to the underlining of a single word that was all he needed, sometimes, to remember a particular passage, to elaborate marginal commentaries. Somehow, Buxtorf's memory held these disparate pieces in suspension and enabled him to connect circuits among them whenever he had to do so.

Buxtorf's methods of compilation changed over time. In his last years, for example, he began to concentrate on a single vast project: a comprehensive dictionary, which must have required collections of data larger in scale and more systematic in approach than anything he had attempted before. Johann Buxtorf the Younger, who actually completed the dictionary, described how this project had sharpened his father's sensibility and changed his working methods. Originally, he had planned to provide only a simple lexicon like the Hebrew one that he had published in 1609. But as he came to see the flaws in earlier lexicons, "he began to make fuller collections of examples, to distinguish more precisely among Targumic, Talmudic and Rabbinic usages . . . to search out the original passages of authors' work and to note them down with their titles and page numbers."[122] Buxtorf

118. See, e.g., Buxtorf, *Thesaurus Grammaticus,* 632 (in Rashi script) and the long section on Yiddish, 652–71.

119. Buxtorf did confide professional ambitions of a very concrete kind in letters to his friend Caspar Waser, professor of Hebrew in Zurich.

120. Wotton to Evelyn, 8 August 1699, quoted in Michael Hunter, "Mapping the Mind of Robert Boyle: The Evidence of the Boyle Papers," in *Archives of the Scientific Revolution: The Formation and Exchange of Ideas in Seventeenth-Century Europe,* ed. Michael Hunter (Woodbridge, 1998), 121–36, at 123.

121. On the Buxtorf papers, see Prijs, *Die hebräischen Handschriften,* 73–85 (by Stephen Burnett).

122. Johann Buxtorf, *Lexicon chaldaicum, talmudicum et rabbinicum* (Basel, 1639), sig. (*) 4v.

had always recorded the provenance of the materials he assembled. Now he used his mastery of bibliography to set texts and extracts into chronological series.

This episode confirms that Buxtorf's surviving notebook is only one foothill from a lost mountain range of notes. But it also helps us understand the difference between filling a notebook with data and filling it with texts. As Buxtorf made his extracts, their variety and richness affected him, changing his views. The components of the canon reacted with one another in unexpected ways, and its borders eroded.

The copybook includes, in Hebrew and Latin translation, a long passage, drawn from a medieval text – the *Alphabet* of pseudo-ben Sira – on the mythical, demonic figure Lilith, who in Jewish tradition was identified as the first wife of Adam.[123] Buxtorf cited this in the *Jüden Schul* to illustrate the superstitious practices with which the Jews warded off demonic attacks on newborn babies.[124] He – or his son – cited it again in the lexicon.[125] But by this time the process of compilation and comparison had taught the Buxtorfs to read it in a new way. Quoting parallel passages from Latin writers – Pliny, Ovid, Serenus Sammonicus – they acknowledged that the Romans too had believed in the existence of blood-sucking birds that flew by night.[126] Compilation and collation had morphed – as they so often did in the late Renaissance – into comparison of beliefs and practices. Jews turned out to resemble Greeks and Romans (just as the Chinese and the Aztecs often did), and to belong with them as a cultured people. Buxtorf's vision of Judaism was less stable than has been thought, and his son, who identified so strongly with him, acknowledged this change in his approach. The explanation for Buxtorf's ability (or that of his son) to change perspective lies partly in the multiple sets of notes that both men assembled, and most of which do not survive: a unique, ordered mass of material in which knowledge grew "unseen and of itself in its cavities and concealed inner spaces."[127]

123. Buxtorf, copybook, 122–23. The text is from the *Alphabet* ascribed to Ben Sira, Question 5, 23a–b. His thickly annotated copy is in Basel, UB, FA VI 10. Among the marginalia is a very conspicuous hand pointing to one statement (25), "all nations are animals." He translates this in the margin as "omnes populi mundi Bestiae sunt pro judaeis," once again taking the statement completely out of context. The text describes a carnivalesque situation in which Nebuchadnezzar asks Ben Sira to marry his daughter, to which Ben Sira answers that he cannot since "all the nations are animals."

124. Buxtorf, *Jüden Schul*, 99–104.

125. There is also a brief note in Buxtorf, *Epitome radicum Hebraicarum et Chaldaicarum* (Basel, 1607), 447.

126. Buxtorf, *Lexicon chaldaicum talmudicum et rabbinicum*, 1141.

127. Hilary Mantel, *Wolf Hall* (London, 2010), 494.

CHAPTER TWO

Compilation and Observation in Buxtorf's *Synagogue of the Jews*

> [H]e hath strange places cramm'd
> With observation, the which he vents
> In mangled forms.
>
> Shakespeare, *As You Like It* 2.7.40–42

In 1603, Buxtorf compiled and published the *Jüden Schul* (The Synagogue of the Jews), a comprehensive, polemical portrait of Ashkenazic Jewry. Twice translated into Latin and expanded by Buxtorf's son Johann, who succeeded to his chair at Basel, this book shaped Christian views of Jewish life and ritual.[1] Buxtorf followed the Jewish male child through the stages of life, from birth and circumcision through education and marriage to divorce and death. His detailed, vivid book described the yearly cycle of Jewish worship in rich, precise detail, making continual digs at the Jews' spiritual failings:

> To let you know in part what they pray, and how they do it, I will translate into German some of the morning prayers and explain them. The first prayer starts: *Adon ólam ásher málach*. It is written in rhyme, like most of their prayers, and is sung and read standing and in a loud voice. It goes: Lord of the world, who reigned before all things were created, at a time when they were created according to his will, he was called King with his name. After everything is gone, he will still remain King, to whom respect and honor will be given. He was always, he is still, and he will be forever in his beauty. He is alone, there is no one that could be compared with him, or could be placed with him. (With this they deny the divine nature of Christ, and regard him

1. See Stephen Burnett, *From Christian Hebraism to Jewish Studies: Johannes Buxtorf (1564–1629) and Hebrew Learning in the Seventeenth Century* (Leiden, 1996), and Rudolf Smend, *Vier Epitaphe: Die Basler Hebraisten-Familie Buxtorf* (Berlin, 2010). Carl Buxtorf-Falkeisen, *Johannes Buxtorf Vater, Prof. ling. hebr. 1564–1629, erkannt aus seinem Briefwechsel* (Basel, 1860), is still valuable.

> only as a bad, common man.) He is without beginning or end; with him is power and dominion. He is my God, and my savior who lives in truth. (Here they scorn our faith that we believe in a savior who died.)[2]

No previous writer had portrayed the lives, works, and worship of Ashkenazic Jews as crisply and precisely as Buxtorf did. An unsatisfactory Latin translation, which appeared in 1604, made his work, retitled *Synagoga Iudaica*, accessible to scholars across Europe.

From its first appearance in Latin, Christian readers have been particularly impressed by the apparently firsthand character of Buxtorf's knowledge. Daniel Tossanus the Younger, a Reformed theologian who became rector of the Basel Gymnasium in 1627, gave the funeral oration for Buxtorf. He stated that Buxtorf achieved his mastery of Jewish sources and traditions by following two paths, that of books and that of the *viva vox*, the living voice:

> he also made use of close association with Jews. That explains why he often supported Jews at his own expense, for some months at a time, in his house. From them he learned the secret and arcane doctrines, and Jewish rituals and ceremonies, with marvelous results. The books that he brought forth, more lasting than bronze, furnish proofs of this that will be famous forever.[3]

Pretty much everyone agreed – for example, the great Hellenist Isaac Casaubon, who noted on the title page of his own copy that Buxtorf "frequented modern rabbis, and to such an extent that he took part in their banquets"[4] (see Fig. 2.1).

More recent scholars have disagreed on this point. Since the 1980s, Jewish historians have examined the works about Jews written by converts and Christians in the sixteenth and seventeenth centuries more and more intensively. They often refer to these books, using a term coined by Ronnie Hsia, as "ethnographies."[5]

2. Johann Buxtorf, *Synagoga judaica: Das ist, Jüden Schul* (Basel: Sebastian Henricpetri, 1603), 204–5; trans. Alan D. Corré, available both as Johannes Buxtorf, *Synagoga Judaica* (https://minds.wisconsin.edu/bitstream/handle/1793/7382/index.html) and as Johannes Buxtorf, *The Jews' Synagogue* (Kindle edition), via Amazon.com. Throughout this chapter, Corré's translation has sometimes been slightly changed.

3. Daniel Tossanus, *Johannis Buxtorfii senioris . . . Vita et mors* (Basel, 1630), 10.

4. Casaubon, note on the title page of his copy of Buxtorf, *Synagoga Iudaica* (Hanau, 1604), London, British Library, 848.b.19: "auctor consuetudinem habuit cum hodiernis Rabb. ut etiam convivijs eorum interesset." Cf. Scaliger's response, in his letter to Buxtorf, 1 June 1606, also stressing Buxtorf's quasi-native fluency and mastery of Jewish matters; Basel, UB, G I 59, 363, printed in *The Correspondence of Joseph Scaliger*, ed. Paul Botley and Dirk van Miert (Geneva, 2012), 6: 436.

5. Ronnie Po-chia Hsia, "Christian Ethnographies of Jews in Early Modern Germany," in *The Expulsion of the Jews: 1492 and After*, ed. Raymond Waddington and A.H. Williamson

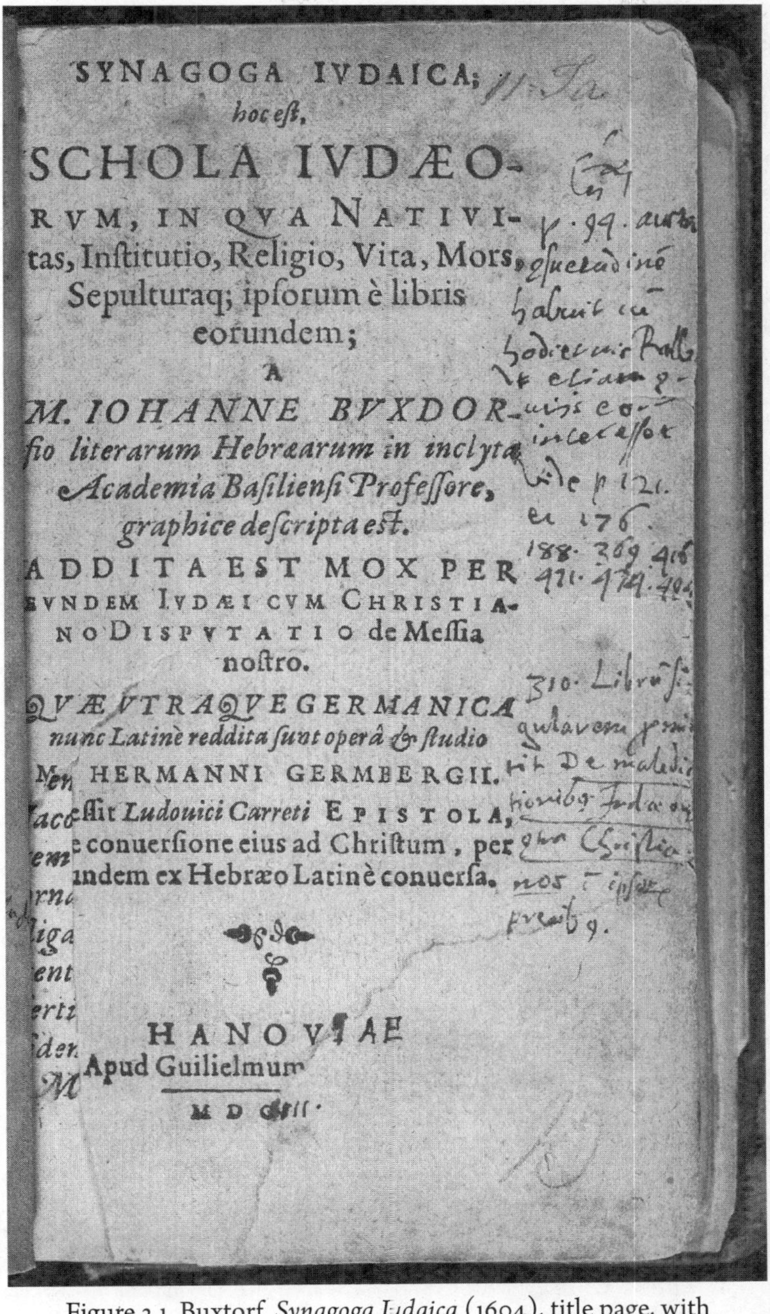

SYNAGOGA IVDAICA;
hoc est,
SCHOLA IVDÆORVM, IN QVA NATIVItas, Institutio, Religio, Vita, Mors, Sepulturaq; ipsorum è libris eorundem;
A
M. IOHANNE BVXDORfio literarum Hebræarum in inclyta Academia Basiliensi Professore, graphice descripta est.
ADDITA EST MOX PER eundem IVDÆI CVM CHRISTIANO DISPVTATIO de Messia nostro.
QVÆ VTRAQVE GERMANICA nunc Latinè reddita sunt operâ & studio M. HERMANNI GERMBERGII.
[illegible] essit Ludouici Carreti EPISTOLA, [illegible] conuersione eius ad Christum, per [illegible] ndem ex Hebræo Latinè conuersa.
HANOVIAE
Apud Guilielmum
M D C[illegible]

Figure 2.1. Buxtorf, *Synagoga Judaica* (1604), title page, with note by Isaac Casaubon. London, British Library, 848.b.19.

Some have argued – as Casaubon, Scaliger, and Tossanus did – that these texts preserve firsthand information on Jewish life and belief. Others – especially Stephen Burnett and Yaacov Deutsch – have noted that most of Buxtorf's material came from texts: Deutsch argues that he and other Christians hoped to show the superiority of their book-based knowledge to the revelations of Jewish converts.[6] We come to the *Synagogue* not with a category but with a question. We seek to fathom what sort of book this actually was, and to work out what Buxtorf meant to achieve, how he went about his work, and what we can learn from it. To clear the ground, we begin by considering whether the term ethnography fits the *Synagogue* at all.

True, in all its forms, the book was designed to suggest that it rested on firsthand inquiry. Buxtorf ordered his book by the life cycle of a Jewish male and the cycle of Jewish holidays. That probably explains why, as we have already seen, he took care to insert a description of his own participation in a Jewish ritual in the chapter on circumcision, which naturally came early.[7] Much later Buxtorf described what took place in the synagogue on Yom Kippur, the day of repentance. In this case he observed that after the opening prayer, *Kol Nidrei*,

> Some stay all night in the shul and pray. Some go home and sleep, some sleep in a corner of the church, distant from the Ark, or in the church for the women [i.e., the women's gallery] after they all have gone home. Some of them who wish to be very pious, and make a full repentance, stand through the whole festival, day and night, erect on their feet, singing and praying continuously. I have seen some that stood erect for twenty-seven hours in one place.[8]

Vivid stories like these, to all appearances drawn from experience, did not make frequent appearances. But they indicated to readers like Casaubon and Scaliger that the book was something like the result of a firsthand inquiry.

Readers also had other reasons to expect firsthand information. One had less to do with Buxtorf himself than with the genre in which he wrote. With its topical organization and its vivid descriptions of customs, the *Synagogue* resembled what its original readers would have called a travel account: a description of a voyage to a generalized Yiddishland, perhaps, rather than to a particular place, but still, the sort of text that reported on travel. Both Scaliger and Casaubon had read widely in such texts – which, they believed, belonged to one of the oldest genres

---

(New York, 1994), 223–235; see further Yaacov Deutsch, *Judaism in Christian Eyes: Ethnographic Descriptions of Jews and Judaism in Early Modern Europe*, trans. Avi Aronsky (Oxford, 2012).

6. Burnett, *From Christian Hebraism to Jewish Studies*; Deutsch, *Judaism in Christian Eyes.*
7. Buxtorf, *Jüden Schul*, chap. 2, 124.
8. Ibid., chap. 21, 529.

of writing. Both men had studied the histories of Herodotus, whom they saw as the founder of the genre, line by line. Their annotated copies, which survive, reveal their shared fascination with the ways in which the Greek writer drew information from informants. Both knew that ancient writers had accused Herodotus of lying.[9] And both believed that his account of his method – which revealed that he was an accurate inquirer and reporter even when he did not believe everything he heard – sufficed to refute these charges. "For myself," Herodotus said, "though it be my business to set down that which is told me, to believe it is none at all of my business; let that saying hold good for the whole of my history."[10] Scaliger underlined this passage and transcribed it on the title page of his copy. On the flyleaf he wrote: "Herodotus's self-defense should be marked."[11] Herodotus explained that he visited Thebes and Heliopolis to learn "if the people of those places would tell me the same tale as the priests at Memphis."[12] "Note," wrote Casaubon beside this passage, "how Herodotus strove to investigate the truth at Heliopolis."[13] Both men read later travel writers just as eagerly: Benjamin of Tudela, for example, and Leo the African. Casaubon's copies of both texts, their margins full of notes, are in the British Library.[14]

Most important of all, both men agreed with the contemporaries who held that firsthand travel accounts formed a distinctively rich body of knowledge about the world. As Justin Stagl, Joan-Pau Rubiés, Nicholas Popper, and others have shown, in the sixteenth and seventeenth centuries, scholars established learned travel literature as a distinctive and vital form of writing.[15] Well-made travel accounts offered readers valuable information about the customs and

9. See Arnaldo Momigliano, "The Place of Herodotus in the History of Historiography," in his *Studies in Historiography* (London, 1966), 127–42.

10. Herodotus, *Histories* 7.152.3; *Herodotus*, trans. A.D. Godley (London and New York, 1920–25), 3: 463 (Greek text at 3: 462).

11. Cambridge, Cambridge University Library, Adv.a.19.2, 274, title page and flyleaf, where Scaliger wrote: "Apologia Herodoti notanda."

12. Herodotus, *Histories* 2.3.1, trans. Godley, 1: 277 (Greek text at 1: 276).

13. Cambridge, Cambridge University Library, Adv.a.3.2.(1.), [56]: "No. Studium Herodoti indagandae veritatis Helopolitae."

14. Casaubon's copy of the 1575 Plantin edition of Benjamin of Tudela's *Itinerarium* is London, British Library, 1046.b.2; for Scaliger's use of this text, see Anthony Grafton, *Joseph Scaliger*, (Oxford, 1983–93), 2: 418–19. Casaubon's copy of Leo Africanus, *De totius Africae descriptione, libri ix* (Antwerp, 1556), is London, British Library 793.d.2. See more generally Anthony Grafton and Joanna Weinberg, with Alastair Hamilton, *"I have always loved the Holy Tongue": Isaac Casaubon, the Jews and a Forgotten Chapter in Renaissance Scholarship* (Cambridge, MA, 2011), 138–40.

15. Justin Stagl, *A History of Curiosity: The Theory of Travel, 1550–1800* (Chur, 1995); Joan-Pau Rubiés, *Travel and Ethnology in the Renaissance: South Asia Through European Eyes, 1250–1625* (Cambridge, 2000); Rubiés, *Travellers and Cosmographers: Studies in the History*

constitutions of other peoples, which could be used for framing both scholarly comparisons and practical policy decisions. They were also meant to train potential travelers by showing them what topics to focus on and how to formulate questions as they visited new lands and cities. Theodor Zwinger, a renowned Basel writer and medical man whose son Jacob was a close friend of Buxtorf's, published in 1577 the formal, and influential, *Method for Travel,* a book that synthesized the experience of humanists and patricians into a set of rules. He laid out the questions that young men in search of political enlightenment on their grand tours must ask as they tried to understand the character of each city they visited: in essence, a questionnaire. To answer it one had to collect information on everything from the city's natural setting and fortifications to its learned institutions.[16] Many other manuals provided similar directions.[17] In-person travel formed part of the education of the political elite in Venice and elsewhere. Gradually it came to be seen as a privileged source of unmediated knowledge.[18] When Tossanus, Casaubon, and Scaliger stressed Buxtorf's use of informants and direct knowledge, they put him into this tradition. Though they did not say it, writing on foreign lands, as distinguished studies have recently confirmed, was likelier to set prejudices and errors in stone for long periods than to portray them dispassionately and precisely.[19] In this respect too Buxtorf worked within a well-established tradition.

Scaliger and Casaubon were far from the only readers who assumed that Buxtorf wrote like a traveler. Early in the seventeenth century, the Basel medical man Thomas Platter the Younger compiled the final version of the wonderful travel diary in which he described his early life as a medical student in Montpellier and grand tourist to London. Platter's lively accounts of the *aiguillette* – the spell used by witches to induce impotence in men – and many other local customs and rituals have enriched the work of French social historians. Platter also visited Avignon,

---

of Early Modern Travel and Ethnology (Aldershot, 2007); Nicholas Popper, *Walter Ralegh's "History of the World" and the Historical Culture of the Late Renaissance* (Chicago, 2012).

16. Theodor Zwinger, *Methodus apodemica, in eorum gratiam, qui cum fructu in quocunq[ue] tandem vitae genere peregrinari cupiunt* (Basel, 1577).

17. For selections from this literature, see *Geschichte der Staatsbeschreibung: Ausgewählte Quellentexte, 1456–1813,* ed. Mohammed Rassem and Justin Stagl (Berlin, 1994). A full bibliography is provided by Justin Stagl et al., *Apodemiken: Eine räsonnierte Bibliographie der reisetheoretischen Literatur des 16., 17. und 18. Jahrhunderts* (Paderborn, 1983).

18. Kathryn Taylor, "Making Statesmen, Writing Culture: Ethnography, Observation and Diplomatic Travel in Early Modern Venice," *Journal of Early Modern History* 22 (2018): 279–98.

19. Carina Johnson, *Cultural Hierarchy in Sixteenth-Century Europe: The Ottomans and Mexicans* (Cambridge, 2011); Surekha Davies, *Renaissance Ethnography and the Invention of the Human: New Worlds, Maps and Monsters* (Cambridge, 2016).

where he investigated the Jewish community. He inserted vivid descriptions of Jewish circumcision and marriage ceremonies into his work.[20] There is only one fly in the ointment. As Stephen Burnett pointed out some time ago, Platter drew the bulk of his material on the Jews not from observation but from Buxtorf's book – even though Buxtorf described Ashkenazic practices largely foreign to the small and confined Avignon community.[21] Evidently, the observant Platter saw Buxtorf as a colleague, a fellow practitioner of sharp-eyed observation, whose vivid descriptions of customs and rituals he appropriated even as he discarded the disparaging comments that had accompanied them.[22]

When Platter copied from Buxtorf, he engaged in another common practice. Observation was always a textual as well as a visual activity. Well-trained travelers prepared to see the world by reading about it and filled out what they saw and recalled with what they read.[23] Evidently, then, the fact that someone set out to write a travel account need not mean that he always or regularly worked with informants – even if his text gave the impression that he did so. And that in turn raises a question about Buxtorf. If everything that looks like direct inquiry in his work was not in fact derived from actual sight and talk, how much of Buxtorf's own work rested on his direct inquiries?

Buxtorf knew and spoke with Jews, some of whom, as we have seen, can be identified. He also bought books from other, unnamed Jews – and complained about still others who refused to sell. Tossanus, in other words, was not wrong to state that Buxtorf conversed with Jews – or that, as we have seen, he asked them about rituals and customs. Buxtorf drew on what he learned from them in the *Synagogue*. How can we assess the nature of these conversations? One way to do so is to follow the hint given by Yaacov Deutsch in the first chapter of his important book, *Judaism in Christian Eyes,* and deprovincialize the study of ethnographies of Judaism. The scholars who have brought these texts out of obscurity in recent years have concentrated for the most part on comparing different texts about the Jews, rather than comparing these with descriptions of other peoples. To juxtapose Buxtorf's reports of his conversations with informants with those of other learned travelers is to become uncomfortably conscious of his limitations.

20. Thomas Platter, *Beschreibung der Reisen durch Frankreich, Spanien, England und die Niederlande 1595–1600*, ed. Rut Keiser (Basel, 1968); *Journal of a Younger Brother: The Life of Thomas Platter as a Medical Student in Montpellier at the Close of the Sixteenth Century*, trans. Sean Jennett (London, 1963).

21. Burnett, *From Christian Humanism to Jewish Studies*, 85.

22. M.A. Katritzky, *Healing, Performance and Ceremony in the Writings of Three Early Modern Physicians: Hippolytus Guarinonius and the Brothers Felix and Thomas Platter* (Farnham, 2012), 75–94, 304–13.

23. Taylor, "Making Statesmen, Writing Culture," 287–92. See also the classic study of Esmond Samuel de Beer in his introduction to *The Diary of John Evelyn* (Oxford, 1955), 1: 1–131.

One traveler whose work everyone knew – Ogier Ghiselin de Busbecq, imperial ambassador to the Ottoman Empire – retailed many things that Ottomans told him in his four chatty, vivid letters about his embassy, which were published in 1581. Busbecq did not pretend that he liked or respected most Turks. At least by origin, his informants were Europeans, with whom he shared languages. Of Ali Pasha, whom he called a friend, he also wrote that "a Dalmatian by birth, [he] is a delightfully intelligent person and (what is surprising in a Turk) by no means lacking in humanity."[24] Like Buxtorf, he sometimes reported Turkish customs and explanations for them in a way that highlighted his own superior attitude. Yet Busbecq also showed, in the body of his text, that he had pursued substantive discussions with his Turkish interlocutors. His account may well, like that of Herodotus, represent a literary performance of ethnographic inquiry rather than an actual one. But Buxtorf and others would have had no way to know that, and probably took it as factual – as others would take Buxtorf's account of the Jews.

Busbecq claimed that he deliberately built his relationships with Turks in order to obtain their explanations of particular customs. Wondering why he saw so many pieces of paper thrust into holes in walls, he asked his acquaintances:

> The Turks made no reply and refused to tell me the reason, either because they were ashamed to tell me a thing which I was unlikely to believe, or else were unwilling to reveal such a mystery to a stranger in religion. Afterwards, when I grew more familiar with the Turks, I learnt from my friends that the Turks have a great respect for paper, because the name of God may be written upon it. . . . With this practice you will probably find no fault.[25]

At least as Busbecq described his inquiries in print, in other words, they reflected multiple conversations with Turkish informants, with whom he claimed relatively close relationships, and for whose customs and intelligence he showed more respect than his occasional derogatory comments might suggest – though it also seems likely that he had most of his conversations with Rumis, Ottoman subjects of European origins like Ali Pasha. Martin Crusius, as we have seen, engaged in much more elaborate conversations with his Greek informants. He recorded these in great detail, drawing on contemporary documents and reports from a

24. Ogier Ghiselin de Busbecq, *The Turkish Letters of Ogier Ghiselin de Busbecq, Imperial Ambassador at Constantinople, 1554–1562*, trans. Edward Seymour Forster (Oxford, 1927; repr., Baton Rouge, 2005), 63; for the original, see Busbecq, *Legationis Turcicae epistolae quatuor* (Paris, 1595), fol. 41v.

25. Busbecq, *Turkish Letters*, 27; *Legationis Turcicae epistolae*, fol. 19r. On Busbecq's cultural ambiguity, his "mixture of self-congratulation and self-scrutiny," see Kaya Şahin, "Political Pragmatism, Humanist Ideals, and Early Modern Orientalism in Busbecq's *Turkish Letters*," in *The Dialectics of Orientalism in Early Modern Europe*, ed. Marcus Keller and Javier Irigoyen-García (London: Palgrave Macmillan, 2018), 33–49; quotation at 43.

Christian friend who accompanied an imperial embassy to Istanbul as well as on what individual Greeks told him.

Even when Busbecq's and Buxtorf's accounts looked similar, their ways of gaining information were strikingly different. Writers of travel literature loved to include details of the languages they encountered – sometimes fantastic (as in Mandeville and More), often genuine. In chapter 10 of the *Jüden Schul*, which deals with the Sabbath, Buxtorf tells the story of three Jews who argued about whether to remain where they were, in a dangerous wood, in order to keep the Sabbath, or to go onward.[26] Here, as elsewhere in the chapter, he took the opportunity to display a bit of Yiddish to his readers. The Jews who want to keep going speak:

> der Weg ist sehr *mesukkan* mit *gazlanim* (das ist, sehr gefehrlich der Räuber halben)... es ist viel besser / dass wir fort gehen / vnnd vnser *chajus matzil* seyn / das ist / vnser Leben erretten (The road is very dangerous with robbers... it is much better for us to depart and save our lives).[27]

Busbecq famously offered his readers samples of Crimean Gothic, which he transcribed from a Greek who had learned to speak the language.[28] Buxtorf's samples of Yiddish reflected his desire to interest Christians in learning Yiddish, but they made up very a small part of a long chapter designed to ridicule Jewish beliefs and practices. Busbecq presented his materials in a set-piece section of his fourth Turkish letter, listing a number of words with definitions and giving part of a song. Above all, he actually learned from a live witness, and the material he collected remains of vital importance for the history of the Gothic language.[29] Buxtorf took his story from the Yiddish story collection, the *Mayse bukh*, that had appeared in Basel in 1602, as his son pointed out in his 1641 Latin edition of the *Synagogue*.[30]

The contrast between these exemplary figures and Buxtorf is clear. Busbecq and Crusius both showed in their publications that inquiry about customs and their explanation could not be a simple, rapid process, and Crusius at least left enough details in his notes to demonstrate that he was telling the truth as he saw it. Busbecq described his informants and evoked multiple conversations; Crusius liked and admired many of his Greek interlocutors. Yet the resemblance

26. Buxtorf calls this account a true story, although the *Mayse bukh* does not claim that, and he slightly changes and omits part of its conclusion.

27. Buxtorf, *Jüden Schul*, chap. 10, 342.

28. Busbecq, *Turkish Letters*, 201–4; *Epistolae*, fols. 135r–137r.

29. D. Gary Miller, *The Oxford Gothic "Grammar"* (Oxford, 2019), 4.

30. Buxtorf, *Synagoga Judaica*, ed. Johann Buxtorf the Younger (Basel 1641), 246 (printed marginal note). The story was number 138 in the 1602 edition of the *Mayse bukh*. For a facsimile and French translation of the original text, see *Un beau livre d'histoires = Eyn schön mayse bukh*, ed. and trans. Astrid Starck (Basel, 2004), 1: 312–15.

between Buxtorf's work and more standard travel accounts remained. It seems clear that Buxtorf saw his claims to have worked with witnesses as giving his work greater credibility, and that learned and less learned contemporaries read him as he wished to be read.

Entering the Book

How then should we approach Buxtorf's German *Synagogue of the Jews*? First, we should be aware of how much artifice and deliberation Buxtorf dedicated to compiling and composing the work. The outer trappings of the book, its paratexts, form a kind of "inclusio": as invitation into the book the reader is offered a small section from Luther's tract on the Jews and for dessert a debate between a Jew and Christian, attributed to Calvin and first printed in Beza's edition of Calvin's correspondence.[31] Buxtorf translated the dialogue from Latin into German fairly faithfully, but made one small but significant change. In the Latin text the debate is set between Calvin and a Jew (Calv. and Jud.). Buxtorf, however, presents the dialogue as a debate between a *Christian* and a Jew.[32] A Reformed German living in Reformed Basel, Buxtorf still esteemed Luther and had no scruples in opening his work with the words of Luther. Purely commercial considerations may have also encouraged Buxtorf or his publisher, Sebastian Heinrich Petri, to remove the name of Calvin from the concluding debate. Within the German-speaking world, Lutheran readers (and purchasers) were numerous. And while they would have found a posthumous endorsement by the Wittenberg reformer attractive, they might have been put off by any mention of the Genevan Calvin.[33] Whatever the case, the strident anti-Jewish pronouncements of these two heroes of the Reformation set their stamp on the book. Any reader who opens the book or simply scrolls to the end gets the message – the Jews distort Scripture and are doomed to perdition.

31. Martin Luther, *Von den Juden und ihren Lügen* (Wittemberg, 1543), sig. P iiiir–v = Buxtorf, *Jüden Schul*, sig. [ ):(viiir–v] (see Burnett, *From Christian Hebraism to Jewish Studies*, 79); John Calvin, "Ad quaestiones et obiecta Iudaei cuiusdam," in Calvin, *Epistolae et Responsa* (Geneva, 1575), 306–14 = Buxtorf, *Jüden Schul*, 665–730.

32. See Mark Cohen, "Leone da Modena's *Riti*: A Seventeenth-Century Plea for Social Toleration of Jews," *Jewish Social Studies*, 34 (1972): 287–321, at 294.

33. See Burnett, *From Christian Hebraism to Jewish Studies*, 79. We are also grateful to Howard Hotson for his advice on this interpretation. See, too, Howard Hotson, "Irenicism in the Confessional Age: The Holy Roman Empire, 1563–1648," in *Conciliation and Confession: Struggling for Unity in the Age of Reform, 1415–1648*, ed. Howard Louthan and Randall Zachman (Notre Dame, IN, 2004), 228–85, and, for the commercial perspective, Ian Maclean, *Scholarship, Commerce, Religion: The Learned Book in the Age of Confessions, 1560–1630* (Cambridge, MA, 2012).

Each chapter, whether long or short, follows a certain pattern, ending with a sting in the tail. This is expressed either by means of apposite verses from the Scriptures or by a clever play of words, which not too subtly subverts the graphic details contained in the description of Jewish life. Even the first chapter, which is devoted to a damning report on Jewish theology, adheres to the same pattern. Buxtorf gathered much of the preparatory material particularly for this first chapter of the *Jüden Schul* in his large copybook and in the margins of his working copies of a wide range of books. From these materials, he constructed an entire vision of rabbinic Judaism in this opening chapter. A prominent place was given at the beginning of the chapter to an exposition of the Maimonidean thirteen principles of the Jewish faith. This is a conspicuous sign, as Burnett already suggested, that we are being introduced to a Judaism in a confessionalized garb. After all, Buxtorf was living in the age of the Christian catechism.[34]

Buxtorf listed the thirteen principles (*ikkarim*), in the form of the *ani ma'amin* (I believe) poem from the prayer book, which, he claimed, had been orally transmitted by the Jews from time immemorial. He then translated Maimonides' commentary on his own set of principles. Although Buxtorf had access to a well-stocked library of Judaica, as we will see, it did not contain Maimonides' Mishnah commentary, the original context of his exposition of the thirteen principles.[35] Instead, the paratexts of Bomberg's first Rabbinic Bible provided Buxtorf with both Maimonides' commentary on his own set of *ikkarim* and a list of the 613 commandments in Abraham ibn Hassan's version.[36] The notion that

34. See Burnett, *From Christian Hebraism to Jewish Studies*, chap. 1 and *passim*, for the role of Buxtorf's confession – as well as that of the wider Republic of Letters – in his scholarly career. On the catechisms of the period and their use see Lee Palmer Wandel, *Reading Catechisms, Teaching Religion* (Leiden, 2016).

35. Later, a 1606 edition of the Mishnah that contained Maimonides' commentary belonged to the younger Buxtorf; it may have been acquired by the elder but shows no signs of use by him (Basel, UB, FA III 15).

36. Both Maimonides' list and explanation of the thirteen principles and the list of 613 commandments are printed at the end of the second volume of the Rabbinic Bible. For a discussion of these Latin formulations of the 613 commandments, see Siegfried Stein, "Philippus Fernandus Polonus, A Sixteenth-Century Hebraist in England," in *Essays in Honour of the Very Rev. J.H. Hertz, Chief Rabbi of the United Hebrew Congregations of the British Empire, on the Occasion of his Seventieth Birthday, September 25, 1942*, ed. I. Epstein, E. Levine, and C. Roth (London, 1942), 397–412, esp. 399–408. It is also clear that Buxtorf used Sebastian Münster's bilingual Hebrew and Latin edition of an abbreviated version of the thirteenth-century Moses of Coucy's *Sefer Mitsvot gadol* (Semag), a compilation of the traditional 613 positive and negative commandments (Basel: Henricus Petri, 1533). His copy (Basel, UB, FA VIII 41a) is dotted with marginal notes on several entries. These indicate his careful perusal of the work in preparation for his first discourse (i.e., chap. 1 of the *Jüden Schul*) on the nature of rabbinic Judaism.

"Taryag Mizvot" (the 613 commandments) was an all-embracing term for rabbinic Judaism or Oral Torah formed a key element in Buxtorf's discourse, as it did in the works of Anthonius Margaritha and Philippus Ferdinandus Polonus. In other words, the two basic elements that underscore Buxtorf's discussion of Judaism were those given prominence in the Rabbinic Bible, a product supervised and printed by a Christian and former Jew. Evidently Judaism had to be treated in terms of dogma and creed.

Buxtorf's discussion of Maimonides' thirteen principles, which he interpreted as a sustained attack on Christianity, then seamlessly merged into his description of the rabbinization of Scripture, as represented by the 248 positive and 365 negative commandments. Buxtorf supported this vision with some carefully chosen quotations from various parts of the Jewish canon and then concluded with a denunciation of the rabbinic sacralization of the Talmud. This, the longest chapter in the book,[37] ends with a carefully chosen set of verses from Jeremiah 8:5–9. "How," the prophet asks, "can you say, 'We are wise, for we have the law of the LORD,' when actually the lying pen of the scribes has handled it falsely. The sages are shamed, they are terrified and caught and have rejected the word of the Lord – what wisdom do they have?" And so the learned homily ends, mirroring the chapter's opening lines taken from Isaiah (29:13–14):

> Because this people approaches me with its mouth and honors me with its lips, but its heart is far from me; and fears me in accordance with commandments which they have learned and done, so shall I also deal wonderfully with this people in the most wonderful and unusual fashion, so that the wisdom of its wise men will degenerate and the understanding of its clever men will become blinded.[38]

This denunciation of Jewish beliefs continues into the beginning of the second chapter. Here Buxtorf set out to describe Jewish customs of childbirth and circumcision. The transition into the narrative section was cleverly achieved by means of a literary jibe. Set in italic type, the Latin line "Nunc lectum admissi risum teneatis Amici" (Having been given this to read could you hold back your laughter, friends), stands out on the German page. Buxtorf's humanistically educated readers would immediately have recognized his clever adaptation, or rather distortion, of a famous line of Horace's *Ars poetica*.[39] This comes from the passage, familiar to every educated Christian, where Horace asks his addressees to imagine

37. It is 93 pages long. The remaining thirty-five chapters take up 665 pages.

38. Buxtorf, *Jüden Schul*, chap. 1, 1; trans. Corré.

39. Horace *Ars poetica* 5. A similar use of this line of the *Ars poetica* is made by Lorenzo of Brindisi in his polemic against the Lutheran Polycarp Leyser, *Hypotyposis Polycarpi Laiseri*, III.ii.4; see Lorenzo of Brindisi, *Opera omnia* (Padua, 1928–56), 2: 162.

a painter who wants to "join a horse's neck to a human head, and then clothe an assemblage of limbs with various feathers, so that what started out on top as a beautiful woman ended below in a vile ugly fish." "If you had been admitted to the spectacle," the poet asks, "could you hold back your laughter, friends?" ("spectatum admissi risum teneatis, amici?"). At this stage in the argument Horace appears to be pointing out that such a grotesque hybrid would prompt derisive laughter.

Here Buxtorf adroitly changed the "spectatum" to "lectum," replacing the act of viewing with reading, and suggesting that the readers were being admitted to the Jewish bedroom (since *lectus* also means bed in Latin). With this resounding allusion to an unreal, distorted world, whether natural or invented, his readers were not left in doubt that they were in for a good laugh at the expense of the Jews. Indeed, Buxtorf highlighted the theatrical nature of his subject on several occasions. In this chapter on circumcision, for example, he described the proceedings in the synagogue and then writes: "And so the first Act ends in the Synagogue. Now follows what you will take as a Comedy when they bring the child home."[40]

No reader of the book saw its character more clearly than a widely read seventeenth-century Jew named Shemaiah, a member of the distinguished Segal Ish Horovitz family. He described his literary projects in a long, elaborate, and somewhat fantastic letter to Buxtorf's son – who, he knew, had edited an edition of the *Jüden Schul* (it is not clear whether he knew that Buxtorf's father had composed the bulk of the text):

> All nations say that you know [literally his honor knows] both the revealed and esoteric. Thus, being of Jacob's root, I, Shemaiah, have also come, to listen to your [his] wisdom and to get to know your [his] greatness in which your [his] humility resides. But I have a small question to ask of you [him] and you [he] should give me a response. What is the purpose and why is it, your [his] honor, that you [he] thought to explain the laws of the Jews to the nations? I noticed that when you [he] entered the domain of Torah with your [his] studies in your [his] *Jüden Schul* you [he] spoke perversely about them [i.e., the Jews]. For even by your [his] words you have [he has] confirmed your [his] power over them. So now I would judge you [him] in accordance with the saying of the sages: It is permitted for a scholar in a place where he is not known to make himself known and claim that he is a worshipper of fire.[41] Does this not explain the praises that the sages and rabbis of Frankfurt and

40. Buxtorf, *Jüden Schul*, chap. 2, 123.

41. This is a conflation of two passages in BT, Nedarim 62a and 62b. "A man may reveal his identity when he is unknown"; "Raba said: A rabbinical scholar may declare, 'I will not pay toll tax. [reference to Ezra 7:24]... Raba also said: A rabbinical scholar may assert, 'I

> their Yeshivah declared at the time of writing the document inserted in the copy that I am sending to you [him] and setting before your [his] eyes?[42] When you see [he sees] it your [his] eyelids will move and your [his] eyes will confront the truth as to who you are and what am I and what I say and how I justify myself, and furthermore, that the Lord has spoken.

In a Hebrew dense with Talmudic citations, Shemaiah made clear that even the revised form of the book, to which Johann the Younger added many quotations of original texts, supported by precise citations, painted a deformed and deceptive portrait of Jewish law and customs.[43]

In his masterpiece, *Legends of the Jews,* Louis Ginzberg assembled a beautiful mosaic of rabbinic Aggadot drawn from multiple sources.[44] The uninformed reader who does not consult the rich and indispensable annotations in volumes 5 and 6 or who is not learned in the complexities of the midrashim and Aggadot will come away with only a vague impression of the rabbinic writings from which Ginzberg gathered his material. For all the differences between the two works, Buxtorf's *Jüden Schul* can leave a similarly distorted impression. A constant appeal to the Jews' "chachamim und Weisen" suggests that they hardly ever engaged in divergent practices or thoughts. As a fluent reader of Hebrew and Yiddish, Buxtorf could dip into Hebrew writings of all periods. As a passionate reader and professional censor of the Hebrew literature, he had privileged access to Jewish religion and thought. Yet what he finally produced was most certainly not a comprehensive overview of Jewish customs.

This is clear from Buxtorf's retrospective account of the way he worked on the book, which he communicated to Matthias Martinius:

> You ask how I managed to collect those Jewish myths. I reply that the Jews themselves furnished them for me. They did so in the first place by providing me with their book of rituals, which is titled *Minhagim,* and is published in Hebrew-German. If you want it, I will send it to you. Then I learned that their whole law, civil, canon, and the rest, can be found stated in *Dinim* [laws], or aphorisms, so to speak, in the book titled *Shulhan arukh.* The Jews themselves also provided this. Later I obtained the Talmud and the further works that I

---

am a servant of fire, and will not pay poll tax." In other words, he was claiming that he was a worshipper of fire, a Zoroastrian priest, and would therefore be exempted from paying the tax.

42. In other words, he seems to be insinuating that Buxtorf is somewhat hypocritical in the *Jüden Schul,* an anti-Jewish work that does not properly represent his true views or behavior toward the Jews. We do not know what was in this document that Shemaiah copied.

43. Basel, UB, A XII 19, no. 28. See Appendix 2.2, below, for a full translation.

44. Lewis Ginzberg, *Legends of the Jews,* trans. Henrietta Szold (Philadelphia, 1909–38).

> cite and explored these for amusement. Whatever I found, I referred to the *Minhagim.*[45]

Buxtorf may have drawn from earlier texts not only the details that he evoked here but the larger structure of his book. A Jew who converted to Christianity and made his living as a Hebrew teacher, Anthonius Margaritha, published a polemical account of Judaism, *Der gantz Jüdisch glaub,* in 1530 and revised it in 1531.[46] Like Buxtorf, and long before, he divided the descriptive part of his work into topical chapters. These dealt, like their counterparts in the *Jüden Schul,* with the religious calendar, important moments in the life cycle of an observant Jew, and customs such as Jewish ways of preparing and eating kosher meat. The two men's order of presentation, interests, and purposes differed sharply, but Margaritha's volume – which had a considerable success in the book market – could have served as one of the lasts to which Buxtorf fitted his often-different material.[47]

Other letters to fellow scholars, and indeed the many extant vestiges of Buxtorf's reading, indicate that he consulted a wider variety of books than he admitted – but also that he was honest when he explained that he compiled his work from other texts. The margins of the *Synagogue* indicate that he went far

45. Buxtorf to Martinius, 4 September 1606, in *Bibliotheca Historico-philologico-theologica,* 4, no. 3 (1721): 598–603, at 602: quoted in Burnett, *From Christian Hebraism to Jewish Studies,* 64 n34. Buxtorf became curious about Jewish books of customs quite early in his career. In an August 1593 letter to Zwinger, he refers explicitly to the 1593 Venetian edition of Simeon Levi Guenzburg's adaptation of Isaak Tyrnau's *Sefer minhagim,* a very popular book of customs that contained references to the usages of the communities of Speyer, Worms, and Mainz, and that had just been brought out by di Gara (Basel, UB, Fr Gr II 9, 89r). On the letters to Zwinger and other scholars about his use of sources, see Grafton and Weinberg, *"I have always loved the Holy Tongue,"* 143.

46. Anthonius Margaritha, *Der gantz Jüdisch glaub* (Augsburg, 1530). In addition to Deutsch, *Judaism in Christian Eyes,* see also Maria Diemling, "Anthonius Margaritha and His 'Der Gantz Jüdisch Glaub,'" in *Jews, Judaism and the Reformation in Sixteenth-Century Germany,* ed. Dean Philip Bell and Stephen Burnett (Leiden, 2006), 303–33; Michael Walton, "Anthonius Margaritha – Honest Reporter?," *Sixteenth Century Journal* 36 (2005): 129–41; Walton, *Anthonius Margaritha and the Jewish Faith: Jewish Life and Conversion in Sixteenth-Century Germany* (Detroit, 2012); Burnett, "Luther's Chief Witness: Anthonius Margarita's *Der gantz jüdisch glaub* (1530/1531)," in *Revealing the Secrets of the Jews: Johannes Pfefferkorn and Christian Writings about Jewish Life and Literature in Early Modern Europe,* ed. Jonathan Adams and Cordelia Heß (Berlin, 2017), 183–200; and the revisionist account by Daniel Lehmann, "'Such an Illumination Cannot Occur': Anthonius Margaritha, the Reformation, and the Polemic against the Jews," *Archiv für Reformationsgeschichte – Archive for Reformation History* 111 (2020): 55–77.

47. See Stephen Burnett, "Distorted Mirrors: Antonius Margaritha, Johann Buxtorf and Christian Ethnographies of the Jews," *Sixteenth Century Journal* 25 (1994): 275–87, and *From Christian Hebraism to Jewish Studies,* chap. 3.

beyond the Yiddish *Minhagim*, the Talmud, and the *Shulhan arukh*. But the letter to Martinius perhaps goes some way to explain the fragmentary or haphazard exposition that we encounter in the book. Buxtorf emphasized certain ceremonies and rituals, while others were totally ignored. Although he sometimes added a chronological note with the date of his source and of its publication, he showed no awareness of the genre or historical context of the texts that he used. Describing the *Minhagim* in his bibliography, Buxtorf pointed out that several authors had written them and that the Venice 1601 edition, which he owned, had illustrations.[48] But he showed no interest in the differences between accounts of customs or visual representations of them in his copybook or in the *Jüden Schul*. This piecemeal form of compilation contrasts with his evident investment in the rhetorical form with which he organized his material.

The Jews certainly provided him with a good up-to-date Ashkenazic syllabus. It is striking that apart from the popular Yiddish *Minhagim*, the moralistic *Brantspigl* and *Frauenbuchlein*, and other texts such as the Yiddish *Shevet Yehudah*, Buxtorf's choice of texts resembled the set of moralistic and pietistic books used and cited by Yuspa Hahn of Nőrdlingen (1570–1637) in his fascinating record of Jewish life and practice in seventeenth-century Frankfurt, the *Sefer Yosif omets*.[49] The inclusion not only of *Sefer hasidim* but also the abbreviated *Reshit hokhmah* of Elijah de Vidas seems surprising[50] – until one sees that Yuspa Hahn also had a copy of a different but also abbreviated form of this important kabbalistic work.[51] Like Hahn, Buxtorf did not cite it in order to communicate kabbalistic ideas. For Buxtorf, the text served as a source of Midrashic extracts, which he could have also procured by

48. Buxtorf, *Bibliotheca Rabbinica*, in his *De abbreviaturis hebraicis liber novus & copiosus* (Basel, 1613), 257–335, at 304: "*Minhagim*, Ritus. Plures sunt, qui Minhagim scripserunt. Sunt autem libri, in quibus ritus describuntur, quibus hodie in Synagogis tempore precum & festorum utuntur. Breviaria quidam vocant. Liber Minhagim excusus Venetijs Hebraeogermanice cum figuris, in 40, et Hebraice Cracoviae." Part of this description is written out in Buxtorf's copy of the 1601 Venice edition, Basel, UB, FA V 47, verso of blank facing title page, probably by his son.

49. Hahn completed this book in 1630. The first edition was printed in Frankfurt am Main by Johann Kellner in 1723.

50. Jacob Poggetto, *Reshit hokhmah ha-katsar* (Basel, 1603). Buxtorf's signed copy is Basel, UB, FA VIII 28:1, available in digital form on e-rara, www.e-rara.ch.

51. On Hahn, see Chava Fränkel Goldschmidt, "Jüdische Religion und Kultur in Franfurt am Main im 16 und 17 Jahrhundert: Yuspa Hahn und sein 'Yosif omez,'" in *Jüdische Kultur in Frankfurt am Main von den Anfängen bis zur Gegenwart*, ed. Karl E. Grőzinger (Wiesbaden, 1997), 101–21; Dean Phillip Bell, *Jewish Identity in Early Modern Germany. Memory, Power and Community* (Ashgate, 2007), 48–56. On the use of kabbalistic sources in seventeenth-century books of Jewish customs, see Jay R. Berkovitz, "Crisis and Authority in Early Modern Ashkenaz," *Jewish History* 26 (2012): 179–99.

reading the classical Midrashim.[52] Jews, in other words, supplied Buxtorf with their own kind of reading matter and he set about to read the texts – without worrying about, for example, whether he was obliged to seek out the original sources that his intermediaries cited. In that sense, his working methods resembled those of Jewish learning more closely than they did the best practices of humanistic philology – though plenty of Christians drew freely on secondary sources and compilations.

Comparison between Buxtorf's notebook and the *Jüden Schul* can help us imagine how he worked back and forth between his notes and other texts that he consulted without transcribing full extracts. In one note he explains the meaning of the *tzitzit* (the fringes worn by male Jews):

> The rule for the *Tzitzit* also includes a device for recalling the 613 precepts. For the word *Tzitzit* amounts to 600 by gematria. When the 8 threads and the 5 knots that hang down are added in, they make up the higher number. *Kad K[H]akke.*, page. 63. cols. 3 and 4.[53] This rule also refers to the resurrection of the human body: and therefore you will find that it includes the whole body. For the *Tzitzit* is threads that hang down like hair, as it is written: "And it took me by the *Tzitzit*, by a lock of my head" (Ezek. 8:3). The holes through which the threads are passed resemble the eyes. The five knots are like the five senses in man. The eight threads indicate the eighth day, on which males are circumcised. At the edges you will also find five knots, which, on four sides, make twenty knots, which refer to the twenty digits of the hands and feet.[54]

In the manuscript, Buxtorf's son has added a cross-reference to chapter 4 of the *Jüden Schul*. A look at that chapter reveals that Buxtorf did use some of this information, though he added to it:

52. E.g., Buxtorf, *Jüden Schul*, chap. 4, 164: "The *author* of the book *Reschis chochmah* writes: 'Page 111. He heard from his elders that it is good to rub tears over your forehead, because some sins might be written on it, and they will be blotted out in this way, as the prophet says: "Draw a sign on the people's forehead."'" Buxtorf is referring to Poggetto, *Reshit hokhmah ha-katsar*, 111a.

53. Buxtorf is referring to Bahya ben Asher's *Kad Ha-kemah* (Venice, 1545), 63, cols. 3, 4.

54. Buxtorf, copybook, Basel, UB, MS A XII 20, 251: "*de zizit*. Praeceptum *zizis* complectitur in se memoriale 613 praeceptorum. Nam vox ipsa ציצית ratione Gematrica facit 600: quibus additis octo filis et quinque nodis dependentibus in ipsis Zisis superiorem numerum exprimunt. Kad Kakke. pag. 63. col. 3. et col. 4. Significat etiam hoc praeceptum Resuscitationem corporis humani: idcirco reperies quod in se complectitur ipsum corpus. Nam ipsae zizis sunt fila dependentia instar capillorum juxta illud: Et accepit me Bezizis, cincinnis capitis mei, Ezech. 8. 3. Foramina per quae dimittuntur fila similitudinem habent oculorum: Quinque nodi similes sunt quinque sensibus in homine: Octo fila indicant diem octavum quo circumciditur homo: In singulis extremitatibus reperies etiam quinque nodos, qui in quatuor oris simul faciunt viginti nodos, quibus indicantur viginti digiti manuum et pedum. etc."

> The *Chachamim* compare the commandment of the *Zizis* to the whole sum of the commandments of the Law and hold it just as high. They write: Whoever keeps this commandment is as if he kept all the commandments. How is that? First, there are five knots on each tassel, which are compared to the five books of Moses. Add eight threads, and this makes thirteen. Now the Hebrew word *Zizis* has the numerical value of six hundred; added up this makes six hundred and thirteen, the same number as there are commandments of God. Ergo, whoever keeps the laws of the *Zizis* correctly has fully observed the six hundred and thirteen commandments. The Rabbis find this proven according to good kabbalistic arithmetic.[55]

But he also provided a much more vivid description of the garment, and a characteristically malign comment:

> They have a foursquare garment or cloth. Some put it on with their other clothes right after they get up, some later when they are getting ready to pray. This little garment consists of two square pieces held together with two ribbons at the top, or sewn together, with an opening large enough for their head; it is made of cloth, taffeta or silk, and both square pieces are therefore hanging from their shoulders covering back and chest. Because of the squares it is called *Arba Canphos* [four corners]. On each corner is a long tassel made of eight white wool threads hanging from a knot approximately four, eight or twelve fingers wide. Such tassels are called *Zizis*. . . . Whoever wants to be very pious and holy wears this little garment every day under the long overcoat in a way that the tassels hang out a little so that he can always see them and have them in sight. Some of them wear it under their coat, some put it on only when they pray. The tassels have to hang down in the front and back so they are surrounded by *Mitzvos* or Commandments (like a fool with bells).[56]

Noting that "there are strange and wonderful things written about it, a few of which I would like to point out," Buxtorf summarized long passages from the Talmud (*Shabbat* and *Baba Batra*), the *Yalkut* and *Sifrei* (in the last case, R. Judah ben Babba's explanation of why women are exempted from the requirement that they wear the *tzitzit*).[57]

In fact, the copybook is not our only guide to the sources for Buxtorf's "Jewish" universe. The Basel University Library possesses many books that he owned and annotated. On occasion they reveal how he read with the *Jüden Schul* in mind. One such book is the Yiddish translation of *Sefer ha-yirah* attributed to

55. Buxtorf, *Jüden Schul,* chap. 4, 178–79; trans. Corré.
56. Ibid., 176–77 (Corré's brackets).
57. Ibid., 176–85.

the thirteenth-century Jonah Gerondi (but now of disputed authorship).[58] This was printed in Zurich in 1546 and bound with the simple, anonymous metrical *Hayyei olam,* printed in Freiburg i. Breisgau. The complications of this prose Yiddish publication, its Hebrew original, and its relation to the rhymed *Hayyei olam* need not detain us here.[59] More important is that Buxtorf used both texts. There is a highly moralistic and prescriptive tone to the work, and his notes show him both learning to read the Yiddish text and picking out some of what he found exemplary in it.[60] He took a serious interest in the work's directives about physical comportment and correct bathroom habits, how to pray and what is correct *kavvanah* or concentration in prayer, and which of his physical parts a person may or may not touch during prayer. The pointing hand that appears at one page (8b; see Fig. 2.2) seems to indicate that Buxtorf felt a particular interest in these points.

In fact, chapter 5 of the *Jüden Schul,* which deals with conduct relating to morning prayers, contains some of the same material, combined with other bits and pieces. Not surprisingly, the eagle-eyed Buxtorf did not overlook the passage where the author advised his readers not to bestow praise on any *goi* in any circumstances. The statement that only in the gentile's presence should you give him the benefit of the doubt "for the sake of peace" would have surely been grist for his anti-Judaic mill.[61] The *N[ota] B[ene]* on another page that exhorts readers to put their heart and soul into their preparations for the Sabbath should not be ignored.[62] Buxtorf's annotations about the eve of Sabbath and nail cutting seem to have been a topic of constant fascination for him (see Fig. 2.3). It is this passage in its verse version (*Hayyei olam* 26–27/13a–b) that Buxtorf transcribed in full in Roman characters in chapter 10 of the *Synagogue,* along with his transcription of the Yiddish adaptation of one of the *Zemirot* sung on Friday night.[63] Coupling these verses with several miracle stories about the Sabbath in the Talmud, Buxtorf assured his readers that the poems are not meant to be taken as poetic fancies, but rather represent true Jewish beliefs and practices. But why did Buxtorf choose to go to all this trouble to transcribe the poems? We will return to this question later.

58. Buxtorf's copy is Basel, UB, FA VI 24: 1.

59. Buxtorf also annotated the *Hayyei olam* (Basel, UB, FA VI 24: 2). On the complicated history of *Sefer ha-yirah,* its Hebrew original, and its relation to the *Hayyei olam,* see Regula Tanner, "Sefer ha-Yira und Sefer Hayye Olam: Zwei jiddische Drucke aus Freiburg i. Br. von 1583," *Judaica* 68 (2012): 105–28.

60. It is noteworthy that Yuspa Hahn cited this work frequently in his *Yosif omets.*

61. *Sefer ha-yirah* (Basel, UB, FA VI 24:1, fol. 22v [*midarkhei shalom*]).

62. Ibid., fol. 25v.

63. The poem is given in Buxtorf's annotated copy of *Hayyei olam,* which is Basel, UB, VI 24:2, fols. 13a–b.

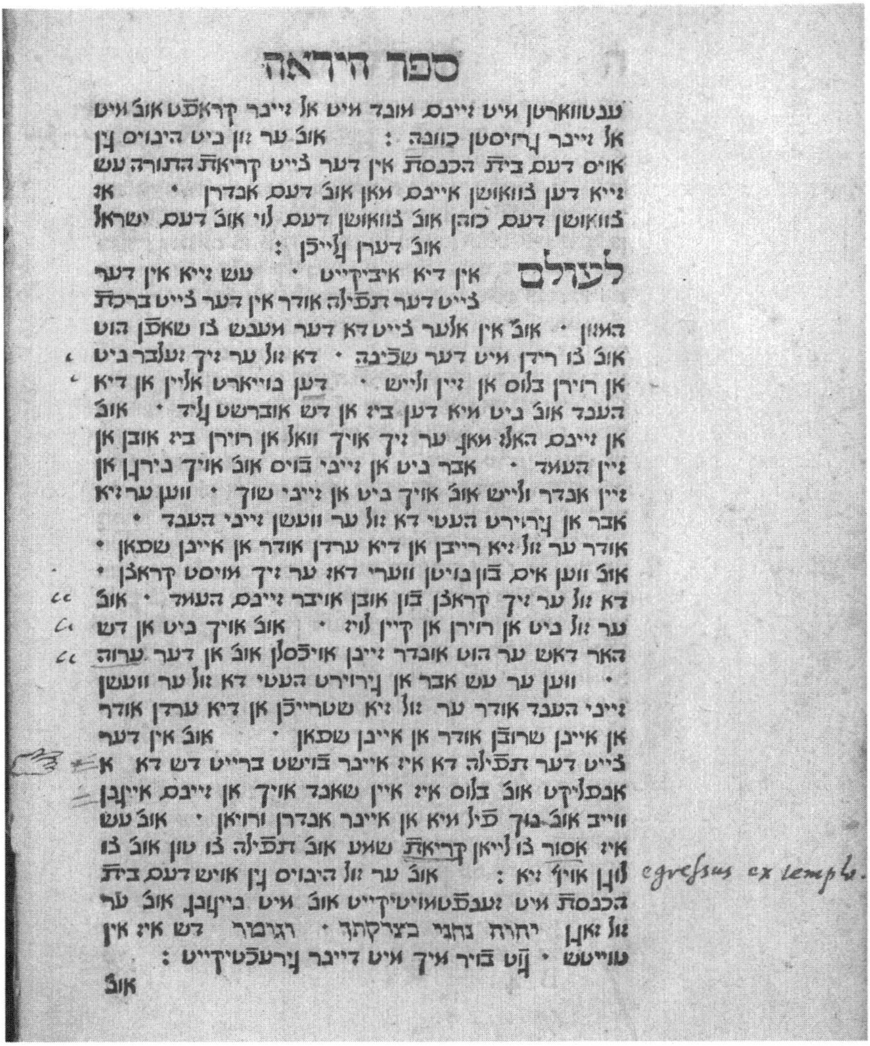
ספר הירׄאה

ענטווארטן מיט זיינם מונד מיט אל זיינר קראפֿט אוב׳ מיט
אל זיינר גרויסטן כוונה : אוב׳ ער זול ניט היגויס נין
אויס דעם בית הכנסת אין דער צייט קריאת התורה עש
זייא דען צווישן איינם מאן אוב׳ דעם אנדרן • אז
צווישן דעם כהן אוב׳ צווישן דעם לוי אוב׳ דעם ישראל
אוב׳ דערן גלייכן :
לעולם אין דיא איביקייט • עש זייא אין דער
צייט דער תפלה אודר אין דער צייט ברכת
המזון • אוב׳ אין אלער צייט דא דער מענש צו שאפֿן הוט
אוב׳ צו רידן מיט דער שכינה • דא זול ער זיך זעלבר ניט
אן רוירן בלוס אן זיין לייש • דען בוייארט אליין אן דיא
העמד אוב׳ ניט מיא דען ביז אן דש אוברשט גליד • אוב׳
אן זיינם האלז מאג ער זיך אויך וואל אן רוירן ביז אובן אן
זיין העמד • אבר ניט אן זייני פֿוס אוב׳ אויך ניט אן
זיין אנדר לייש אוב׳ אויך ניט אן זייני שוך • ווען ער זיא
אבר אן גירוירט העטי דא זול ער וועשן זייני הענד •
אודר ער זול זיא רייבן אן דיא ערדן אודר אן איינן שטאן •
אוב׳ ווען אים פֿון בויטן ווערי דאז ער זיך מויסט קראצן •
דא זול ער זיך קראצן פֿון אובן אויבר זיינם העמד • אוב׳
ער זול ניט אן רוירן אן קיין לויז • אוב׳ אויך ניט אן דש
האר דאש ער הוט אובדר זייבן אויכסלן אוב׳ אן דער ערוה
• ווען ער עש אבר אן גירוירט העטי דא זול ער וועשן
זייני הענד אודר ער זול זיא שטרייכן אן דיא ערדן אודר
אן איינן שרובן אודר אן איינן שטאן • אוב׳ אין דער
צייט דער תפלה דא איז איינר פֿוישט ברייט דש דא א
אנבליקט אוב׳ בלוס איז איין שאנד אויך אן זיינם איינגן
וייב אוב׳ גוך פֿיל מיא אן איינר אנדרן ורויאן • אוב׳ עש
איז אסור צו לייאן קריאת שמע אוב׳ תפלה צו טון אוב׳ צו
לונן אויף זיא : אוב׳ ער זול היגויס נין אויש דעם בית
הכנסת מיט זענפֿטמויטיקייט אוב׳ מיט בייגנן אוב׳ ער
זול זאגן יחזה נחזי בצדקתך • וגומר דש איז אין
טוייטש • גאט פֿויר איך מיט דיינר גירעכטיקייט :
אוב׳

Figure 2.2. *Sefer ha-yirah* (1546) with notes by Buxtorf.
Basel, Universitätsbibliothek, FA VI 24: 1.

Occasionally he misunderstood a passage. In one case, the text justifies the reading of the Aramaic paraphrase (Targum) together with the weekly reading from the Torah for those who are not schooled in Hebrew. Buxtorf seems to understand the text as stating that "the Bible must not be read in any language other than the holy one" – that is, Hebrew.[64] But earlier on the page the details

64. Buxtorf's note is in *Sefer ha-yirah* (Basel, UB, FA VI 24: 1), fol. 25r: "Biblia non legenda in alia lingua, quam sancta."

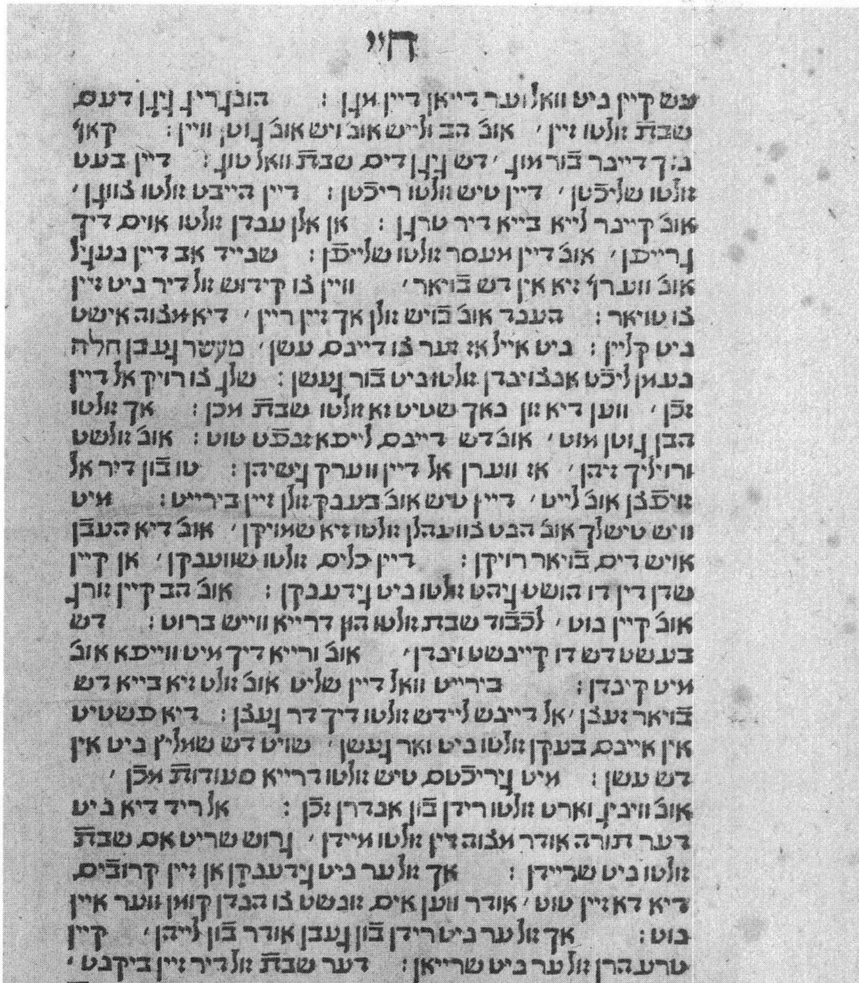

Figure 2.3. The source of Buxtorf's verses on nail cutting: *Hayyei olam*. Basel, Universitätsbibliothek, FA VI 24: 2.

of intercourse have also attracted his attention. The Yiddish translator seems a little prudish about the material. When he writes "and this suffices for the *mevin* [the one who understands]," Buxtorf gives the apposite Latin phrase "intelligenti satis" in the margin (see Fig. 2.4).

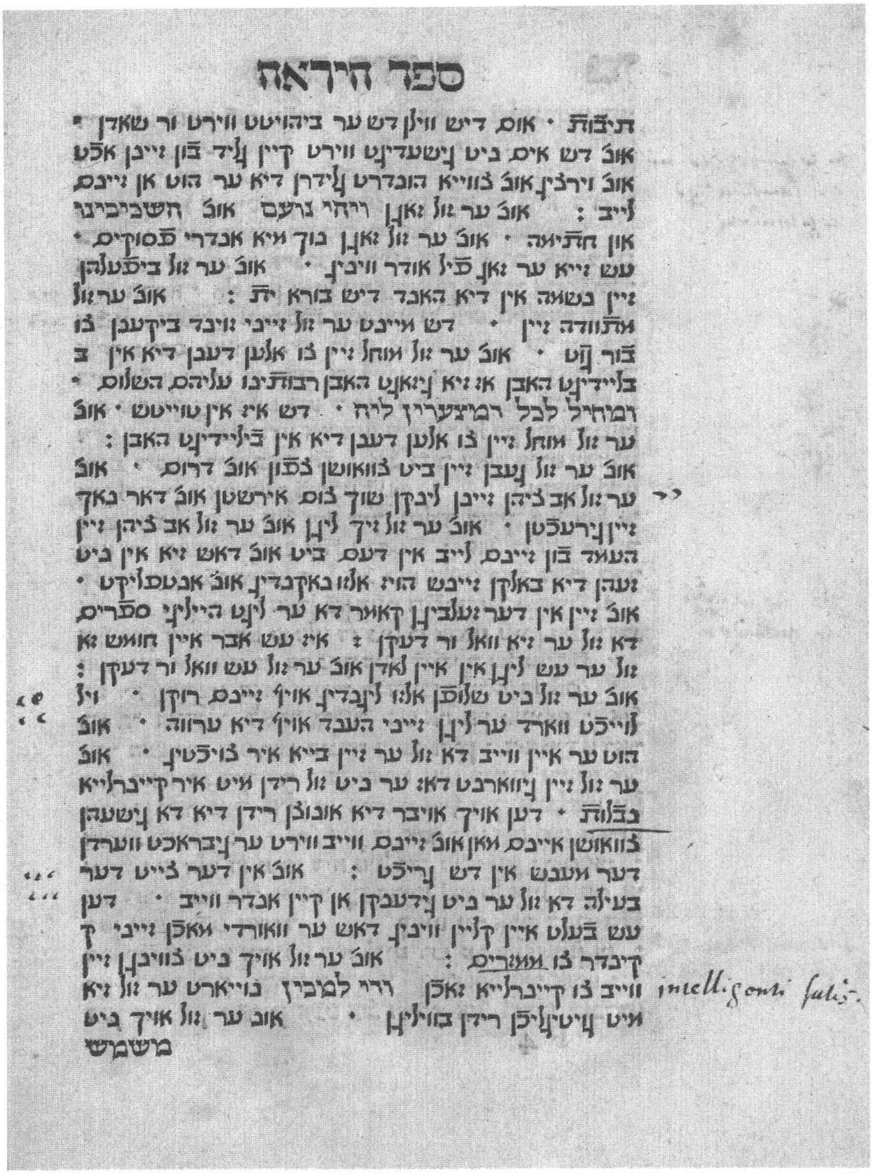
ספר הירא֗ה

תיבות · אום דיש ווילן דש ער ביהויטט ווירט ור שאדן ·
אונ׳ דש אים ניט גישעדיגט ווירט קיין גליד פון זיינן אלט
אונ׳ וירצין אונ׳ צווייא הונדרט גלידרן דיא ער הוט אן זיינם
לייב : אונ׳ ער זול זאגן ויהי נועם אונ׳ השכיבינו
אין חתימה · אונ׳ ער זול זאגן נוך מיא אנדרי פסוקים ·
עש זייא ער זאגן פיל אודר ווינגן · אונ׳ ער זול ביפעלהן
זיין נשמה אין דיא האנד דיש בורא ית׳ : אונ׳ ער זול
מתוודה זיין · דש מיינט ער זול זייני זינד ביקענן צו
צור גוט · אונ׳ ער זול מוחל זיין צו אלין דענן דיא אין צ
בליידיגט האבן אז זיא גיזאגט האבן רבותינו עליהם השלום ·
ומוחיל לכל המיצערין ליה · דש איז אין טוייטש · אונ׳
ער זול מוחל זיין צו אלין דענן דיא אין בליידיגט האבן :
אונ׳ ער זול נעבן זיין ביט צוואושן צפון אונ׳ דרום · אונ׳
ער זול אב ציהן זייבן לינקן שוך פום אירשטן אונ׳ דאר נאך
זיין גירעכטן · אונ׳ ער זול זיך לייגן אונ׳ ער זול אב ציהן זיין
העמד פון זיינם לייב אין דעם ביט אונ׳ דאש זיא אין ביט
זעהן דיא באלקן זיינש הויז אלזו נאקנדיג אונ׳ אנטפלויקט ·
אונ׳ זיין אין דער זעלבגן קאמר דא ער לינט היילגי ספרים
דא זול ער זיא וואל ור דעקן : איז עש אבר איין חומש זא
זול ער עש לינן אין לאדן אונ׳ ער זול עש וואל ור דעקן :
אונ׳ ער זול ניט שלופן אלזו ליגנדיג אויף זיינם רוקן · ויל
לייכט ווארד ער לינן זייני העבד אויף דיא ערווה · אונ׳
הוט ער איין ווייב דא זול ער זיין ביא איר בוליטין · אונ׳
ער זול זיין גיווארנט דאז ער ניט זול רידן מיט איר קיינר לייא
נבלות · דען אויך אויבר דיא אובזן רידן דיא דא גישעהן
צוואושן איינם מאן אונ׳ זיינם ווייב ווירט ער גיבראכט ווערדן
דער מענש אין דש גריכט : אונ׳ אין דער צייט דער
בעילה דא זול ער ניט גידענקן אן קיין אנדר ווייב · דען
עש פעלט איין קליין ווינג דאש ער וואורדי מאכן זייני ק
קינדר צו ממזרים : אונ׳ ער זול אויך ניט צווינגן זיין
ווייב צו קיינר לייא זאכן ודי למבין נוייארט ער זול זיא
מיט גיטיגליכן רידן צווינגן · אונ ער זול אויך ניט
משמש

Figure 2.4. *Sefer ha-yirah* (1546) with notes by Buxtorf.
Basel, Universitätsbibliothek, FA VI 24:1.

This note seems to be reflected in the end of chapter 8 of the *Synagogue*. Here Buxtorf writes:

> The Jews have also a very beautiful bedroom ethics, particularly how a man should behave toward his wife. It is discussed and disputed in a very subtle manner, in accordance with their customs, but I would not report it to cultivated ears. Whoever has the desire can read the book Orah Hayyim. Here I will let the Jews go to sleep, and I shall describe further their holiness and piety.[65]

Once again, the form of Buxtorf's portrayal reinforces the attitudes that inspired it, and that he meant to transmit. After all, as he must have known, humanists had regularly discussed the proper way to behave in the bedroom. Unlike the poets of ancient Rome, Johannes Secundus and his fellow Christians wrote epithalamia that took the bride and groom into the bedroom and gave the groom instructions: As Secundus wrote, "Touch now her thighs and now her waist / And now what is beneath it placed."[66] In the words of Leonard Forster, later neo-Latin poets treated the wedding "as a sexual experience."[67] It was something close to Tartuffery to reproach the Jews for doing the same in a different genre.[68]

From time to time, Buxtorf did his best to suggest that he had studied the material world of the Jews as an antiquarian would have studied that of ancient Greeks or Romans.[69] In his description of ritual slaughter, for example, Buxtorf argued that in order to become a ritual slaughterer, a Jew did not undergo a practical training but studied a manual over and over again. To prove this he

65. Buxtorf, *Jüden Schul*, chap. 8, 289. In the margins of this passage Buxtorf refers to the *Brandspiegel* (chap. 38) and *Reshit hokhmah* (*schaar hakkeduscha* c. 16), but there also seems to be an echo of this passage from *Sefer ha-yirah*, which exhorts readers to engage in similarly pietistic behavior. In his copy of *Hayyei olam*, Buxtorf enters a related note on 13r: "modeste cum uxore in lecto loquendum: et caste vivendum."

66. "Nunc per crura tenella perque ventrem / Et quae proxima sunt et huic et illis"; quoted, with the translation by F.A. Wright, in Leonard Forster, *The Icy Fire: Five Studies in European Petrarchism* (Cambridge, 1969), 111–12.

67. Ibid., 108.

68. Christian literature on conduct – e.g., Leon Battista Alberti's *On the Family* – also gave detailed advice on comportment in the bedroom.

69. On the connections between antiquarianism and travel literature, see John Howland Rowe, "Ethnography and Ethnology in the Sixteenth Century," *Kroeber Anthropological Society Papers* 30 (1964): 1–19; David Lupher, *Romans in a New World: Classical Models in Sixteenth-Century Spanish America* (Ann Arbor, 2003); Jonathan Sheehan, "Sacred and Profane: Idolatry, Antiquarianism and the Polemics of Distinction in the Seventeenth Century," *Past & Present* 192 (2006): 37–66; Guy Stroumsa, *A New Science: The Discovery of Religion in the Age of Reason* (Cambridge, MA, 2010); Kathryn Taylor, "Ancients and Moderns in Sixteenth-Century Ethnography," *History of European Ideas* 46 (2020): 113–30.

quoted what he described as a "Testimonium" that he had recently read, and which he translated into German: a certificate, signed by a rabbi, that granted a man permission to slaughter animals so long as he continued to study the textbook regularly, once a week for the first year, once a month for the second, and then once a year for the rest of his life.[70] Here, in the terms of early modern antiquarianism, was an exemplary "proof": a document, rather than a literary text, which supported Buxtorf's description of customs. In fact, though, Buxtorf did little if any ransacking of archives and notarial collections. In his copybook, he recorded that he had found the rabbinical certificate, copied out, "at the end of the book that describes the form and ritual of slaughter."[71] His careful deployment of the document in his own text masked the fact that he found it in the very book whose authority it confirmed.[72]

Questions of Style

The true key to Buxtorf's practice of description is more elusive – perhaps because he hid it in plain sight. Even when he did not claim to have witnessed a scene – and he usually did not – Buxtorf showed a remarkable gift for plain, unemphatic description, which evidently seemed to reflect observation. Consider, for example, what he had to say about a Jewish household to which a son has been born:

> If the woman recovers well and gives birth to a son, then there is great joy in the whole house, and the father will set out to look around for good geese, chicken, fish, meat and good drinks for the *Bris milah,* the circumcision feast. For God ordained that the male be circumcised on the eighth day, as everybody knows from the books of Moses. In the interval, the guests are invited to the circumcision-feast. At least a *Minjan,* that is, ten men, all of them over thirteen, must be invited.
>
> On the seventh night some of the invited guests come, and others sit with the mother, they have a good meal and stay up all night and do many

70. Buxtorf, *Jüden Schul,* chap. 27, 566–67.

71. Buxtorf, copybook, 324: "Testimonium, quo licentia Iudaeo data ad mactandum, scriptum in fine libri, in quo forma et ritus mactandi describitur." As transcribed in the copybook, the name of the person receiving the certificate is given as Todros, and that of the person granting it as Menachem ben Jacob. Buxtorf did not include these names in his German translation.

72. In fact, Buxtorf had unwittingly come across the widespread practice of writing a license to slaughter into the textbook that the student had to master. There are many copies of Jacob Weil's manual, the *Hilkhot shehitah u-vedikah,* containing handwritten rabbinic licenses for their pupils. See, e.g., Oxford, Bodleian Libraries Opp.add. 4°iii. 57 (Cracow, 1600).

> entertaining things, they play cards, throw dice, sing, and tell tales. The men drink themselves into a stupor, and they entertain and amuse the young mother so that she should not worry too much about the forthcoming circumcision. However, the learned and pious Jews add some devotions and prayers, and they warn the man who will perform the circumcision not to drink too much, so that his hands do not tremble when he must circumcise the child.[73]

The scene seems eminently plausible. Its details, from the Jewish father being sent out to do the shopping to the staging of a party to distract the mother, are quotidian and realistic. The narrative as a whole moves seamlessly, without interruption for accusations of superstition, through the eight days after the birth until the climactic night.

Yet vital details about how Buxtorf observed these practices are not clarified. Did Buxtorf in fact gain entrance to a Jewish home and observe the card playing and dicing? He does not explicitly set the scene. As soon as one wonders about its location, one realizes how unlikely it is that Buxtorf actually spent the eve of a circumcision with a Jewish family. And in fact, the evidence suggests that he did not – at least in the early years when he was preparing the *Synagogue*. As Stephen Burnett pointed out years ago, Buxtorf's vivid narrative distills a longer series of details and actions that he had read about rather than seen.[74] The convert Ernst F. Hess published his *Flagellum Iudaeorum* or *Juden-Geissel* only a couple of years before Buxtorf published the *Synagogue*.[75] And he described the preparation for a circumcision in the following terms:

> If the mother bears a little boy, then the father sets himself to work to obtain a fine cow, or sheep, or calves, or something else of the sort, as well as geese, chickens, doves, fish and meat. Once he has ordered all these, he starts writing invitations to his Jewish neighbors, giving them the happy news and asking them to come to him on such or such a day (for he is required to summon ten Jews who must all be more than thirteen years old). Above all, he orders a Mohel, that is, a circumciser: for they have no books from which to learn how to circumcise a child and don't like to hire someone who has not practiced this art beforehand. Rather, they say, I don't want him practicing on my beard.... When the seventh day comes and it is near evening, the father of the child gives two young men their orders: one brings a pitcher with wine,

73. Buxtorf, *Jüden Schul*, chap. 7, 105–6.

74. Burnett, *From Christian Hebraism to Jewish Studies*, 66.

75. One copy (London, British Library, 4033.de.7.(3.)) is bound with Buxtorf's *Jüden Schul* and Georg Nigrin's *Jüden Feind* (Frankfurt am Main, 1605).

> the other candies, apples, pears and nuts. They go to everyone who has been invited to the circumcision and share out the wine and candies. This happens in Friedberg, Frankfurt, Worms, Prague, Posen, Thoren, Cracow – wherever many blind and godless Jews live together in a city or a village. When the evening begins, close friends arrive and the party begins, they begin soaking and joking and drink themselves full. They also make the mother happy, so that she isn't afraid of the circumcision. They sit up and pray all night near the child. Experience shows that many children have trouble on this night, for many of them die then.... Accordingly, many friends come on this night, they read and pray and hang lots of charms for the mother on the bed, so that the baby doesn't die. In addition, the father of the child asks the Mohel not to drink too much so that he doesn't hurt the child when his hand trembles. For he might cut too deep and spoil the whole outfit.[76]

All but a couple of the details in Buxtorf's account – the fact that the watchers play cards, for example – come from Hess. And that point comes from another convert, Anthonius Margaritha, who explained that in some places, the Jews "gambled with the baby's mother to console her, so that she would not feel sorry on the child's account, because of the circumcision."[77] Buxtorf's seamless, lucid narrative is in fact a patchwork of older and newer texts, most of them not named. In his account of circumcision, for example, he states that the mohel must be a man and not a woman.[78] Nothing in his account or its margins informs the reader that he drew this information from Moses Isserles's gloss on the laws of circumcision in the *Shulhan arukh*, which he sharpened and misrepresented even as he summarized it.[79] These written sources provide him not only with the content he deploys (and sometimes distorts), but also with the matter-of-fact form of his description and his responses to particular questions.

The style that Buxtorf imported, with the details, into his text, matters greatly. Hess – like Johann Pfefferkorn[80] and other Jewish converts – produced not a literary text, elaborated with fine writing, but a concrete description, which set the scene and described preparations for and the progress of the ceremony. These narratives are couched in a language whose syntax is loose even by period

76. Ernest Ferdinand Hess, *Flagellum iudeorum: Juden Geissel* (Erfurt, 1600), 2: 2, sigs. H iiiiv– h vv.

77. Margaritha, *Der gantz jüdisch glaub*, sig. hv.

78. Buxtorf, *Jüden Schul*, chap. 2, 106.

79. The passage is hilkhot millah, par. 261. Isserles wrote that "a woman is not obliged [permitted] to circumcise her son." In his copy, Basel, UB, FA IV 13, Buxtorf underlined these words and wrote: "Mulieri non est permissum circuncidere filium suum ullo modo."

80. For a sample of Pfefferkorn's descriptions of Jewish life and worship see Johann Pfefferkorn, *The Jews' Mirror*, trans. Ruth Cape, int. Maria Diemling (Tempe, AZ, 2011).

standards. It is possible that their authors did not write them at all, in the sense of formal composition. Rather, they may have spoken them, perhaps in answer to a series of questions, while an amanuensis – Elisheva Carlebach speaks of "the role of Dominicans as ghost authors" – recorded the result.[81] The free, vivid style of Buxtorf's ethnographic narratives – and the odd decisions about which ritual details to include, and which to omit, a feature of his work that is shared, as Carlebach and Deutsch have shown, with the works of some of the converts – reflects not his deadpan attendance at Jewish rituals but the untutored language of the modestly educated Jews and converts, recalling the rituals of their childhoods in response to batteries of questions, whose works he pillaged.

Buxtorf's performance as an observer is admirably consistent: he achieves a high level of verisimilitude, what the rhetoricians would have called *enargeia*, while giving the impression that the customs in question were uniform and unchanging. But like his presentation of the work as a whole, his accomplishment has more to do with compilation than with observation: it is a bookish work successfully designed to look less bookish than it really is. At times, indeed, something that sounds like a genuine oral component breaks in, but it does not always make his exposition more profound or reliable. In chapter 27 of the *Jüden Schul*, for example, Buxtorf states that Jewish butchers had their children urinate on the meat they sold to Christians, and "blessed" (cursed) it. "Perhaps," he commented, "they are not all, and in all places, this evil, but in general they are by nature very similar in their opposition to Christians."[82] Here Buxtorf transforms what can only have been individual reports, whose sources he does not identify, into a general description of the malevolence of all – or almost all – Jews.

The Act of Compilation

Buxtorf, as we have seen, did not have to sketch all his stories of Jewish religious life from scratch. But unlike his predecessors, he prided himself on his scholarship, which he exhibited in its external form by means of citation and reference to multiple sources, both within and on the margins of the pages of the book. All this apparatus can be deceptive. Buxtorf's choices of sources and citations of them are often arbitrary. He used what was to hand. But this was not always the rule – and we sometimes can detect a method in both the selection and handling of the raw material.

81. Elisheva Carlebach, *Divided Souls: Converts from Judaism in Germany, 1500–1750* (New Haven, 2001), 178. See also Maria Diemling, "Patronage, Representation and Conversion: Victor von Carben (1423–1515) and His Social Networks," in *Revealing the Secrets of the Jews*, ed. Adams and Heß, 157–81, at 177–78.

82. Buxtorf, *Jüden Schul*, chap. 27, 572–73.

Exemplary in this respect is the chapter on Tabernacles. Between the tidbits of information about the building of the *sukkah* (booth) and the biblical foundations of the four species connected with this festival, Buxtorf directs his readers to a specific element in the service – the *hakafot*: the processions in the synagogue performed with the four species – palm branch, myrtle, willow, and citron – in hand. Before taking up his pen to write the chapter, Buxtorf had already done his homework. He had discovered the interpretation for all this ceremonial in the thirteenth-century exegete Bahya ben Asher's *Kad ha-kemach*.[83] According to Bahya, in Buxtorf's reading, all these processions pointed to one symbol: the Jewish hope for the ultimate downfall of Edom, that is, Christian domination. His books were often ruthlessly censored. It was not difficult for Buxtorf to spot the relevant passages. Buxtorf offers his readers a long extract and then comments:

> You see how much good they also wish the Christians on this festival. But in their prayer books they turn out differently, they pray and wish that God may strike us, as he did to the firstborn in Egypt as we read in one of the prayers, which starts *Ana hoschia na* [save us!]. They say [and here there is a reference to the Cracow *Mahzor* (1585) in the margin] specifically *Takkeh ojebhenu*, strike our enemies, as you struck the firstborn in Egypt.... By enemies, they mean none other than the ones who keep them prisoners.[84]

Buxtorf did not reveal to his readers that he had collected this information about the prayer when he was rifling through Jewish liturgies, probably in his role as censor, hunting the anti-Christian passages.[85] In his notes, which take up several pages of his copybook, not only had he transcribed this bit of the liturgical poem, but he had also noted the reading "your enemies" (i.e., God's enemies) in the Venice *Mahzor*.[86] Apparently, the *Jüden Schul* was not the place to inform readers that not all prayer books articulated anti-Christian sentiments, at least not in this particular poem; nor was this the platform to discuss textual variants in an effective way.

Negative accounts of foreign religious practices regularly appeared in ethnographic works. When the Calvinist Jean de Léry described the observances of the Tupinamba that he had witnessed in Brazil, he admitted, with some hesitation, that these Americans shared the sense of a divinity that Cicero had held to be universal. But their rituals horrified him. When he watched them howl and leap, he could

83. Buxtorf, copybook, 15: "Kad [Hakkemach pag. 51. col. 4]. Quare septies circumeant altare in festo tabernaculorum."

84. Buxtorf, *Jüden Schul*, chap. 16, 460.

85. See below, Chapter 5.

86. Buxtorf, copybook, 259: "In Veneto legitur אויביך."

only conclude that "the devil entered their bodies."[87] And when he described the *caraïbes* who led their dances, shaking maracas, he compared them to "the bell-ringers that accompany those impostors who, exploiting the credulity of our simple folk over here, carry from place to place the reliquaries of Saint Anthony or Saint Bernard, and other such instruments of idolatry."[88] Jews and Catholics made natural terms of comparison for him.

Hugo Blotius made even sharper observations about the Catholic cult of the House of the Virgin at Loreto:

> All the sorts of people who live here occupy themselves with three sorts of work. For either they sell wreaths of little stones or *patinostra,* like leaden bulls, or they work as shopkeepers, or they manage churchy matters. Here you encounter a vast quantity of Jesuits, friars, and monks: the order of the ecclesiastics. . . . Around 25 inns here are always open. At Easter, though, all the shrines are hostels, or rather, the other shrines besides the Palace of the Priests and the inns serve as shops for those petty merchants I mentioned. . . . The inside walls of the church are covered with votive tablets from almost everywhere. In the middle appears the shrine of the Virgin mother, which is said to have been carried by angels, first from Palestine, then from Illyria, from a nearby wood, and finally from the property of Laura, a certain woman with that name, into the place where it stands today. The history that the natives told me about this needed an Iliad, not this brief and trivial description.[89]

87. Jean de Léry, *History of a Voyage to the Land of Brazil, Otherwise Called America,* trans. Janet Whatley (Berkeley, 1990), 141; *Voyage au Brésil,* ed. Michel Contat (Lausanne, 1972), 195.

88. De Léry, *History of a Voyage,* 142; *Voyage au Brésil,* 196.

89. Blotius, *Odeporicon Italiae,* Vienna, Österreichische Nationalbibliothek, MS 6070, fols. 136r–137r: "Omne hominum genus, qui hic habitant omnino in tribus exercitiorum generibus sunt occupati. Aut enim serta lapilllorum seu Patinostra, ut Plumbeas bullas vendunt, aut rem cauponariam exercent; aut res templarias administrant, magna hic Jesuitarum, fratrum [MS saucs] et monachorum est ubertas. Ordo Ecclesiasticorum. Pallatium quoddam habent Instar arcis, in quo plus quam 100 homines quotidie aluntur. Diversoria hic fere semper patent 25. Sub Pascha vero omnes aedes sunt diversoria, alias reliquae aedes praeter Palatium Sacerdotorum et cauponas sunt tabernae illorum quos dixi marcatorculorum. Templum illum magnum et ingens, non tantum ad rerum sacrarum usum, verum etiam ad bellum et hostem sustinendum est accommodatum. Ingredienti hiantibus faucibus ad portam temple in rotis adstant magnae aliquot [MS aliquod] machinae bellicae. Templum intus totum tabulis votivis paene undique coopertum est, in medio visitur aedicula Virginia matris quae primum ex Palestina, deinde ex Illyrico, ex proxima Sylva, atque postremo ex agro Laurae, faeminae cuiusdam ita vocatae, in hunc usque locum ubi nunc stat ab angelis translata dicitur. Historia quam mihi indiginae narrarunt [MS narrarum] hac de re Iliadem, non hanc postulabat brevem descriptiunculam." See Paola Molino, "Istruzioni per un viaggio

Both Léry and Blotius did their best to show that they were describing what they had seen and heard. Buxtorf indicated at times that he was reporting his own experiences, but he was far more reliant than the others on the books that all three consulted from time to time.

From textual symbolism Buxtorf returned to the synagogue and the regulations compiled in Joseph Karo's Sephardic code, *Shulhan arukh* (Set Table). But his is an Ashkenazic endeavor: he hardly ever quoted the *Shulhan arukh,* as a rule drawing his information instead from Moses Isserles's Ashkenazic annotations in his *ha-mappah (Tablecloth).*[90] Even then Buxtorf set his own stamp on the material he extracted from the text. Speaking about the ceremonies on Hoshana Rabba and the custom of taking the seven Torah scrolls out of the ark, he wrote: "In Poland they take out all the books; even if there were seven thousand, they would have to come out."[91] In fact, Isserles had stated: "There are some places where they bring out all the scrolls and that is what is done in our land."[92] In Buxtorf's hands Isserles's reference to local custom was reduced to absurdity, and with it the symbolism of the number seven, connected with the festival. Here the compiler becomes the inquisitorial commentator.

The Compiler and Humanist

Although Buxtorf's portrayal of the Jews is highly charged and tendentious, reading between the lines we can see, next to the polemicist, the scholar at work – the Buxtorf who was to manifest himself in full philological grandeur in his later writings, the *Hebrew and Chaldaic Lexicon* and, above all, the *Tiberias.* An inkling – just an inkling – of this Buxtorf appears *in potentia* at the end of chapter 25, which comprises his short overview of Jewish fast days. Buxtorf writes:

> In conclusion to this chapter, I do not want to omit reporting how the highly learned Rabbis believe that a devout fast should be conducted. You can read about it in *Medrasch rabba,* in the chapter that starts *Vezos habberáchah.*[93] It reads thus in German: As soon as Moses saw that God had condemned him

in Italia: 'Hodoeporicum Hugoni Blotii earum rerum quas in Italia vidit et observavit,'" *Biblos* (2006): 115–37, at 133, 136.

90. Buxtorf's copy of Isserles's *Shulhan arukh* (Venice, 1593–94) contains sporadic annotations (Basel, UB, FA IV 13).

91. Buxtorf, *Jüden Schul,* chap. 16, 462.

92. Isserles, *Shulhan arukh, Orah hayyim,* par. 660, 1.

93. See Midrash Deuteronomy Rabbah, vezot ha-berakhah, 11, 10. The text reads: "He drew a small circle."

> to punishment he undertook a fast and made a pasty or cookie as big as his body, put on sackcloth, strewed ashes on his head, and stood in the middle of the cookie fasting and praying incessantly. He had no intention of moving until God changed and canceled the decree, which is what happened.[94]

Buxtorf begins with a passage from a late Midrash on Deuteronomy, compiled between the sixth and the ninth century, as cited in an early modern kabbalistic text, to which he refers in a marginal note. Here as in the following passage he translates the word *ugah* as "cookie," although undoubtedly, in this context, the entire expression *ag ugah* means "he drew a circle." Buxtorf then continues his satirical discourse:

> Without any doubt this comment derives from the source that spins all the Jewish fables, that is, the holy Talmud. You can read in the tractate about fasts [in the margin: "Taanis chap. 3. 19. 23"] that a very holy and pious Jew, named *Chone hammaagal*, behaved in the same way when he wanted to beg God for something of pressing importance. Since it had not rained for the whole month of February, and the rain was essential for the coming harvest, he fasted and placed himself, as though imprisoned, in one such cookie. And he said: Lord of the world, your children lift up their eyes to me because I am beloved to you and am a member of your household, just as a child is dear to his father. I swear by your holy name that I will not move from this place until you show mercy to your children. Immediately it rained. What's more, it is also reported [in margin: a reference to Rashi's comment ad loc.] that the prophet Habakkuk did a similar thing, for it is written (2.1): "Here I stand on guard and stand still and look to see what answer I will get, that is, I will stand here in my pasty or cookie and wait until you answer me as to why the godless live so long."[95] ... The stubborn Jews find themselves in such a pitiable, perverted state of mind that they are not ashamed, and do not have a bad conscience about speaking so vulgarly and nonsensically about God's word, and about interpreting Scripture according to their own fancy and liking.[96]

What Buxtorf is presenting here is a free rendering of a notable passage about Honi, the circle drawer, who makes his first appearance in the Mishnah (Ta'anit 3:5). The passage describes a time of extreme drought. When all prayers fail, the charismatic Honi draws a circle (*ag ugah*), stands in the middle, and insists that

94. Buxtorf, *Jüden Schul*, chap. 25, 556–57.

95. Buxtorf uses Rashi's gloss on Ta'anit 23a. He cites the expression "as in a prison" to gloss the verse from Habakkuk quoted in the Talmud but inserts it into his retelling of the story.

96. Buxtorf, *Jüden Schul*, chap. 25, 557–58.

God produce the "right kind of rain to ensure that all water supplies are provided for." Honi's provocative action and insolent behavior toward God are successful and the "right" kind of rain falls. Though the Mishnah contains an implicit criticism of Honi's demanding stance toward God, which appears to verge on the blasphemous, it also endorses him as "God's beloved and spoiled son." It would appear that simply by including the gripping narrative within the confines of the legal discussion, the nonrabbinic figure Honi has been given a legitimate place within rabbinic society.

Buxtorf's assessment of these notable and much-loved passages is predictable: Talmudic blasphemy of the first order. More interesting and significant, however, is the way he went about presenting his glossed version of the stories. Particularly noteworthy is his ability to surf the rabbinic web, making connections between texts and detecting tropes within the rabbinic corpus. First, we should note that the word *ugah,* which also means "cake," clearly refers to a circle in this context.[97] Buxtorf deliberately chose to translate the word as "cookie," garnishing it with his own caustic addition, the "pasty or cookie of human dimension," into which the fasting protagonist inserts himself. The freedom and authority with which Buxtorf spun his own fantasies, as he fashioned his rendering of these tales, are certainly impressive. He used the Talmudic expansion of the mishnaic story – which introduces Honi's action of placing himself in a circle as comparable to that of the prophet Habakkuk waiting on his post for an answer from God – and combines it with Rashi's gloss on the verse, which describes Habakkuk's position "as one who is in a prison." Though he combined all these texts and glosses into his retelling of the story of the "cookie," we can appreciate Buxtorf's ability to detect the circulating topos of the "ugah," which he traces back to the "holy Talmud."

At the end of this chapter, Buxtorf gave his own justification for writing the book:

> This is enough talk about the ceremonies of the Jews' holy days, and feast days, and the customs of their Jewish faith, from which we can conclude that their religion is no longer based on Moses and the prophets but rather on lies and the false orders of the Rabbis and scribes, as I undertook to prove in the beginning of this book.[98]

But we may well ask whether this is a satisfactory reason for all the compilation and industry that Buxtorf seems to have expended on composing the book. The

97. As Rashi explains, the word *ugah* indicates a form like a cake that is circular in shape (Rashi ad Ta'anit 23a).

98. Buxtorf, *Jüden Schul,* chap. 25, 559–60. In the margin there is a reference to page 10 of the abbreviated *Reshit hokhmah.* In other words, Buxtorf is not citing Deuteronomy Rabbah but is working directly from chap. 3 of "sha'ar teshuvah" in Poggetto, *Reshit hokhmah ha-katsar,* 10r.

last chapter of the *Jüden Schul*, on the messiah, is no exception to the rules that govern Buxtorf's writing. Though he offers much correct information about the vague concept of the messiah in Jewish tradition, he nevertheless weaves his usual hyperboles and admixture of literary allusion and rhetoric into the narrative.

We have to admit then that this is not an ethnography. The compiler is not simply a compiler: the texts are often not left intact. But as we have shown, there is another Buxtorf. He appears rarely, but he does appear. The act of translating Yiddish Sabbath poems into German may be seen to belong to that other Buxtorf – the humanist who produced booklets on Hebrew epistolography, taught the rules of Hebrew prosody, and collected Hebrew epitaphs.[99] But in these activities he was no more an ethnographer than the Buxtorf we have encountered here, the writer of the *Jüden Schul*.

99. On Buxtorf's interest in epitaphs, see Michela Andreatta, "Collecting Hebrew Epitaphs in the Early Modern Age: The Christian Hebraist as Antiquarian," in *Jewish Books and Their Readers: Aspects of the Intellectual Life of Christians and Jews in Early Modern Europe*, ed. Scott Mandelbrote and Joanna Weinberg (Leiden, 2016), 260–86, at 264–66, and Appendix 2.1 below.

CHAPTER THREE

Guides to Reading Jewish Texts: Buxtorf the Bibliographer

In April 1614 Thomas Erpenius, the renowned Arabic scholar of Leiden University, sent Buxtorf a letter. He regaled his colleague with a list of his newly acquired Hebrew books and promised to send him a copy of the Zohar printed in Mantua. The younger scholar could not hold himself back from insinuating a note of criticism of the more senior Hebraist, who had just published two compendia, one on Jewish abbreviations and one on rabbinical bibliography:[1] "Your book *De abbreviaturis* gives me enormous pleasure, as does the *Bibliotheca Rabbinica* although there are quite some titles missing which you would not deny; but then it is never possible to begin something and immediately bring it to completion."[2]

Buxtorf himself would have been the first to acknowledge the deficiencies of his bibliography.[3] After all, in his dedicatory letter to Philippe du Plessis-Mornay he admitted that both the *Brevis recensio operis Talmudis*, his separate account

1. Johann Buxtorf, *De abbreviaturis hebraicis liber novus & copiosus* (Basel, 1613). The *Bibliotheca Rabbinica* appears in the same volume (257–335), without a separate title page; unless otherwise specified, this is the edition cited.

2. Erpenius to Buxtorf, 4 April 1614, Basel, UB, MS G 2 II 24, fols. 19r–21v: "Liber tuus de abbreviaturis mirum in modum mihi arridet, ut et Bibliotheca Rabbinica in qua tamen non pauca desiderantur quod nec ipse ignoras: nimirum inchoari simul et perfici nihil unquam solet [cf. Cicero, *Brutus* 71: 'nihil est enim simul et inventum et perfectum'][.]" On the letter, see Arnoud Vrolijk and Joanna Weinberg, "Thomas Erpenius, Oriental Scholarship and the Art of Persuasion," in *Scholarship between Europe and the Levant: Essays in Honour of Alastair Hamilton*, ed. Jan Loop and Jill Kraye (Leiden, 2020), 34–59, esp. 54–57.

3. Not only is the bibliography lacunose, but it also contains several errors: e.g., the wrong place of publication for the classic mystical work, *Sefer yetsirah* (Venice instead of Mantua), and the ascription of *Or adonai* to Obadiah Sforno. The actual title of Sforno's philosophical treatise is *Or ammim*, which title is then given without author in the following entry.

of the Talmud, and the *Bibliotheca Rabbinica* were afterthoughts. The basic purpose of the volume had been to teach his readers about the omnipresence of abbreviations in Hebrew writings.[4] In the course of his work on this enterprise he realized how widespread was the use of abbreviations as references to Hebrew books or their authors or rabbinic authorities as well as to chapters or sections of the Talmud. He envisioned his two little books as practical handbooks for Christian scholars who wished to enlarge their Hebraic horizons, by reading their way into the vast range of what he called "Rabbinic" literature. As he warned readers, "to read and understand the books of the Hebrews is not for just anyone."[5] Abbreviations for everything from individual names to holidays and set phrases to book titles confronted everyone who tried to read Jewish literature – including Christians who had not mastered them by immersion in a Jewish primary school (heder) or higher academy (yeshivah).

> The Hebrews rejoice so much in this form of abbreviation, that hardly any book, any writing, any letter, any sheet or note lacks it. These are rough spots. The rash run into them and become stuck. The prudent work their way across them, shrewdly and with difficulty. But those who jump over them often miss the true and complete sense of the passage. They were brought into being partly by speed in writing, partly by the frequency of their use.[6]

In Buxtorf's first text, accordingly, he explained how Jews formed abbreviations and then provided a massive collection of them, in alphabetical order, with explanations. His book offered newer entrants to the world of Hebrew letters the practical guidance of a generous senior colleague.[7]

Though Buxtorf must have known these characteristics of Jewish writing well before he undertook the *De abbreviaturis*, he had apparently not conceived of the idea of adding a Hebrew bibliography to a list of Hebrew abbreviations. The final product, in which the bibliography appeared as an independent work, was therefore in several instances inextricably linked to the *De abbreviaturis*

4. Buxtorf, *De abbreviaturis*, sig):(3r.
5. Buxtorf, *De abbreviaturis*, 2.
6. Ibid.
7. Buxtorf's gathering of bibliographical information on Jewish texts began early. In the margins of his Latin translation of Elijah Levita's *Massoret ha-massoret* (Copenhagen, Det Kongelike Bibliotek, MS E Don. Var. 13. 8° (a 1629 copy by Hans Jensen Svaning), which he completed by 1593, a few notes already reveal his interest: p. 9 on "Iter semitae scientiae": "[est Grammatica R. Mose Kimchi qui sic inscribitur]"; p. 12 on "Sed ex libro Achla veachla": "[אכלה 1. Sam. 1.9. ואכלה Et comede Gen 27.19. Nomen hoc a primis verbis libro inditum. Ex eo creduntur descripta, quae in fine magnorum Bibl. Venetorum titulantur magnae Masorae nomine.]"; p. 18 on "liber Ietzirah": "[de creatione ubi multa arcana proponuntur]."

both in content and in method. The *De abbreviaturis* is, however, much more comprehensive than the booklist, even in one case containing an abbreviation that Buxtorf had never encountered![8] By contrast, as his son's dedicatory epistle to the second printing of 1640 indicates, Buxtorf had every intention to add to the list of books.[9]

The nature of Buxtorf's immediate inspiration for these linked projects is not clear. One of his models for the collection of abbreviations seems to have been Elijah Levita's *Tishbi* (1541), which offered 712 definitions, often extended, for words that appeared in the Talmud or Midrash. Levita's Christian associate Paul Fagius translated this book into Latin, in which form it became an essential tool for Christians mastering Hebrew. Like Levita, Buxtorf offered definitions, in his case for standard abbreviations of many kinds: the conventional use of acronyms, such as רש״י (Rashi), for Rabbi Salomon Iarchi (Solomon ben Isaac);[10] for titles, such as אב״ד (ABD) for *Av bet din*, the president of the law court; and for religious terms and phrases, such as אגלא (AGLA), the abbreviation for "Thou art powerful in eternity, Lord," which Buxtorf defined as a "kabbalistic divine name" – and which figures in later medieval Jewish magical practices.[11]

Some of the materials that Buxtorf used he had compiled years before. He explicated the terms שתי (warp) and ערב (weft), terms that originated in the language of weavers, as references to the Cross. Bahya, he pointed out, had interpreted a phrase in Isaiah 66:17, "They that sanctify themselves and purify themselves to go unto the gardens, behind one in the midst," as a reference to Christians, who sanctified themselves by making the sign of the cross ("behind one in the midst"), and to Turks, who purified themselves by washing their hands and feet. Kimhi, commenting on the same passage, had noted that "many commentators" interpreted the phrase at issue in the same way.[12] Buxtorf summarized these discussions, very briefly and without referring explicitly to Bahya and Kimhi, in his copybook.[13] There he also made clear that he had discussed the passage with

8. Buxtorf, *De abbreviaturis*, 125–26, s.v.: ס״הב: "ספר הבהיר Liber Cabalisticus, qui manuscriptus tantum inter judaeos habetur.... Sic abbreviate citari, non memini me legere."

9. Buxtorf, *De abbreviaturis Hebraicis liber novus & copiosus: cui accesserunt operis Talmudici brevis recensio. Item Bibliotheca rabbinica nova*, ed. Johann Buxtorf the Younger (Basel, 1640), Epistola dedicatoria, sig.):(4v.

10. Buxtorf, *De abbreviaturis*, 169–70. As he does for many other important Jewish authorities, Buxtorf provides a potted biography for Rashi, the data for which he culled from his three favorite Jewish historians: David Gans, Abraham Zacuto, and Gedaliah ibn Yahya. It is therefore remarkable that he nevertheless still used "Iarchi," the Christian name for Rashi, which he would not have come across in the Jewish sources.

11. Ibid., 169–70, 17–18, 20.

12. Ibid., 172–73.

13. Johann Buxtorf, copybook, Basel, UB, MS A XII 20, 29: "Sequitur in Iesiae testimonio אַחַד אַחַת בַּתָּוֶךְ id est unam (vel unum) in ipso medio ירמוז לשתי וערב id est, haec

a Jew: "When I myself asked a certain Jew, on 1 July 1596, what they understood by those words, he answered, a *tselem*, that is, an idol, that is, of the cross."[14] It seems highly likely that in other references as well he drew on conversations with informants, but sources to identify them are lacking.

Buxtorf's entries varied greatly in length, density, and tone. Sometimes, as in his substantial discussions of the terms *Rav* and *Rabban*, he compiled large amounts of historical material that clarified their meanings and development – including the connection of the latter to Christianity through Paul's teacher, Rabban Gamaliel.[15] Sometimes he covered material that he would also discuss in his bibliography: for example, the dates of Onkelos the translator, which he considered in both works.[16]

And sometimes, he let his attitudes toward the Jews appear. In a long entry on Jewish titles, Buxtorf drew on Isaac Abravanel to argue that the Jews had created the term the term *Morenu* (our teacher) relatively recently, as their own counterpart to the Christian *Doctor*. Their effort to declare in public the wisdom and erudition of their teachers was only natural: "there is no more boastful people under heaven, than the Jews."[17] Sometimes his mask slipped more visibly. In discussing the Jews' practice of inserting blank spaces into the beginnings or middles of lines in the Torah, he concluded by observing that "this superstitious method of writing is used only in the holy manuscript of the Torah, which they use when they perform rituals in their Synagogues."[18] Discussing the abbreviation קק for קהלה קדושה (holy synagogue), Buxtorf described how he had seen it written in the Frankfurt synagogue and how it figured in the title of the head of the Cracow synagogue. "They always call their synagogues, schools, and churches this," he noted, "though they are rather pits filled with every sort of impiety."[19]

---

indicant stamen et subtegmen, (id est Crucem quem sic appellant et circumscribunt Iudaei ex pertinaci ejus odio, quod ista fila in modum crucis se intersecant) quo Edomitae sese sanctificant. או ירמוז ליום אידם. Aut indicant diem ruinae seu perditionis eorum (hoc est Diem domini cum suis solis) qui illis institutus et ordinatus est etc. Verba hebraea sunt הקבוע לה' יום שישי ואחד בשבת ואינו בתוך."

14. Ibid.: "Ipse enim 1 Julii 1596, cum quaererem a quodam Iudaeo quod per ista verba intelligerent: respondit unum צֶלֶם id est idolum, scilicet crucis." See Chapter 1, above, for a discussion of this example.

15. Buxtorf, *De abbreviaturis*, 148–53.

16. Ibid., 105; cf. the less elaborate discussion in *Bibliotheca Rabbinica*, 261–62.

17. Buxtorf, *De abbreviaturis*, 99–101, at 100.

18. Ibid., 140–43, at 143. For a contemporary Jewish description and explanation of this practice, see Marvin Heller, *Printing the Talmud: A History of the Earliest Printed Editions of the Talmud* (Brooklyn, NY, 1992), 361.

19. Buxtorf, *De abbreviaturis*, 146: "Sic semper suas synagogas, scholas & ecclesias appellant, cum potius omnis impietatis lacunae sint."

Buxtorf's close friend Waser also wrote a treatise *De abbreviaturis Hebraeorum.*[20] The manuscript is not dated but in many respects, though far shorter, it closely resembles Buxtorf's tract. The two men had similar goals in drawing up their lists of Hebrew abbreviations. Both hoped to assist the beginner as he or she attempted to enter the closed textual world of the Jews.[21] Talmudic and Masoretic abbreviations are unraveled, many acronyms of rabbinic authorities are revealed, whilst kabbalistic ideas in abbreviation usually drawn from Reuchlin are also displayed. Waser employed the same categories as Buxtorf to distinguish different kinds of abbreviations: *propriae/impropriae* and *privatae/communes*. Yet he did not refer to Buxtorf, not even in the entry ופגין ("whoso breaketh an hedge, a serpent shall bite him": Ecclesiastes 10:8) where he mentioned epistolary use of this abbreviation; Buxtorf himself alluded to his own publication: "Vide nostram Institutionem epistolarem Hebraicam cap. 6."[22] The similarity between the two tracts is also in evidence in the entry for זה"ל צ"יב where both scholars referred to the inscription of the biblical phrase "This is the gate of the Lord: the righteous will enter it" (Ps. 118:20) on the lintels of synagogues, and mentioned the Frankfurt synagogue as a case in point. Buxtorf also noted that the formula sometimes appeared on the title pages of Hebrew Bibles, and that it was part of the title of the head of the yeshiva of the Cracow synagogue.[23] Both recognized the abbreviation י"ש ("may his name be eradicated") as a curse that Jews used against Christians. Waser simply conveyed its meaning and usage.[24] Buxtorf, typically, devoted a long passage to the vituperations that the Jews used against Christians and Jesus in particular, and that Christians failed to recognize.[25] He also mentioned that he owned "handwritten examples of that thing by the Jews," and that he had discussed the abbreviated curse with a Jew, who informed him that some resolved it in an even more frightful way. "That Jew still lives," he remarked, "and I know where, and he is quite well

20. Caspar Waser, *De abbreviaturis Hebraeorum*, Zurich, ZB, MS F 172c, fols. 863r–891v. The manuscript appears to have been written in preparation for publication. The Hebrew words are sometimes incorrectly written, and it seems that Waser's Hebrew script is not as fluent as Buxtorf's.

21. Waser, *De abbreviaturis*, fol. 863r: "Abbreviaturarum Hebraicarum, non in Biblijs, sed in Rabbinorum scriptis usus est maximus; ac proinde illarum cognitio necessaria illi est, qui in eorum scriptis cum fructu cupit versari."

22. Buxtorf, *De abbreviaturis*, 73.

23. Waser, *De abbreviaturis*, fol. 875r: "Habetur Francofurti ad Moenum In superliminari portae Synagogae Judaeorum"; Buxtorf, *De abbreviaturis*, 146–47: "Cum servilibus . . . in synagoga sancta Francofurtensi: . . . Caput academiae synagogae sanctae Cracoviensis."

24. Waser, *De abbreviaturis*, fol. 877r: "Deleatur nomen ejus. Sic Christianis imprecari solent."

25. Buxtorf, *De abbreviaturis*, 87–89.

known at Frankfurt."[26] Despite the almost identical wording in some cases, Waser and Buxtorf sometimes expanded abbreviations in quite different ways. Unlike Buxtorf, Waser had travelled in Italy and elsewhere. He explained that "in letters written by Italian Jews," the abbreviation מי׳ "means מיסיר Missir, which the French say Monsieur."[27] Buxtorf, by contrast, pointed out that it stood both for the phrase "from whence do we know?' in the Talmud and for the title of Levi ben Gershom's treatise *The Wars of the Lord*. On other occasions, the two scholars spoke with one voice. Under the entry for ק׳ק ("holy community"), Waser claimed, as Buxtorf did, that the Jews used this phrase to indicate that their synagogues were holy. In fact, he insisted, in the same words Buxtorf chose, they were *impietatis lacunae*, "pits of impiety."[28]

The main point of both enterprises, however, was not ideological but practical. Buxtorf nourished many ambitions – personal, professional, and pedagogical. He felt none of these more strongly than the will to induce other Christians to move from reading only the Hebrew Bible to studying the full range of Hebrew, Aramaic, and Yiddish texts. The *De abbreviaturis* was avowedly a pragmatic work, designed to be handy for Christians exploring anything from biblical commentary to Kabbalah. It could not in the nature of things be complete or rigorous.

Yet it was more elaborate than its companion. Here too he was out to create a book that, though both erudite and original, was not a massive work of scholarship but a modest aid for Christian readers – one full and novel enough that they welcomed it. Buxtorf presented his bibliography as provisional and unsystematic. His eloquence warmed up as he warned the reader not to expect too much:

> If someone had given me '"the wings of a dove," as the Psalmist speaks [55:7], so that I could have flown swiftly and searched through the public libraries of cities and princes (for private ones on this subject are very rare), I could have attained a deeper and more thorough knowledge of Hebrew books. Who

26. Buxtorf, *De abbreviaturis*, 87: "Maledictio judaica est, qua non tantum inter se in suae gentis malitiosos & improbos utuntur, sed multo maxime in Christianos. Ejus rei manuscripta apud me sunt judaeorum exempla": 88: "Mihi ante paucos annos hac de re cum quodam judaeo conferenti, immo, respondit, non sic tantum explicatur, sed & ישו שקר ותועבה

Quis ad haec toto pectore non exhorreat? & tamen haec dicuntur inter recutitos istos, & feruntur, nemine Christianorum ad id advertente. Vivit iste Judaeus adhuc, & scio ubi, estque Francofurti satis notus." He goes on to say that the Jew was considering conversion when they spoke.

27. Waser, *De abbreviaturis*, 882r: "In Epistolis ab Italis Judaeis scriptis, valet מיסיר Missir, quod Galli dicunt Monsieur."

28. Buxtorf, *De abbreviaturis*, 146: "Sic semper suas synagogas, scholas et ecclesias appellant, cum potius omnis impietatis lacunae sint"; Waser, *De abbreviaturis*, 887r: "Sic synagogas suas appellant, cum tamen omnis impietatis lacunae sint."

> would have stayed me with spice powders or pomegranate juice well enough that I could have penetrated the smoky and evil-smelling dens of the Jews? And yet anyone who considered putting his hand to this task would have had to do something of the sort, or to find a trustworthy Jew who had thought about this project. Since I lacked these means at this time, it would perhaps have been better to remain silent about this thing, which was so unknown, rather than to speak about it.[29]

Of what one cannot speak, Buxtorf advised, thereof one must be silent. Perhaps a truly serious Hebrew bibliographer would have chosen that path.

But Buxtorf did not take his own advice. Instead, he made clear from near the outset that he saw the *Bibliotheca Rabbinica* as provisional and incomplete. He urged readers to supplement their copies. Other lexicographers prominently stated the cataloguing rules they had followed. Instead, Buxtorf set out the rules that he wanted others to follow as they filled out his own sketchy descriptions:

> Therefore, kindly and learned reader, you yourself add and explicate the things that are not recorded or explicated here (I know there are a great many of them). For the books that you have or have seen, name them, state their subject matter, add the place of printing, and either share them with me, if you are willing to do so, or add them to my beginnings and publish them under your name.[30]

To his readers, Buxtorf's invitation to others to collaborate with him seemed to be tinged with sincerity. In February 1614, Buxtorf's friend Walter Keuchen wrote to him from Hanau. He made clear that he was doing his best to follow Buxtorf's instructions:

> As to your instructions that I complete your little book, the *Bibliotheca Rabbinica*, I have enlarged it with roughly a hundred new titles, which I have turned over with my own hands, and learned their subjects, in addition to

29. Ibid., 257–58: "Et quis dedisset mihi אברי כיונה *alas columbinas* (ut Psaltes loquitur [55:7]) ut perniciter avolans & bibliothecas publicas (nam privatae ad hanc rem rarissimae) Principum vel urbium perquirens, libros Hebraeos sagacius & penitius cognovissem? quis me pulvere aromatario aut succo malogranatorum satis confirmasset, ut fumosas & maleolentes judaeorum caveas permeare potuissem? Et tamen tale aliquid fieri necesse fuisset, aut certe fidi alicujus Judaei opera utendum illi, qui huic operi manum adjicere cogitasset. His medijs quum destitutus hoc tempore fuerim, praestitisset forte de re adeo ignota tacere quam loqui."

30. Ibid., 259: "Ergo, Lector benevole & docte, quae hic nec annotata, nec explicata (quae scio plurima esse) tu adde & explica: tu libros quos habes aut videris, nominato, argumentum dicito, locum impressionis su[b]jicito, eaque vel mecum, si placet, communicato, vel hisce meis initiijs apposita, sub tuo nomine publicato."

many titles which you entered after copying them from other sources [as we will see, this description of Buxtorf's practices was accurate].[31]

Keuchen pointed out that even his collections and Buxtorf's, when combined, would not add up to half of the printed books in Hebrew, "to say nothing of the manuscripts. I have collected a good many of the more secret ones, many of them with difficulty, but I will perhaps be able to gather them more easily than any other Christian in the course of time."[32] Keuchen counseled his friend to await his new material patiently, so that they could produce a perfect version of the *Bibliotheca*, one that would be useful to Jews as well as Christians.[33] More expert readers than Keuchen responded to the book with a similar mixture of enthusiasm and criticism. According to Johann the Younger, the learned Constantinople rabbi Jacob Roman, "as soon as he saw my father's *Bibliotheca Rabbinica*, was so delighted by his find that he decided to enlarge it and publish it, translated into Hebrew, for the use of his people" (Johann printed Roman's list of additional titles, in Latin, in his 1640 edition of the *Bibliotheca*).[34] Deep into the seventeenth century, Christian scholars used Buxtorf's bibliography as a standard reference work (see Fig. 3.1).

Buxtorf's technical practices – from his rules for bibliographical description to his request for collaboration – were not innovative. When he offered full lists of the segments and chapters of the Babylonian Talmud, with brief characterizations of both, he followed well-established precedents in the tradition of Christian Hebrew scholarship. The Franciscan Pietro Galatino had analyzed the Talmud in detail in 1516, in his *De arcanis Catholicae veritatis*.[35] The Dominican Sixtus of Siena did the same in his *Bibliotheca sancta* in 1566.[36] What Buxtorf added to their approach was a massive descriptive introduction to the work, designed both to

31. Keuchen to Buxtorf, 7 February 1614, Basel, UB, MS G I 60, 319: "Quod jussisti, ut libellus tuus, Bibliotheca Rabbinica, a me completur; auxi illum plus minus centum titulis novis, quos ipse manibus meis pervolvi, atque argumentum cognovi, praeter plurimos titulos, quos tu aliunde descriptos, unde posuisti."

32. Ibid.: "Neque tamen si & tui & mei simul jungantur, crediderim nos dimidiam partem librorum impressorum, nedum manuscriptorum, quorum ego secretiorum non pauca habeo, plurima tamen difficulter quidem, facilius tamen quam Christianorum quisquam habere temporis successu fortassis potero, congessisse."

33. Ibid.

34. Johann Buxtorf the Younger in *De abbreviaturis* (1640), 436. The list appears in ibid., 437–72. For discussion, see Joseph Prijs, *Die Basler hebräischen Drucke, 1492–1866*, ed. Bernhard Prijs (Olten, 1964), 371; Stephen Burnett, *From Christian Hebraism to Jewish Studies: Johannes Buxtorf (1564–1629) and Hebrew Learning in the Seventeenth Century* (Leiden, 1996), 3, 166.

35. Pietro Galatino, *De arcanis Catholicae veritatis* (Frankfurt, 1612), book 1, chap. 5, cols. 13–20.

36. Sixtus of Siena, *Bibliotheca sancta* (Cologne, 1586), 126–28.

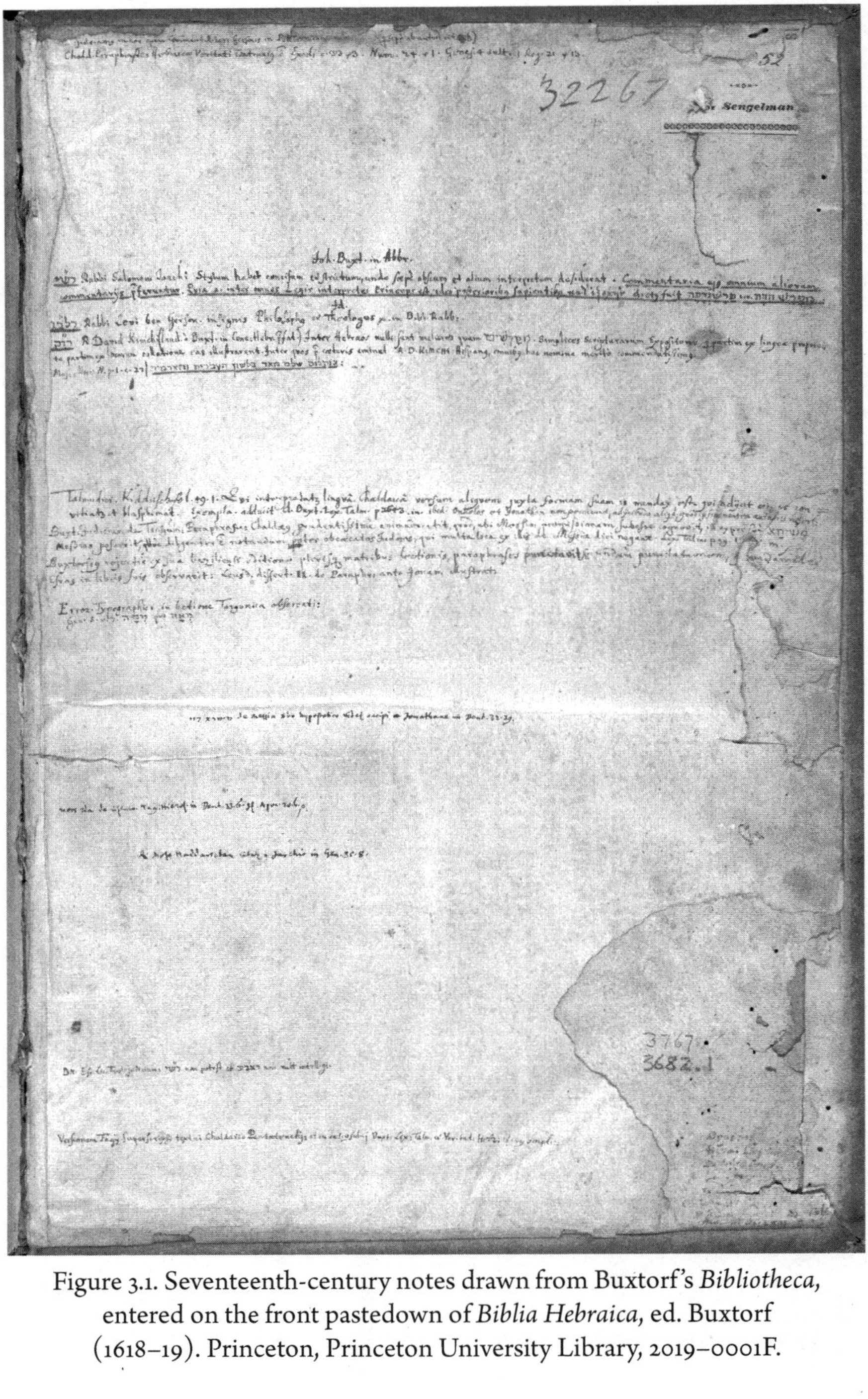

Figure 3.1. Seventeenth-century notes drawn from Buxtorf's *Bibliotheca*, entered on the front pastedown of *Biblia Hebraica*, ed. Buxtorf (1618–19). Princeton, Princeton University Library, 2019–0001F.

trace its origins and to show how Jews used it. He made clear again and again that the text had evolved over time. Buxtorf described it as "a doctrinal work, or a great corpus of doctrine, compiled by a variety of very learned rabbis."[37] He followed its development, from its original oral form to its canonization. Discussing the Mishnah, for example, he explained that Rabbi Judah ha-Nasi, working around 150 CE, worried that as the Jewish people were scattered, their oral law might disappear as well:

> He therefore gathered the notes and papers of all the Jews of his time, which each of them had annotated as helps to memory, and made a book from them, which was called *Sefer mischnajoth, The Book of Traditions*. This book received public authority and credit, and was received by all the Jews, living both in the land of Israel and in Babylon, as an authentic *corpus juris*, and explicated in the academies.[38]

Buxtorf's terminology, with its emphasis on codification and reception, shows that he had the growth and acceptance of the Roman law in mind as he traced that of the Talmud. Joseph Scaliger had taught him to see the Talmud in these terms, as a Jewish counterpart to the Roman law.[39] They and other Christian scholars held that the rabbinical opinions in the Mishnah, the compilation ascribed to Judah haNasi, and the rabbinical interpretations and debates in the later Gemara were comparable to the opinions and debates of Roman jurists collected in the Digest.[40] Buxtorf traced the creation of the Gemara as well as the Mishnah, and described the two different versions of the Talmud, Palestinian and Babylonian.

More important than the history of the Talmud was its status. Using his familiar technique of commonplacing, Buxtorf collected a long series of passages from rabbis and the Talmud that stressed the perfection and authority of the Oral Law. The rabbis, he emphasized, "cite the books of the Talmud everywhere

37. Buxtorf, *Bibliotheca Rabbinica*, 191.

38. Ibid., 201.

39. Anthony Grafton, "'Pandects of the Jews': A French, Swiss and Italian Prelude to John Selden," in *Jewish Books and Their Readers: Aspects of the Intellectual Life of Christians and Jews in Early Modern Europe*, ed. Scott Mandelbrote and Joanna Weinberg (Leiden, 2016), 169–88.

40. Recent scholarship has tended to bear out the view of Scaliger, Buxtorf, and many of their contemporaries that these corpora can usefully be compared. See, e.g., Catherine Heszer, "The Codification of Legal Knowledge in Late Antiquity: The Talmud Yerushalmi and Roman Law Codes," in *The Talmud Yerushalmi and Graeco-Roman Culture*, ed. Peter Schäfer (Tübingen, 1998), 1: 581–641; Nicholas de Lange, "Jews in the Age of Justinian," in *The Cambridge Companion to the Age of Justinian*, ed. Michael Maas (Cambridge, 2005), 401–26, at 405–6; and more generally Mark Letteney, *The Christianization of Knowledge in Late Antiquity: Intellectual and Material Transformations* (Cambridge, 2023).

in their works and confirm their entire doctrine from the Talmud, as if from an established code of law."[41] Or, as he put it in less laudatory fashion elsewhere: "in all their writings the sages and the rabbis babble continually about the Talmud alone." These "most impudent and impious praises" of the text confirmed that the Jews were a blind and obstinate people.[42] Christians needed to read it not for its own value but as a key to the vast world of rabbinical literature, and occasionally to a passage in the New Testament. True, Buxtorf was more enthusiastic about the work than Sixtus of Siena. After describing the burnings of the Talmud on papal command in 1553 and 1559, Sixtus expressed his regret that these books, "so often destroyed, came back to life like the heads of a hydra." He laid out the structure of the Talmud, he explained, chiefly to give Christians ammunition if they encountered "the vilenesses, impieties, and blasphemies" of the text or if further copies of it came to Europe from the copious stocks that still existed in "Asia, Africa, Greece, and Germany."[43] Yet Buxtorf's evaluation of the work was also largely negative.

Though Buxtorf insisted that his students move beyond the study of biblical Hebrew, he had always seen it as important for them to learn about some of the ways in which Jews read and referred to the Bible.[44] As early as 1593, he asked his friend Jacob Zwinger, then in Venice, to buy him four or even six copies of the *Humash*: "the five books of Moses with the readings from the prophets, printed in Hebrew in a very small format, without vowel points."[45] These editions of the Bible, which he presumably wanted for some of his first students, contained the Torah, divided up into the fifty-four weekly reading portions for synagogue use, or *parashiyyot*, and the weekly reading portions from the prophets, or *haftarot*. In the *Bibliotheca* Buxtorf described the system of weekly readings in detail. He stated, without hesitation, that "this division into sections is from Ezra, though they commonly ascribe it to Moses."[46] The practice of reading these passages aloud during the synagogue service, he argued, was also ancient, as James made clear in Acts 15:21: "For Moses of old time hath in every city them that preach

41. Buxtorf, *Bibliotheca Rabbinica*, 191.

42. Ibid., 199.

43. Sixtus of Siena, *Bibliotheca sancta*, 125.

44. See the passage from a 1593 lecture by Buxtorf quoted in the Introduction, above: *Encomion Hebraeae Linguae*, Basel, UB, Ki Ar 189, pt. 4, 1r–8r, at fol. 7r.

45. Buxtorf to Jacob Zwinger, 4 August 1593; Basel, UB, MS Fr Gr II 9, 89, previously quoted by Stephen Burnett, "The Christian Humanism of Johann Buxtorf (1564–1629)" (PhD diss., University of Wisconsin, Madison, 1990), 148 n7; Burnett, *From Christian Hebraism to Jewish Studies*, 252; and Anthony Grafton and Joanna Weinberg, with Alastair Hamilton, *"I have always loved the Holy Tongue": Isaac Casaubon, the Jews and a Forgotten Chapter in Renaissance Scholarship* (Cambridge, MA, 2011), 143.

46. Buxtorf, *Bibliotheca Rabbinica*, 251.

him, being read in the synagogues every sabbath day."[47] Following Elijah Levita, Buxtorf dated the reading of the *haftarot* to the time of Antiochus Epiphanes in the second century BCE. After Antiochus prohibited the reading of the Law and burned the copies of it, he explained, the pious began to meet and read aloud sections from the prophets that corresponded to the weekly Torah portions.[48] Buxtorf also explained how the yearly cycle of readings fit the Jewish liturgical calendar.[49] Always concerned with the practical needs of Christian readers, he noted that Jewish scholars normally used these divisions, which were unfamiliar to Christians, to identify biblical citations. Accordingly, he provided a table, with their familiar equivalents, for use by his readers.[50]

But the core of his work dealt not with canonical texts but with later rabbinical works. Here he found himself compelled to work in a different tradition: to adopt a set of bio-bibliographical methods and practices that humanist lexicographers and bibliographers had created for much less polemical purposes.[51] Buxtorf imitated them, for example, when he treated his work as a temporary resource that others had his permission to reuse and edit as they wished. The Zurich scholar Conrad Gessner, who had produced his massive *Bibliotheca universalis* in 1545, made clear that his own work had engulfed the more limited bibliographies of older compilers by listing "the catalogues of writers which I inserted in full in my own work."[52] On a separate sheet, he explained that one of his chief sources, the Benedictine Joannes Trithemius, whose name he abbreviated as T. in citations, had done the same.[53] In a final note, handwritten on this new leaf, Gessner revealed just how systematically he had gone about exploiting the work of his predecessor: "Sometimes I have set the letter T. twice before a name, when Trithemius wrote

47. Ibid.

48. Ibid., 251–52, following Elijah Levita, *Sefer Tishbi* (1541), s.v. "*patar*" (dismiss or exempt); Levita, *Lexicon Hebraicum utilissimum*, ed. and trans. Paul Fagius (Basel, 1557, but mostly made up of pages from the original edition: Isny: Fagius, 1541), 169–70.

49. Buxtorf, *Bibliotheca Rabbinica*, 253.

50. Ibid., 249, 254–55. It should be noted, however, that because he set the lectionary sections in alphabetical order, readers would have to have had sufficient proficiency in reading Hebrew texts to discern when the shorthand references to each section were being cited.

51. On the standard practices of humanist bio-bibliography, see Frederic Clark, "Reading the Life Cycle: History, Antiquity, and *fides* in Lambarde's *Perambulation* and Beyond," *Journal of the Warburg and Courtauld Institutes* 81 (2018): 191–208; Clark, *The First Pagan Historian: The Fortunes of a Fraud from Antiquity to the Enlightenment* (Oxford, 2020); and Clark, "Writing about Reading: On Early Modern Annotation Practice and the Future of Book History," in *Gabriel Harvey and the History of Reading: Essays by Lisa Jardine and others*, ed. Anthony Grafton, Nicholas Popper, and William Sherman (London, 2024), 307–24.

52. Conrad Gessner, *Bibliotheca universalis* (Zurich, 1545), sig. [* 6v]: "Catalogi scriptorum, quos integros inseruimus nostro."

53. Gessner, ibid., separate leaf inserted between sig. [*6v] and sig. ar.

about the same writer in two places: that is, in his ecclesiastical catalogue and in his German writers."[54] The *Bibliotheca universalis* gave readers many peeps into Gessner's workshop, allowing them to watch him combine derivative materials from different sources.

Gessner clearly saw his *Bibliotheca*, as printed in 1545, as a work of temporary scholarship. He filled his working copy of the book, preserved in Zurich, with corrections and additions, in the hope of bringing out a new and more perfect edition. Buxtorf did the same with his *Bibliotheca*. In 1640, Johann the Younger brought out a new edition of Buxtorf's compendia. He noted that his father and he had entered corrections and additions in two copies of the original edition, and that he had used both of them in his revisions of the book.[55] One annotated copy of the *Bibliotheca*, signed by one J. Buxtorf, survives in Basel. Material from its margins appears in the 1640 edition. It appears to be the copy annotated by Buxtorf the Younger.[56]

When Buxtorf the Elder urged his colleagues to add places and dates of printing to their notices, he also followed Gessner's model. As Gessner explained,

> The time at which particular books appeared also matters. For you can hardly find many of those that were printed some time ago any longer. And the later ones are generally better than the earlier ones, so long as careful printers produce the new edition. For the careless ones – of whom there are many – always put out worse ones. Knowledge of the places will also be useful. That informs us where we should look for each of them, and how to find them if all copies have now been sold separately, if that is necessary. It is also likely that copies of the book will survive for a long time in public and private libraries where they were printed.[57]

54. Gessner, ibid., handwritten note in his working copy: Zurich, Zentralbibliothek, ZB DrM 3 (available online on e-rara, https://www.e-rara.ch/): "Aliquando binariis nomina T. literae praeposuimus, cum de eodem scriptore Tritemius duobus in locis scripsit, nempe in catalogo ecclesiastico et in scriptoribus Germanis."

55. Johann Buxtorf the Younger, ep. ed. to *De abbreviaturis* (1640), sig.):(2v: "Additiones illae [to the Bibliotheca] profectae sunt partim ex paternis reliquiis, partim ex propriis scriniis. Reliquit enim ille exemplar, cuius margini (ut quivis authores facere consueverunt) adscripserat, quae ipsmet adiicienda observarat. Tale exemplar habebam & ego, in quo nonnulla adnotaram, quae mihi progressu temporis sese obtulerunt."

56. Buxtorf, *De abbreviaturis*, Basel, UB, FA II 24b, signed "I. Buxtorfj" on the title page.

57. Gessner, *Bibliotheca*, sig. *4v: "Non parum quoque refert quo tempore quaeque prodierint. Multa enim olim excusa, vix amplius invenias: & posteriora fere prioribus meliora sunt, modo a diligentibus chalcographis editio repetatur: indiligentes enim, quales plurimi sunt, semper emittunt depravatoriora, Locorum quoque notitia conducet ut sciatur unde quaeque petenda sint, & ut olim distractis iam cunctis exemplaribus, si opus erit, inveniri

Buxtorf was anything but consistent in his own efforts to follow these rules. Many of his listings included only author and title, for reasons that will become clear.

But he did take over other practices from Gessner. For example, he sometimes commented on typography and paper, especially when he approved of them.

> Hebrew Bibles, bare, in folio and quarto, published at Pesaro in 1494, very correctly. It was printed there also in octavo, in an elegant minuscule type, with points.... [On a Venetian Bomberg 1546 edition of Bahya ben Asher] printed in elegant type and clean paper, with the passages from Scripture noted in the margin.... I have also seen a quarto printed at Riva in very small type.[58]

Gessner, similarly, noted that printers differed greatly: "for some print in a polished and corrected form, with handsome characters, and on good paper. Others don't."[59]

Even a sharply and characteristically evaluative passage in Buxtorf's prefatory matter – one that seems to express with great directness and precision his feelings about Jewish literature and its uses – may have more to do with Gessner and Latin scholarly traditions than it seems. Once the literature of the Jews had been fully listed and collected, Buxtorf promised, "this web will finally be rewoven, and our schools will know the wisdom that the Synagogue of the Jews professed on the past and the doctrine that it professes today. It will be possible to gather gold from the dung, and to find the precious pearls hidden in the excrement."[60] Here he seems to echo the views of the thirteenth-century Dominican Ramon Martí. In his *Pugio fidei,* Martí stated, he had used "certain traditions that I found in the Talmud and midrashim, that is glosses and traditions of the ancient Jews. I removed them like pearls from an immense dunghill."[61]

---

possint. Verisimile est enim in publicis privatisque bibliothecis ibi quam diutissime superstitem fore librorum copiam, ubi excusi fuerint."

58. Buxtorf, *Bibliotheca Rabbinica,* 268: "Biblia Hebraea nuda in fol. & in quarto, impressa *Pisauri,* Anno 1494, correctissime. Ibidem quoque in 8° excusa, typo minusculo eleganti, cum punctis"; 279: "... typo eleganti & charta munda, cum locis Scripturae in margine annotatis.... Vidi & in quarto excusum Rivae, typo minutiore" [i.e., Riva di Trento, 1559].

59. Gessner, *Bibliotheca,* sig. *4v: "Multum interest inter typographos, cum alij polite, emendate, pulchris characteribus, & bonis chartis imprimant: alij contra."

60. Buxtorf, *Bibliotheca Rabbinica,* 259: "Sic tandem tela haec capta pertexetur, & scient nostrae quoque scholae, quam Synagoga judaica sapientiam olim professa sit, & quam hodie doctrinam profiteatur. Licebit & hic aurum ex stercore colligere, & margaritas pretiosas fimo reconditas reperire." Also quoted by Burnett, *From Christian Hebraism to Jewish Studies,* 163 and n140.

61. Ramon Martí, *Pugio fidei,* ed. Benedict Carpzov (Leipzig, 1687), 3: "... quaedam traditiones, quas in Talmud & *midraschim,* id est glossis, & traditionibus antiquorum Judaeorum reperi, & tanquam margaritas quasdam de maximo fimario sustuli[.]"

The two men agreed that the proper way for a Christian to use the works of the Jews was by careful selection. One must pan for gold, filtering out the limited number of valuable pieces concealed in a highly unappealing setting. Both clearly saw Jewish texts in a very different light than Galatino. In his *De arcanis Christianae veritatis,* he had offered a less expert but much more positive judgment: "It scarcely ever happens in the Talmud that one finds some Gemara – that, is a tradition or conclusion that those ancient fathers taught, accepted, and determined – that does not express some secret of Catholic truth, either explicitly and openly, or implicitly and figuratively."[62] Though Buxtorf cited Galatino regularly for information on the Kabbalah, he did not trust the compiler's assessments of the texts that he cited. Of one of Galatino's principal sources for the Kabbalah, the *Gale razaya,* Buxtorf wrote: "Galatino often praises and cites this book. Many doubt its reliability."[63] No wonder, it seems, that he took such a different stance on the Talmud.

In fact, though, Buxtorf's use of scatological expressions to describe the texts he had worked through grew from deep roots in the Latin scholarly tradition. Cassiodorus recalled in his *Institutes* that Virgil was once asked what he was doing while he was reading the archaic poet Ennius. He replied, "I am looking for gold in the dung." (1.1.8). As to the pearls, Jerome wrote to the widow Furia, advising her to read the Bible and the works of Christian sages: "you do not need to seek for gold in the muck: use many pearls to buy the one pearl of price."[64] Any scholar might decide, after assembling a compilation, that much of the material he had sifted was not of great value. Gessner quoted Vergil's line about Ennius in the course of a diatribe of his own about how much time he had wasted taking notes on the prefaces of the books he listed.[65] Perhaps Buxtorf only meant to nail his bibliographical flag to the mast.

Whether or not Buxtorf had Gessner's language in mind when he described the writings of the Jews, Gessner's testimony helps us understand how Buxtorf compiled his *Bibliotheca.* Though Gessner described books he had seen, he also drew information together from a wide range of other sources, highly varied in value and precision. He described these ill-assorted sources with an engaging lack of embarrassment:

62. Galatino, *De arcanis Catholicae veritatis,* book 1, chap. 6, cols. 19–20: "Nam vix fere in ipso Talmud contingit, aliquam reperire גמרא *ghemara,* id est traditionem seu conclusionem, ab antiquis illis patribus traditam, conclusam, ac determinatam, quae aliquod Catholicae veritatis arcanum aut explicite & aperte, aut implicite & per figuras non exprimat."

63. Buxtorf, *Bibliotheca Rabbinica,* 283: "Galatino saepissime hic liber laudatus & citatus, de cujus fide multi dubitant."

64. Jerome, *Ep.* 54.11.1 (*Sancti Eusebii Hieronymi Epistulae, Pars 1: Epistulae I–LXX,* ed. Isidor Hilberg, Corpus scriptorum ecclesiasticorum Latinorum 54 [Vienna, 1910], 478): "non necesse habes aurum in luto quaerere: multis margaritis unam redime margaritam."

65. Gessner, *Bibliotheca,* sig. *3v.

> I scraped up the material of my work from everywhere: from the catalogues of printers, a good many of which I obtained from various regions; from the catalogues of libraries, and then from libraries themselves, everywhere, both public and private, in Germany and Italy, which I examined with care; from the letters of my friends; from what learned men told me ...[66]

In passing, Buxtorf similarly identified some of the many ways in which he collected his materials. He too described the active life that he led in search of books to add to his collection and his listings.

One source of the *Bibliotheca*, as Prijs and Burnett have pointed out, was autopsy – exactly as Gessner claimed for his vastly larger project. Buxtorf worked from his own collection, for a start. Copies of some of the books that he described in the *Bibliotheca* are listed in the eighteenth-century inventory of the family library.[67] Several remain in the Basel university library, frequently bearing bibliographical information, some of which at least seems to have been entered by Buxtorf the Younger.[68]

Buxtorf paid special attention to books that had played a role in his own literary production. Thus, for example, he added a long entry on *Avkat rohel*, a

66. Ibid., sig. 3* recto: "Materiam operis undecunque corrasi: ex catalogis typographorum, quorum non paucos diversis e regionibus conquisivi: ex Bibliothecarum elenchis, tum Bibliothecis ipsis passim, et publicis & privatis in Germania Italiaque diligenter inspectis, ex litteris amicorum, ex narratione doctorum hominum, denique ex Catalogis scriptorum, quos paulo post nominabo."

67. Joseph Prijs, *Die Handschriften der Universitätsbibliothek Basel: Die Hebräischen Handschriften: Katalog auf Grund der Beschreibungen von Joseph Prijs*, ed. Bernhard Prijs and David Prijs, with Stephen G. Burnett and Thomas Willi (Basel, 1994), 324; Burnett, *From Christian Hebraism to Jewish Studies*, 159–60.

68. Fairly typical is Buxtorf's *Sammelband*, containing three books: Jacob ben Isaac Luzzatto, *Kaftor va-ferah* (Basel, 1581); Judah ben Samuel he-hasid, *Sefer Hasidim* (Basel, 1581); and Menahem Recanati, *Sefer Ta'ame ha-mitsvot* (Basel, 1581); Basel, UB, FA V 1:1–3, all described in *Bibliotheca Rabbinica*, 298–99, 290, 291. A description of the first volume appears on the verso of the front flyleaf: "כפתר ופרח Kaphtor uperach. Liber est, continens historias Talmudicas, Agados, Medraschos et similes sermones allegoricos et hyperbolicos. Basilea: apud Ambrosium Frobenium. 1581" (298–99 contains a more extended version of the same description). A copy of the *Yosippon* (Basel, 1541), Basel, UB, Frey-Gryn A II 52, belonged to the elder Buxtorf, as a signature records: "Ex libris Ioannis Buxdorfij Westphali. Basil. An. 1590." But it bears bibliographical annotations by his son. A copy of the *Kuzari* (Venice, 1594), Basel, UB, FA V 11, presumably belonged to Johann but bears bibliographical notes in the hand of Johann the Younger, For Buxtorf's curiosity about bibliographical detail, see his copy of Issahar Moses ben Betsalel Katz's popular introduction to the concepts of the Zohar, *Imrei binah* (Prague, 1610), Basel, UB, FA V 37, which bears Buxtorf's signature and the date 1611. On fol. 4a the author states that he will refer to the Cremona edition of the Zohar. In a marginal note, Buxtorf writes: "Cremonensis editio Zohar." See also Basel, UB, FA VI 8 (another *Sammelband* with brief bibliographical notes).

work attributed to Jacob Machir. Buxtorf had used this anonymous booklet, a popular anthology of messianic tales and omens, to disparage Jewish messianism in the last chapter of his *Jüden Schul*. In the *Bibliotheca* he did not cast judgment on the book; nevertheless, he provided a full description of its contents. He referred to the Venetian print of 1597 [*sic*], adding that he owned a manuscript of the work.[69] The authoritative legal compendium of Isaac Alfasi was also treated to a fairly lengthy description.[70] Buxtorf briefly outlined the structure of the work, drawing attention to the section on the biblical duty for men to don fringes that he had already discussed in the *Jüden Schul*. He mentioned the latest edition (Basel, 1602), which Jews had put up for sale in Frankfurt. In addition, he described the Cracow edition of 1597, which, unlike the Basel print, contained all the standard commentaries and was the version he himself owned.[71]

Sometimes, Buxtorf gave no indication that he was familiar with a book that figured in his collection. He described the book *Zevah Pesah* as a work written by Isaac Abravanel, without further elucidation. Given that he had made extracts from the work in his copybook, it is puzzling that he did not proffer more information about this classic commentary on the Passover Haggadah.[72] On the other hand, he devoted a long entry to the *Alphabet* of Ben Sira, in which he flaunted his ownership of the old Hebrew print published in 1519 in Constantinople that Sebastian Münster had used in his *Cosmographia*, and described its miscellaneous contents.[73]

Buxtorf continued both acquiring new Hebrew books and compiling lists of them after the *Bibliotheca* appeared. In a letter of 25 March 1615 to Erpenius, Buxtorf excused himself for being slow to send a document he had promised – evidently, a handwritten copy of his own library catalogue: "I have not yet copied the full catalogue of my rabbinic books. At this [book] fair I expect three or four more. You will have it, as well as that of Waser, at the next opportunity."[74]

69. The work was published in Venice by Giorgio di Cabali in 1566 (not 1597); Buxtorf's copy is Basel, UB, FA VIII 28:4. His manuscript of *Avkat rohel* is Basel, UB, R III 3 (see Prijs, *Die Hebräischen Handschriften*, 53).

70. Buxtorf, *Bibliotheca Rabbinica*, 273–74.

71. Buxtorf's copy of Alfasi's *Sefer ha-halakhot* (Cracow, 1597) is Basel, UB, FA III 3a–c. It contains his annotations, pointing hands, and underlinings.

72. Buxtorf's copy of Isaac Abravanel's *Zevah Pesah* (Cremona, 1557) is Basel, UB, FA V 3. He refers to the work and Abravanel's use of the biblical phrase "[ha-]mayim ha-zedonim" (the arrogant waters) for baptism in his copybook, 40.

73. He referred to the Latin translations of the work by Paulus Fagius (Basel, 1542) and Johannes Drusius (Franeker, 1597). He omitted the date of the Constantinople edition (1519). His thickly annotated copy is in Basel, UB, FA VI 10. Buxtorf also owned a copy of Fagius's Ben Sira, Basel, UB, FB III 10/2.

74. Buxtorf to Erpenius, 25 March 1615, Munich, BSB, clm 10359, 215: "Catalogum librorum meorum Rabbinicorum nondum totum descripsi. Exspecto his nundinis tres vel quatuor. Habebis eum proxima alia occasione, ut et D. Waseri."

The picture that emerges is lively: a form of sociable knowledge, in the words of Elizabeth Yale, in which the collecting of books played an essential role – and in which giving books to distinguished friends, as well as receiving them or information about them as gifts and circulating lists of private library holdings, deserved prominent mention in correspondence and in published works of scholarship.[75] Bibliography and sociability had been organically connected since the early fifteenth century, when Niccolò Niccoli directed Poggio Bracciolini's "deliberate and focused" hunts for the Latin classics.[76] Gessner, closer to Buxtorf in space, time, and method, stuffed his *Bibliotheca universalis* with personal anecdotes, brief accounts of information that friends had provided, and the like.[77] Here too Buxtorf worked in a humanist tradition.

In the simplest cases, Buxtorf's friends provided him with information about particularly glamorous acquisitions. When they were distinguished Christian scholars, he took the occasion to engage in a bit of book chat, to offer them some warm personal acknowledgment, and sometimes even to pose a question in his printed text, as one might do today on social media. "I recently learned," he wrote in the *Bibliotheca*, "that my friend Walther Keuchen, Rector of the school at Hanau, has obtained a complete Abravanel printed in Turkey in two great volumes. If this is true, he should take pride in it, for the book is warmly praised."[78] After Keuchen obtained Buxtorf's book, he replied with a note that offered some corrections and made clear that he had no very high opinion of Abravanel – and incidentally revealed how he and Buxtorf had tried to exchange information in this instance:

> In the beginning you mention my Abravanel. It is true that I have him in complete form, but not printed in Turkey in two volumes, as that Jew lied: but the commentaries on the prior and posterior prophets, [written] at Naples in Italy. The rest are printed, as you write in your little book. In the

75. Elizabeth Yale, *Sociable Knowledge: Natural History and the Nation in Early Modern England* (Philadelphia, 2016).

76. See Julia Haig Gaisser, "Poggio and Other Book Hunters," in *Poggio Bracciolini and the Re(dis)covery of Antiquity: Textual and Material Conditions: Proceedings of the Symposium Held at Bryn Mawr College on April 8–9, 2016*, ed. Roberta Ricci (Florence, 2020), 173–88, at 174, and *Two Renaissance Book Hunters: The Letters of Poggius Bracciolini to Nicolaus de Niccolis*, ed. Phyllis Walter Goodhart Gordan (New York, 1974). For the larger background, see L.D. Reynolds and N.G. Wilson, *Scribes and Scholars: A Guide to the Transmission of Greek and Latin Literature* (Oxford, 2013), and Christopher de Hamel, *The Posthumous Papers of the Manuscripts Club* (London, 2022).

77. E.g., Gessner, *Bibliotheca*, fol. 518r; fol. 275v; fol. 294r–v.

78. Buxtorf, *Bibliotheca Rabbinica*, 263: "Intellexi nuper, integrum Abarbinelem in Turcia duob. magnis voluminibus excusum, nactum esse amicum meum, Gulaterum Keuchenium, Rectorem Scholae Hannovianae. Quod si verum, est quod sibi gratuletur. Nam liber est summe commendatus."

> Prophets his aim is to refute the Christians expressly. Accordingly, I had written some very sharp things to you before this, but I have the feeling that my letter went astray.[79]

In similar passages, Buxtorf recalled that he had given a copy of Benjamin of Tudela's account of his travels to Joseph Scaliger, "who has left his very valuable Hebrew Bible to the University of Leiden," and that he had given *Sepher hamussar* to his friend Ioannes Drusius.[80]

In other cases, Buxtorf recalled that he had seen a particular book "with the Jews" or, more precisely, "at Frankfurt with the Jews."[81] In one instance, he inserted a reference to the work *Kene binah,* which he was given by "a certain Jew," but noted that he did not have time to scrutinize it properly.[82] When Jews provided books, he named no names and did not praise their learning and humanity. It seems likely that these exchanges involved purchases rather than gifts. Probably he visited the same display of Jewish books for sale in Frankfurt where, as we will see, Franciscus Gomarus bought a copy of the Zohar for Buxtorf. Still, it seems clear that his transactions with Jews sometimes involved book talk.

Manuscripts occasionally provoked it as well. He noted the "enormous and complete manuscript of Rashi's commentaries, three hundred years old," that Scaliger had left to the University of Leiden Library; also the two-volume manuscript of the Jerusalem Talmud that Scaliger had owned and bequeathed to Leiden.[83] As we will see, Buxtorf listed many books that had not yet been printed. But he seems to have had relatively little interest in describing manuscripts. His son Johann had a keener eye for them. In one note in his working copy of the *Bibliotheca,* he added a reference to a manuscript titled *Even shetiyyah* (Foundation Stone). He described its contents precisely, noting that he had seen it in the library

79. Keuchen to Buxtorf, 7 February 1614; Basel, UB, MS G I 60, 319: "In principio facis mentionem Barbenalli mei, quem totum me habere verum est, sed non in Turcia duobus voluminibus impressum, ut Judaeus ille mentitus fuit: sed Neapoli Italiae commentarii in Prophetas priores & posteriores. Reliqui sunt impressi, ut in libello tu scribis. In Prophetis hoc ipsum propositum est, ut Christianos ex professo refutet: unde ante haec ad te verba quaedam mire contumeliosa perscripseram; sed sentio literas meas aberrasse."

80. Buxtorf, *Bibliotheca Rabbinica,* 304–5. There were various printings of Judah Kalats's ethical work, *Sefer ha-mussar*: Constantinople, 1536; Mantua, 1560; Cracow, 1598. Drusius referred to it several times in his *Annotationum in totum Jesu Christi Testamentum, sive praeteritorum libri decem* (Franeker, 1612).

81. Buxtorf, *Bibliotheca Rabbinica,* 299, 273–74, 283.

82. Ibid., 322: "Mihi, dum hunc ipsum catalogum scriberem, a Judaeo quodam fuit donatus, sed prae festinatione praeli non licuit accuratius perlustrare." Elkana ben Gadur Altschuler was the author of *Keneh binah* (Prague, 1609), a kabbalistic work on the Shema.

83. Buxtorf, *Bibliotheca Rabbinica,* 263, 332. The manuscripts are now respectively Leiden, Leiden University Library, MS Or. 4718 (Scal. 1) and MS Or. 4720 (Scal. 3).

at Heidelberg.[84] This information was incorporated into his revised edition of 1640: "Once upon a time I saw this manuscript in the Heidelberg library."[85] With regard to the *Sefer amana* (The book of faith), which was printed in Hebrew and Latin by Fagius in 1542, Buxtorf the Younger stated in the revised edition that he owned a copy written in a "Jewish hand."[86]

Historians often claim that Buxtorf's *Bibliotheca Rabbinica*, though deficient in so many ways, constituted an important contribution to the study of Hebrew bibliography.[87] Yet, even though subsequent bibliographers modeled their own catalogues of Hebrew books on Buxtorf's *Bibliotheca*, there is no denying the rather idiosyncratic and lacunose nature of his enterprise. Close examination of the books listed in the *Bibliotheca* reveals that Buxtorf himself had never actually seen the vast majority of the items listed. Rather, as Keuchen pointed out, he derived his information from secondary or even tertiary sources.[88] His system was not so much to itemize his own library as rather to list names and titles as they appeared in his favorite Jewish history textbooks, in particular Abraham Zacuto's *Sefer yuhasin*, which he read in the second edition printed in Cracow in 1581;[89] Gedaliah ibn Yahya's *Shalshelet ha-kabbalah* (Venice, 1587);[90] and David Gans's *Tsemah David* (Prague, 1592).[91] He rifled these works systematically for

84. Basel, UB, FA II 24b, "Vidi M in Bib. Acad, H. in fol.3." The younger Buxtorf spent some years in Heidelberg after taking his MA at Basel in 1615. The MS is now Vatican City, BAV, MS ebr. 250 (part of the great collection taken from Heidelberg in the seventeenth century). Its author is Abraham ben Judah of Candia. As Buxtorf says, it is divided into four parts (for a description of the MS see *Hebrew Manuscripts in the Vatican Library: Catalogue*, ed. Benjamin Richler, palaeographical and codicological descriptions by Malachi Beit-Arié in collaboration with Nurit Pasternak [Città del Vaticano, 2008], 187). See also Buxtorf the Younger to Hottinger, 8 September 1647, Zurich, ZB, MS F 51, 194, where, regarding the manuscript in Heidelberg, Buxtorf the Younger commented: "Heidelbergense aliquando vidi, et manibus meis tractavi; cuius autem antiquitatis fuerit, certo asseverare nequeo."

85. Buxtorf, *Bibliotheca Rabbinica* (1640), 304.

86. Ibid., 314: "Habeo etiam exemplar manu Judaica exaratum." This is probably to be identified as Basel, UB, R IV 3 (see Prijs, *Die Hebräischen Handschriften*, 56).

87. See Burnett, *From Christian Hebraism to Jewish Studies*, 164, and Shimeon Brisman, *A History and Guide to Judaic Bibliography* (Cincinnati; New York: Ktav, 1977), 4–5.

88. Menahem Slatkine had already alerted readers to Buxtorf's thorough use of Gedaliah ibn Yahya's *Shalshelet ha-kabbalah* in his *Reishit bikkurei ha-bibliyografyah ba-sifrut ha-'Ivrit: ha-sefer Siftei yeshenim me-R. Shabbetai Meshorer Bass* (Tel Aviv, 1957/8), 23. Nearly half of the items in the catalogue were derived from secondary sources.

89. His own copy of Zacuto is Basel, UB, FA VI 28:2. In the *De abbreviaturis* under the entry אש"ש he refers to the first edition printed in Constantinople (1566).

90. We have not located Buxtorf's copy of *Shalshelet ha-kabbalah*, which was clearly a key source for him. He often refers the reader to precise page numbers in the work.

91. Buxtorf's copy of the 1592 Prague edition of Gans is Basel, UB, FA V 12:1–2. Burnett notes that Buxtorf did not consistently work from books he saw (*From Christian Hebraism*

information about authors and their books, some of which had not been printed, and still today remain in manuscript.[92] In his copies of *Zemah David* and *Sefer yuhasin*, both of which survive, he underlined and sometimes annotated the bibliographical material that he later included in the *Bibliotheca*.

In Buxtorf's day, and until fairly recently (1997), the very first book on his list, *Even bohan*, an anti-Christian polemic by Shemtov ibn Isaac Shaprut, existed only in a number of manuscript versions.[93] Buxtorf had no knowledge of the work's content and had simply derived his minimal information about ibn Shaprut – "he lived in 1375" –from the *Tsemah David*.[94] His entry for sermons, "derashot," refers to the sermons of Rabbenu Nissim, a disciple of Rabbi Moses ben Nahman who died in 1268 – a fact drawn from Gans and Gedaliah ibn Yahya. In the same entry Buxtorf spoke of "Gedaliah, who wrote a book of one hundred and eighty sermons which he preached in various places throughout Italy." This particular piece of information was taken from Gedaliah ibn Yahya's own testimony.[95] Yet when Buxtorf came to write the entry for *Shalshelet ha-kabbalah*, he referred to the page where ibn Yahya claimed to have preached and to have written several books. Here, he assumed a more critical stance, casting doubt on the existence of any of these books and dubbing ibn Yahya "a strikingly ambitious man."[96]

Even the entry for one of the most important codes, the *Arba'ah turim* of Jacob ben Asher, was mostly derived from the *Tsemah David*. When Buxtorf gave the date of the code (1240) according to the information given "at the end of the section on the laws of the new moon, paragraph 427," he was simply

---

to Jewish Studies, 160–63), but does not make clear how indebted Buxtorf was to these three historical sources.

92. He also occasionally inserted references to books that he came across in other works; e.g., in the entry for *Derech emuna* he states that this work is often cited by the author of *Kaftor* (i.e., Jacob Luzzatto).

93. José-Vicente Niclós, *Šem tob ibn saprut = 'La Piedra de toque' (Eben bohan): una obra de controversia judeo-cristiana; introducción, edición crítica, traducción y notas al libro I* (Madrid: 1997).

94. David Gans, *Sefer Tsemah David*, ed. Mordechai Breuer (Jerusalem, 1983), fol. 59r. Buxtorf the Younger knew that the work was anti-Christian. See *Bibliotheca Rabbinica* (1640), s.v. *Even bochan*.

95. Gedaliah ibn Yahya, *Shalshelet ha-kabbalah*, fol. 67r. Buxtorf's statement, "R. Gedalia composuit *Sepher hadraschos*, librum concionum centum octoginta, quas in Italia in diversis locis habuit" (*Bibliotheca Rabbinica*, 284), seems to take for granted that his readers would understand that he was referring to ibn Yahya and his statement in *Shalshelet ha-kabbalah*. Oddly, in the *De abbreviaturis*, under the entry ש"ה (172) Buxtorf shows no knowledge of this work, which was so essential for the entire undertaking: "*Catena kabalae*. Nomen est libri historici qui in Tzemach David saepe citatur."

96. Buxtorf, *De abbreviaturis*, 327: "R. Gedaliah ben Jechaiah qui fol. 66 enumerat unum et viginti libros a se conscriptos . . . An omnes isti libri editi sint, ignoro. Praecipui illorum in hoc catalogo suis locis propositi sint. Videtur mire ambitiosus homo." See also 284.

transmitting the facts that he had discovered in Gans's chronicle.[97] In the entry for *She'elot* ("Quaestiones et Responsiones"), the first reference to *She'iltot R. Ahai* derived from Abraham ibn Daud's chronicle *Sefer ha-Kabbalah*. The list of authors of Responsa was culled from Gans's *Tsemah David*. Buxtorf's system was simply to troll through the first part of Gans's chronicle, identifying, examining, and excerpting named authorities whom the author had described as authors of Responsa.[98] Buxtorf mentioned two unpublished works by Johanan Alemanno, who, he noted, had been Pico della Mirandola's teacher.[99] He drew all of this information, including Alemanno's connection to the Christian scholar, from the *Shalshelet ha-kabbalah*.[100]

These examples could be multiplied, but they suffice to indicate the nature of Buxtorf's method of assembling his material. The limitations of this procedure and similarly its pedagogical usefulness are self-evident. In fact, the entire *Bibliotheca Rabbinica* is something of a mixed bag its records of specific books combine bibliographical information, personal anecdotes, and mini historical disquisitions. In this respect above all it resembles Gessner's first massive compilation of 1545.

At times, to be sure, Buxtorf devoted special energy to the description of a particular book. Buxtorf was no admirer of Kabbalah or mystical Jewish literature. Yet not only did he own several works in this category, but he used these books where it suited his scholarly purpose. As we will see, the antiquity of the Hebrew vowel points was of central importance to Buxtorf's appraisal of the Masoretic Hebrew Bible.[101] He made every effort to endorse the antiquity of a kabbalistic work, the Book of Bahir, which he thought supported his view. Taking on trust the attributed authorship to Nehuniah ben ha-Kanah, a first-century Mishnaic sage, he reproduced notable sayings ascribed to Nehuniah in the Mishnah and then proffered another piece of information: the work testified to the antiquity of the vowel points that grammarians and he himself had demonstrated in his *Thesaurus*.[102] In the entry for *Zohar hadash*, a compilation of texts that were lacking

97. Gans, *Tsemah David*, fol. 58v (ed. Breuer, p. 131); Buxtorf, *Bibliotheca Rabbinica*, 276: "Scripsit librum anno Christi 1340, ut ex ipso libro apparet, in iuribus ראש חודש *Rosch chodesch* id est, novilunii, num. 427."

98. Buxtorf, *Bibliotheca Rabbinica*, 324–25. Following the same method, he also included an entry for *Teshuvot*, Responsiones (ibid., 335).

99. Ibid., 290–91 (on Alemanno's *Heshec shlomo*, a commentary on the Song of Songs); 316 (on Alemanno's *Einei ha-edah*, a commentary on the first five chapters of Genesis). At 291 Buxtorf says of Alemanno: "qui fuit praeceptor Johannis Pici Mirandulae Comitis."

100. Ibn Yahya, *Shalshelet ha-kabbalah*, fol. 63v.

101. On Buxtorf's treatment of the topic, see Chapter 6, below.

102. Buxtorf, *Bibliotheca Rabbinica*, 279: "... Ex libro Bahir testimonium Grammatici adducunt pro antiquitate punctorum vocalium astruenda, quod & ego in Thesauro meo Grammatico ex Cephalerio descripsi[.]"

in printed editions of the Zohar, which was first printed in Cracow in 1603, not only did he refer to his *Thesaurus* but he also quoted several significant passages that contained references to the names of the vowel points. Here he adopted a more cautious tone in his assessment, although he was clearly intent on using these texts as evidence for the antiquity of the vowel points.[103]

Buxtorf devoted substantial attention to the Zohar even though he had no real interest in the mystical speculations it contained.[104] In the abbreviations, in the entry for רשבי (Rashbi, Rabbi Simeon bar Yohai), he confronted some of the central scholarly issues relating to the Zohar and its purported author.[105] Beside a series of passages cited in the original and in Latin translation, Buxtorf probed the person of Simeon bar Yohai as portrayed in the Talmud. On the basis of the evidence compiled by Zacuto and ibn Yahya, he questioned the story of Simeon's composition of the Zohar while in hiding in a cave during the persecutions of Emperor Hadrian. He also followed his Jewish sources in underlining the pseudepigraphical nature of the Zohar – it had been ascribed to Simeon bar Yohai, whose teachings had been transmitted orally over time, but who had never put it into writing.[106] In the shorter entry for the Zohar in the *Bibliotheca*, he referred the reader to the entry for "Rashbi" in the *Abbreviations*, and pointed to the question of Simeon's authorship of the extant text of the Zohar and its authority by means of a long passage from Gedaliah ibn Yahya's *Shalshelet ha-kabbalah*. Ibn Yahya had reported the view, also recorded in Zacuto's history, that the Zohar did not contradict the Gemara (Talmud) and that when it gave an explicit ruling on a matter not expounded in the Gemara, the zoharic ruling was followed.[107] The Zohar, according to the Jews, was not a recent production but, like many other Jewish books, had remained buried in Egypt and "those eastern regions" for centuries.[108]

In this case, Buxtorf had taken the task of bibliographer far beyond conventional standards. In these two entries, Buxtorf aired the questions about the

103. Ibid., 287–88, at 287: "In eo Medrasch multa de punctis vocalium & accentuum habentur, quae si essent eius antiquitatis, ut judaei volunt, multum haberent ponderis[.]"

104. See Chapter 6 for further discussion of Buxtorf's evolving views.

105. Buxtorf, *De abbreviaturis*, 164–68.

106. Buxtorf follows Zacuto in comparing the Zohar – which was attributed to Simeon bar Yohai but was actually put together by his students and his students' students – to the Mishnah, Sifra, Sifrei, and Tosefta, which originated from R. Akiba.

107. Ibn Yahya, *Shalshelet ha-Kabbalah*, fol. 31v; Abraham Zacuto, *Sefer Yuhasin* (Constantinople, 1566), fols. 42r–v. On the question of the Zohar as a legal authority, see Leon Modena's critique, discussed by Yaacob Dweck, *The Scandal of the Kabbalah* (Princeton, 2011), 76–78

108. Buxtorf, *Bibliotheca Rabbinica*, 286–87, at 287: "Quod tamdiu delituerit, id aiunt non esse novum inter judaicos libros, quorum multi adhuc hodie extent, qui similiter per multos centenos annos in Aegypto & istis orientalibus locis occultati fuerint. Sed satis."

Zohar's antiquity and authorship that had vexed both Jews and Christians and that continued to be the subject of fierce debate. So engrossed was he in this issue that he failed to give information about the printing of the text in question.[109] Gessner, similarly, had quoted extensively from and speculated searchingly about texts and authors that left him and other readers uncertain: for example, the works of the humanist, engineer, alchemist, and magus Heinrich Cornelius Agrippa.[110]

As we noted above, the *Bibliotheca Rabbinica* is inextricably linked to the *De abbreviaturis*. This certainly comes to the fore in the case of the Zohar, where Buxtorf referred the reader back to his entry on Simeon bar Yohai in the *De abbreviaturis*.[111] Yet on other occasions he failed to supply the reader with such directions. In the entry for *Yad hazakah* the reader is given a full breakdown of the fourteen books of Maimonides's code, the *Mishneh Torah* or *Yad hazakah*. But it is only in the *De Abbreviaturis*, in the entry for Rambam (the acronym for Moses ben Maimon or Maimonides), that we are told that the code is titled either *Yad* [*hazakah*] or *Mishneh Torah*, a vital piece of information for those wishing to consult the work.

Personal anecdotes or asides crop up regularly throughout Buxtorf's *Bibliotheca*, as they did in many early works of bio-bibliography. For example, in addition to providing details of the contents and publication of *Neve shalom* written by the fifteenth-century philosophical homilist Abraham Shalom, Buxtorf felt the need to explain to his readers that his friend, Erpenius, had given him a copy of the book. He ended the entry with the prayer that God should bless Erpenius's studies and labors.[112] Erpenius had aided Buxtorf in his quest for Hebrew books. Presumably, Buxtorf felt it politic to mention his debt to Erpenius in print.[113]

109. His son added the requisite information in the 1640 edition. Interestingly, he changed the text, claiming that the Zohar contradicts the Gemara in many legal points ("in multis iuribus repugnat doctrinae Gemarae"; Buxtorf, *Bibliotheca Rabbinica* [1640], 337), whereas the senior Buxtorf had originally written "in nullis juribus repugnat doctrinae Gemarae" (*Bibliotheca Rabbinica*, 287).

110. Gessner's article on Agrippa appears in *Bibliotheca*, fols. 307r–309v. At the end he writes: "Hic author in plaerisque Romanae ecclesiae consensit, in multis dissensit, authoritati ipsius derogans, nuptias sacerdotum suadens, imaginum omnem usum taxans, cucullam diaboli inventum asserens, recentiorum theologorum, ut Scoti, Thomae, scripta reijiciens, &c. In quibusdam parum constans, nec ausus animi sententiam proferre" (fol. 309v).

111. Buxtorf, *Bibliotheca Rabbinica*, 286: "Sohar . . . plane Cabalisticus, de quo vide in Abbreviatura רש״בי."

112. Buxtorf, *Bibliotheca Rabbinica*, 310: "Hoc me nuper donavit M. Thomas Erpenius, amicus meus è primis, Arabicae linguae & aliarum Orientalium in Academia Leidensi Professor, cujus studiis & laboribus Deus benedicat."

113. On Buxtorf's relationship with Erpenius, see Vrolijk and Weinberg, "Thomas Erpenius, Oriental Scholarship and the Art of Persuasion," 54–57.

Buxtorf evinced appreciation of certain literary works, especially elegant and rhetorically complex books such as Isaac Obadiah ben Jacob's *Iggeret dofi ha-zeman* (Letter on the disgrace of the time).[114] Buxtorf's recommendation of Yedaiah Bedersi's *Behinat olam, Probatio, examen mundi* – "splendid both in content and in linguistic style" – is particularly noteworthy.[115] He enhanced his own endorsement of the didactic poem by citing that of his distinguished predecessor in Basel, Sebastian Münster.[116] In a prefatory epistle to his edition of Abraham bar Hiyya's *Sphaera mundi*, Münster had scanned a list of Hebrew books that Andreas Masius had sent to him, and declared: "I would like to read it [i.e., *Behinat olam*] more than all the other [books on Masius's list], if I could get hold of a copy."[117] Buxtorf's extraordinary promotion of this popular, but very difficult and sometimes obscure, philosophical and moralistic text is significant. It brings to the fore what has already been noted: Buxtorf's positive assessment, and indeed ruling passion, for certain kinds of Ḥebrew literature. Buxtorf annotated his copy of the book very thoroughly. For the most part he added biblical allusions, but he was especially drawn to a line in chapter 4 about the consequences of following human desires.[118] It is telling that such a moralistic poem attracted him, as it would many generations of readers, including Goethe and Lessing.

It also demonstrates his awareness of working within a tradition of Hebrew studies. Sebastian Münster was one of the most distinguished Hebraists of the sixteenth century – as professor of Hebrew in Basel, Buxtorf was as it were taking on his mantel, and he felt obliged to take Münster's views into account. This is certainly the case in the entry for *Mikneh Abraham* (Peculium Abrahamae), written by the physician and grammarian Abraham de Balmes and printed in

114. Isaac Obadiah (of Patras), *Iggeret dofi ha-zeman* (Venice, 1598), is as Buxtorf wrote, a highly ornate monograph on the disgrace of the time. The work ends with a series of letters, which Buxtorf would have put to his own use in his work on Hebrew epistolography.

115. Buxtorf, *Bibliotheca Rabbinica*, 279–80 at 279.

116. Yedaiah ben Abraham Bedersi wrote the *Behinat ha-olam* (Examination of the world) in Provence in the thirteenth century. It was first published in Mantua in 1474, and many times thereafter. Buxtorf mentioned the prints of Cracow (1591) and Prague (1598). He corrected David Gans's mistake in referring to the author as Jacob Bedersi, but also transmitted Gans's incorrect information that Beḍersi lived in Barcelona.

117. Münster, in Abraham bar Hiyya, *Sphaera mundi* (Basel, 1546), sig. *r: "Item ספר בחינת העולם quem prae omnibus legere cuperem si mihi copia contingere posset." On Masius and Münster, see Theodor Dunkelgrün, "The Hebrew Library of a Renaissance Humanist: Andreas Masius and the Bibliography to His *Iosuae Imperatoris Historia* (1574), with a Latin Edition and an Annotated English Translation," *Studia Rosenthaliana* 42–43 (2010–11): 197–252, esp. 200–201.

118. Buxtorf's copy of *Behinat olam* (Prague, 1598) is Basel, UB, FA VI 17.

Venice by Bomberg in both a Hebrew and a bilingual Hebrew/Latin edition in 1523.[119] Buxtorf wrote out Münster's highly negative and scathing assessment of the work. According to Münster, de Balmes was more intent on disparaging the views of his predecessors than on teaching. Where he should have conveyed grammatical rules, he gave physical explanations or metaphysical propositions. The bilingual text enabled the student to learn Hebrew phrases, but other than that, the work was of scant use for the reader.[120] Buxtorf made no comment on Münster's assessment and simply (and rightly) added that the Latin translation was too literal and that the Hebrew text was more intelligible than the Latin.[121] It would appear that he needed to air Münster's opinion of the book, but withheld himself from either endorsing or negating the somewhat damning appraisal of his illustrious predecessor.

We have noted how extensively and intently Buxtorf had read some of the major sixteenth-century Jewish historiographical sources. His interest in these texts was not solely to excavate the bio-bibliographical references they contained, but also to gain an entry into the poorly charted course of Jewish history, which, as we will see, was vital for his reconstruction of the historical foundation of the Masorah. This historical preoccupation spilled over into the bibliography. In one of the very last entries he listed *Tikkun shtaros* (Ordinatio contractuum), which he ascribed to "Juda Barcellonius." He then proceeded to give a potted history of the Jewish mode of dating and, in particular, the *minyan shtarot* ("era of documents," the standard term for the Seleucid era, 312–63 BCE).[122] It is not clear how Buxtorf

119. The title of the work is *Mikneh Abram* (Peculium Abrae), based on the phrase in Genesis 13:7.

120. Münster's attack occurs in his edition of Elijah Levita's *Grammatica Hebraica absolutissima* (Basel, 1525), sig. a, fol. 2v: "Nam autor operis Abraham a Balmis nihil aliud agere mihi visus est, quam veterum doctrinam perpetuo convellere atque impugnare, magis in insectando occupatus quam in docendo. At in dubium tantum vocare priscorum praeceptiones, cum interim nihil certi statuas, non docere est sed ridere. Quin saepe ubi grammaticae regulae tradendae erant, ipse prorsus ἀπροσδιονύσως & velut oblitus argumenti, physicas rationes et metaphysica theoremata affert, ut praeter Hebraismi phrasim, quam ex opposita versione latina degustare licet, parum utilitatis alioqui ad lectorem sit rediturum." De Balmes's work was a new venture in Hebrew grammar. He applied logical and particularly Aristotelian categories, possibly modeled on the Latin speculative grammarians, and intertwined his arguments with a mystical conception of the Hebrew language.

121. Buxtorf, *Bibliotheca Rabbinica*, 378 s.v. "Mikne Abraham": "Versio Latina, literam rigide persequens, adeo obscura & inepta est, ut longe facilius Hebraea intelligas, quam Latina."

122. In the 1640 *Bibliotheca Rabbinica* Buxtorf's son added a reference to Azariah de' Rossi's *Light of the Eyes* (chap. 24), a vital source for treatment of the use of the *minyan shtarot* or era of documents.

came across this title. He claimed that R. Juda Barsellonius had written this work around 1070, probably relying on David Gans's report that Judah ibn Bargelloni had written the *Sefer ittim* in 1070.[123] Ibn Bargelloni had indeed written a *Sefer shtarot*, part of a larger work on cases of civil law. Buxtorf clearly had no access to a manuscript of this book, but simply used the opportunity to educate his readers on deciphering dates in Hebrew books. In contrast, Buxtorf the critical reader comes to the fore in the entry for *Seder olam S[Z]utta* (The small order of the world), the medieval chronicle that lists generations from Adam to Jehoiachin and sets out to authenticate the authority of the Babylonian exilarchs. Basing his claims on the absence of references to some of the most distinguished Geonim, such as Sherira Gaon, he noted that "the Hebrews" believed that the date given at the beginning of the text, 804, might have been a marginal note inserted into the text by "an imprudent historian," who followed the custom of the discipline and brought his text up to date.[124]

Buxtorf's *Bibliotheca Rabbinica* did not emerge as a *creatio ex nihilo*; yet, previous attempts at listing Jewish books had been far less ambitious in scope and detail. On close examination it bears all the characteristics of a copybook, into which he had deposited all manner of interesting information and facts that he had collected over decades. Perhaps the most revealing section of the work comes at the end. A short text addressed to the reader explains how to pronounce the Latin transliterations Buxtorf had offered for Hebrew titles. Buxtorf had crafted them, the text explained, "according to the form of the Jewish pronunciation." He had done so "to the end that if anyone asks for them from the Jews, he will know how to name them in their way."[125] At the end as at the beginning, practicality

123. Gans, *Tsemah David*, fol. 53r (ed. Breuer, 117). Buxtorf had given an entry for *Ittim* vel Sepher *haittim*: "Auctor ר' יהודה בר ברזילי [Judah bar Barzilai] qui simpliciter apud authores vocatur אלברזלוני Albarcelonius" (*Bibliotheca Rabbinica*, 318." He referred to another work by the same author in the entry for *Jichus basar*: "Author R. Judah Barcellonius qui apud פוסקים Iure consultos simpliciter vocatur רב אלברצלוני" (*Bibliotheca Rabbinica*, 295). Thus, it would appear that Buxtorf was not aware that all three works were written by one and the same author.

124. Buxtorf, *Bibliotheca Rabbinica*, 312–13: "Est autem moris historicorum, ut historiam scribant usque ad sua tempora. Videtur ergo Hebraeis, ista verba in principio libri ab aliquo istius temporis, ad marginem ascripta fuisse, quae postea per imprudentem historicum in textum relata."

125. Buxtorf, *Bibliotheca Rabbinica*, 335: "Lectori S. Noveris primo nomina librorum Latine scripta esse ad formam pronunciationis judaicae, quibus Thau lene sive raphatum semper ut S effertur. Id eo fine, ut si quis eos a judaeis petat, sciat eos ipsorum more nominare. Deinde notandum, iudaeos voces polysyllabas fere semper efferre cum accentu in penultima. Tertio Beth lene quandoque per V consonam, quandoque per F expressam: in harum enim potestatem incidit. Denique quae hic desiderantur, tu benevolus addito. Vale. Vade."

reigned: Buxtorf's guidance was designed for Christians in northern Europe who might have occasion to talk about books with or buy them from Ashkenazic Jews.[126] Within those limits, his compendia were remarkable works of shrewdly packaged erudition.

126. On the whole, early modern Christians transliterated Hebrew according to the "Sephardi" pronunciation of Hebrew. Buxtorf's use of Germanic pronunciation was therefore remarkable and required explanation. But his transliteration is not at all consistent – a problem that is still encountered today.

CHAPTER FOUR

In the Printing House: Buxtorf the Bookman

In August 1600, Buxtorf sent his friend Caspar Waser in Zurich a freshly printed copy of Waser's Hebrew grammar, which Buxtorf had just seen through the press at the printing house of Conrad Waldkirch.[1] His letters to Waser went into considerable detail about the production of the book. He even went so far as to ask for a tip for the workman who had set the type:

> The compositor was quite diligent and hardworking, so that everything came out in good order. He is a good man, and in this line he is a very dedicated craftsman. He has a wife who will very soon give birth. She is quite pious, industrious, and hardworking as well. But they must live from day to day by his work. You would do well, not only because it is customary, but rather for the sake of piety, if you would send a frank, or however much you prefer, as a tip, or rather as charity, so that the pains of childbirth don't weigh so heavily on the wife. This will be an act of piety for you.[2]

Buxtorf's plea was successful. In a second letter, he thanked his friend and provided him with further details about the compositor and his wife, to prove that he was helping the worthy poor:

> Your letters delighted me, gentle Waser. My heartfelt thanks for the tip that you provided for the compositor. I shall give it not to him, so that he can wet his whistle (the custom of just about all those who hang about in printers' shops),

1. Caspar Waser, *Archetypus grammaticae Hebraeae duabus praecipuis partibus, etymologia & syntaxi, absolutus* (Basel, 1600).

2. Buxtorf to Waser, 6 August 1600; Zurich, ZB, MS S-155, 39: "Qui literas composuit, satis fuit diligens et industrius, ut omnia ordine starent. Bonus vir est, et in hac arte strenuus artifex. Habet uxorem partui vicinam, piam alioqui, industriam et laboriosam. In diem tantum ex labore vivunt. Feceris bene, non tantum pro consuetudine, sed magis pro pietate, si francum aut quantum libuerit, pro Trinckgelt, vel eleemosyna potius miseris, ut puerperii incommoda uxori sint leviora. Hoc tuae erit pietatis."

> but to his hardworking wife, who comes from my homeland, to support the expenses of childbirth, or rather to restore them. In this you show the sense of duty and the generosity of a pious man.[3]

It all seems smooth and seamless: a world of scholarly publication, hard to recognize now, in which craftsmanship on the part of workers was recompensed with charity on the part of managers and authors.

But the publication of Waser's volume was not so easy – and relations between artisans and authors were not consistently so pleasant – as this exchange suggests on its own. Buxtorf and Waser had corresponded since the early 1590s. Living in Zurich, Waser lacked direct access to a press that could handle Hebrew.[4] Buxtorf urged him to complete ambitious projects, such as a bilingual edition, Latin and Hebrew, of David Kimhi, and asked for opinions on his own projects.[5] Evidently he had ambitious plans for Waldkirch, who eventually became Basel's first major producer of Hebrew books. But though Basel had a tradition of printing in Hebrew and Aramaic that went back to Ambrosius Froben, Buxtorf could not be altogether encouraging about prospects there for the moment either. In 1595, he told Waser that "I would speak to Waldkirch now, but at the moment his Hebrew type is in disorder."[6] In September 1599, he asked Waldkirch about the chances for printing a Hebrew grammar that Waser had just completed. But the printer was fully occupied completing jobs for the impending Frankfurt book fair. Printing would begin, at the soonest, in the winter.[7]

3. Buxtorf to Waser, Basel, 7 September 1600; Zurich, ZB, MS S-155, 40: "Literas tuas, honorande D. Wasere accepi, quae me plurimum exhilararunt. Pro dono illo vinario, compositori exhibito, gratias ago maximas. Illud non ipsi, ad humectandum guttur (ut mos fere est omnium in typographiis versantium) sed uxori ejus, mulieri industriae, et meae compatriotae, ad sublevandos sumptus aut restaurationem potius Accubitus puerperalis, dabo. Officium et beneficium illud pium est. Quod industria in Tuam Gram. collocata Tibi placet, gaudeo."

4. On Hebrew printing in Basel, see in general the classic work of Joseph Prijs, *Die Basler Hebräischen Drucke, 1492–1866*, ed. Bernhard Prijs (Olten, 1964), as well as Marvin Heller, *The Sixteenth-Century Hebrew Book: An Abridged Thesaurus* (Leiden, 2004); Heller, *The Seventeenth-Century Hebrew Book: An Abridged Thesaurus* (Leiden, 2011); and Clemens Sidorko, *Basel und der jiddische Buchdruck (1557–1612): Kulturexport in der frühen Neuzeit* (Basel, 2014). For the larger context, see Stephen Burnett, *Christian Hebraism in the Reformation Era (1500–1600): Authors, Books and the Transmission of Jewish Learning* (Leiden, 2012).

5. Buxtorf to Waser, 28 January 1595; Zurich, ZB, MS S-150, 96: "Tu vero, qui per Dei gratiam multa potes, fac ut res hebraica promoveatur. Commentarium Kimchi, nisi fallor, fac ne diu premas. Commodum mihi videretur hebraea cum latinis conjungere."

6. Ibid.: "Waldkirchius noster suos typos hebraeos confusos adhuc habet, alias illum hoc alloquerer hoc tempore."

7. Buxtorf to Waser, 1 September 1599, Zurich, ZB, MS S-154, 56: "Cum Waldkirchio de tua Gram. locutus sum, sed hactenus nihil responsi certi obtinere potui, quod ille ad has

Waser sent his grammar to Basel at the end of February 1600. He was worried on more than one account. Uncertain that Waldkirch had actually committed to printing the book, he asked Amandus Polanus, professor of the Old Testament, to see to it that his hard work did not go to waste.[8] Even the prospect of publication was not altogether happy. In Hebrew printing, as he remarked to Polanus, "a slip in one small mark often causes a serious error." He suggested that Polanus might intercede with Buxtorf, "so that he does not lose heart in the course of correcting it, as he took on the job."[9] Jacob Zwinger, he urged, should also help by insisting that Buxtorf produce a correct and faithful version of his book.[10]

As it turned out, Buxtorf did not abandon his commitment to the project. But Waldkirch took his time. In March, as Buxtorf reported, the printing house was idle, awaiting the master's return from Frankfurt.[11] To show his commitment to the project, perhaps, he sent Waser both a poem by a Silesian "studiosus" and a two-line Hebrew poem of his own:

אשר יבדוק או שדקדק בלי דקדוק
את דברי תורה:
הלוא סבור נער ובור בהדבור
מסתרי תורה:

One who examines or parses the words of Torah without grammar, he is to be regarded a mere child and an illiterate in matters of Torah's secrets.[12]

---

nundinas occupatissimus esset. Ad instantem hyemem spem fecit, de quo quamprimum potero te monebo."

8. Waser to Polanus, 1 March 1600, Basel, UB, MS G II 12, 807: "Per eundem [sc. one Ionas Libingus, whom he recommends] misi ad Dn. Buxtorfium Archetypum meae Grammaticae Hebraeae, ut apud vos per vestrum Waldkirchium imprimatur: quam impressionem si tu quoque tua intercessione, ut potes, potes promovere, quaeso facias pro tuo in me amore. Nam caussae plurimae sunt, cur nolim eum supprimi, et non [nunc?] tandem, post maximos exhaustos labores, in lucem proferri."

9. Ibid.: "Praeterea, ut in ea corrigenda, siquidem impressurus eam sit Waldkirchius, non defatigetur, sicut in se recepit, Dn. Buxtorfius, velim te quoque ipsum hortari. Nam in Hebraicis lapsus unius apicis magnum saepe errorem parit, ut optime nosti."

10. Waser to Jacob Zwinger, 27 February 1600, Basel, UB, MS Fr Gr II 14, 125r: "Vt item correcte & juxta exemplar meum, quod diligentissime exaravi, imprimatur, instabis non tantum apud typografum, sed inprimis apud Dn. Buxtorfium, qui ejus corrigendae curam suscipiet."

11. Buxtorf to Waser, 18 March 1600, Zurich, ZB, MS S-154, 113: "Waldkirchii typographia ab opere cessat nunc: ubi ipse Francofurto redierit, causam tuam apud illum urgebo et promovebo."

12. Ibid. Buxtorf's own Latin version of the two verses reads: "Ita profecto est: Grammaticus bonus sit, necesse est, qui de scripturae sensu penitus judicare vult."

Buxtorf commented: "If you recognize that this is true, is good, I shall also add it to your Grammar."[13] It was standard practice, in the sixteenth and seventeenth centuries, to preface books of all kinds with celebratory verses. By sending candidates for inclusion, Buxtorf could give an impression of editorial activity even as nothing substantive was happening.

When Waldkirch returned, moreover, other jobs intervened. On 30 May, the news was mixed. Buxtorf reported that only four quires of copy, two weeks' work, remained to be set and run off before printing of Waser's book could start.[14] But he also admitted that one of the prefatory poems meant to adorn the *Grammar* "has been lost, or should I say misplaced, so that it can't be found, by me or by the printer."[15] Buxtorf hoped that a replacement could be sent. Two weeks later he promised that he and Zwinger would both add verses of their own.[16] Yet the main news was still not uniformly good. Buxtorf had to inform his friend that more Hebrew copy had arrived unexpectedly, and it had to be printed before the *Grammar* could reach the head of the queue.[17] Finally, work began. In July, Buxtorf was sending printed pages of the book to Waser as they were completed. Less than a month later, the whole book was done. Even then, Buxtorf explained why he and Waldkirch had altered the title; set Waser's tables of the eleven servile Hebrew letters, unappealingly, in the same small octavo format as the book itself; and added a short treatise, in Hebrew and Latin translation, on Hebrew versification.[18]

13. Ibid.: "Hoc si verum, si bonum agnoscis, Grammaticae tuae quoque affigam."

14. Buxtorf to Waser, 30 May 1600, Zurich, ZB, MS S-155, 14, 3: "Hactenus vero, quod neque tuo, neque meo voto satisfactum, scribere nolui. Ingrata tibi scribere nolui, ideo tacui. Quatuor folia, vier Bögen, excudenda restant, antequam tuus liber praelo subijci possit. & duabus septimanis absolvi poterunt, si res rite peragatur. Utut sit, ad nundinas autumnales primum habebis. Aliud polliceri non possum. Feci quod potui: idem D. Amandus et Zwingerus fecerunt."

15. Ibid.: "Sed audi: Carmen illud a praeceptore nobilium conscriptum, amissum: non quidem periit, sed inter chartas abjectum latet. Si adhuc exemplar habes, remittere poteris. De reliquis non sis sollicitus."

16. Buxtorf to Waser, 13 June 1600, Zurich, ZB, MS S-155, 14, 1: "Nescio an antehac scripserim, quod carmen praeceptoris istius nobilium vel a me vel a typographo amissum dicam an sepositum sit, ut reperiri non possit. Tu aliud exemplar mitte, si habes. Zwingerus carmen conscripsit. Meum quoque addetur." The lost poem was recovered and the others added: Waser, *Archetypus grammaticae Hebreae*, sigs. α4v–α5v [α7r, α7v].

17. Ibid.: "Quod vero editio tui libri hactenus cessat, id praeter spem fit et meam et typographi. Acceperat quinque folia hebraea excudenda, quibus libellum absolutum putabat, interea missa sunt plura, a quibus excudendis cessare non potuit. Hic liber post octiduum finem habebit, et is tuo initium dabit."

18. Buxtorf to Waser, 6 August 1600, Zurich, ZB, MS S-155, 39: "Miraberis forte, qua temeritate Titulus libri alius est, quam Tu voluisti. Causam scribam. Tipographus Syntaxeos, quam plenius quam quisquam alius tractasti, mentionem in fronte fieri voluit, ut libro major

Like many editors, before and since, Buxtorf did his best to shift the blame for any changes that displeased his friend. Practical needs had dictated a number of editorial interventions: "Some things had to be cut severely because of excess length. Here is another reason for changes. At the end certain additions were made because there were empty pages."[19] Waldkirch had made these decisions, just as he had decided to change the title "so that it would sell more handily": if Waser felt that these changes were excessive, he should "blame the printer."[20] Buxtorf also confessed that he himself had probably missed some errors, since the hot summer weather had made him sleepy and inattentive.[21] Waser's book, Buxtorf admitted, had reached the world "not as it should have been made, but as it could be."[22] It was in this context, in fact, that he called Waser's attention to the poor but diligent compositor – possibly as much to distract his author as to serve the needs of Christian charity.

The details of these transactions are not distinctive: dozens of books underwent delays in early modern printing houses, as their owners switched tasks to keep their presses busy, only to be both cut and stretched once they reached the printer's Procrustean bed.[23] Changes in titles and mix-ups about prefatory material happened in the best-regulated printing houses.[24] But the role that Buxtorf

esset gratia. Hinc prima illa mutatio nata. Finis Libri cum in medium folium caderet, ac primo intuitu impossibile videretur, Tabulas Literarum Servilium in forma 8ª libro annectere: inde vacaturae videbantur aliquot paginae, de iis complendis a typographo monebar. Nam conclusum erat Tabulas non aliter quam a te erant scriptae excudi posse, nempe in forma quam Quarta complicat. Forte tum in libellum quendam hebraeum incidi, in quo Tractatulum istum de Ratione Carminum hebraicorum inveni: eum statim descriptum, Latinum feci, et punctis vocalibus, ad subsidium studiosorum, ornavi, eoque vacantes paginas complere commodum visum fuit. Haec dum excuduntur, de Tabulis in 8° excudendis consultatio initur ad majorem libri commoditatem. Utcunque res in effectum deducta."

19. Ibid.: "Quasdam enim propter prolixitatem lacerare necesse fuit. En alterius mutationis causam. In fine rursus quaedam adjecta sunt propter vacantes paginas."

20. Ibid.: "Haec si bono animo accipias, ut te facturum spero, est quod laeter. Certe bono fine animoque a me factum est: non quod Tuae Grammaticae, quae studiosis sufficiens est, deesse aliquid putarim, sed ut Typographo potius gratificarer, qui eo modo Titulum Liberaliorem fore existimavit, ut commodius venderetur. Audacia itaque si hic in vitio est, Typographum accuses."

21. Ibid.: "In corrigendo, qua potui diligentia usus sum. Errata si quae inciderunt, imbecillitati meae attribues, quam aestus magnus inducto somno saepe debiliorem fecit."

22. Ibid.: "Utcunque Res in effectum deducta: non ut debuit, sed ut fieri potuit."

23. See the classic study by Donald McKenzie, "Printers of the Mind: Some Notes on Bibliographical Theories and Printing-House Practices," *Studies in Bibliography* 22 (1969): 1–75, repr. in McKenzie, *Making Meaning: "Printers of the Mind" and Other Essays*, ed. Peter McDonald and Michael Suarez (Amherst, 2002), 13–85.

24. Percy Simpson, *Proof-Reading in the Sixteenth, Seventeenth and Eighteenth Centuries* (London, 1935); Adam Smyth, *Material Texts in Early Modern England* (Cambridge, 2018), chap. 3, 75–136.

chose to play is remarkable. Though his professorship was poorly paid, he earned substantial sums in other ways – for example, by providing room and board to students. Printers' correctors, in Basel as elsewhere, were ghostly, Bartleby-like figures. They wore out their eyes deciphering illegible copy, took abuse from compositors and pressmen more anxious to get on with their work than to ensure that the results were free from error – and usually received less pay per hour than the workmen.[25] Yet Buxtorf worked as a corrector of books in Hebrew, Aramaic, and Yiddish, first for Waldkirch and then for Ludwig König.[26] Like Walter Keuchen, a former pupil who corrected Hebrew books in Hanau, he combined this post and its activities with closely related ones: that of the bibliographer who set out to establish control over the mysterious treasures of the Jewish book world and that of the official censor who helped the city magistrates determine which of these Jewish books might be printed in their cities, in complete or corrected form.[27] Like Keuchen, too, Buxtorf described his work in these capacities in both detailed letters and crisp official documents. These rich materials enable us to follow him into his printers' shops, to see the messy stacks of letters in which a manuscript page could easily get lost, and to meet the compositors who did skilled work many hours a day in return for very modest wages. In this chapter, we will watch Buxtorf work as corrector. Then we will turn to his related work as censor of Jewish books.

Like many correctors, Buxtorf carried out a wide range of tasks. When Theodor Zwinger, father of Jacob, drew up a table of organization for the printer's

25. See Simpson, *Proof-Reading*; Anthony Grafton, *The Culture of Correction in Renaissance Europe* (London, 2011); and, for the practices used in Yiddish printing in Basel, Sidorko, *Basel und der jiddische Buchdruck*, 321–39.

26. Buxtorf's first recorded meeting with a printer took place at the spring book fair in Frankfurt in 1588, as a page of his *Album amicorum*, Basel, UB, MS AN VI 26z (no foliation or pagination), reveals: "Sine tuo numine, Nihil est in homine. Nam Ex te uno, Deus, omnia. Christophorus Corvinus typographus Herbornensis scribebat haec Frankoforti in nundinis vernalibus anni 1588." Christoph Rab, a highly educated printer (note his use of the imperfect tense in signing his album entry), called himself Corvinus (raven) in Latin. He and Buxtorf both arrived in Herborn in 1585 and may well have known one another before this dated meeting. In a letter written from Basel some years later to Johann Piscator, Buxtorf mentioned that a particular compositor had been one of the most competent of Corvinus's workers at Herborn, which suggests direct knowledge: Buxtorf to Piscator, 10 September 1607, in *Clarorum virorum epistolae cxvii: E bibliothecae Gothanae authographis*, ed. Ern. Sal. Cyprianus (Leizpig, 1714), 154–55, at 154.

27. On Keuchen, see Stephen Burnett, "Hebrew Censorship in Hanau: A Mirror of Jewish-Christian Coexistence in Seventeenth-Century Germany," in *The Expulsion of the Jews: 1492 and After*, ed. Raymond Waddington and Arthur Williamson (New York, 1994), 199–222. He was a useful source of information for Buxtorf about ongoing and new Hebrew publication projects. See, e.g., Keuchen to Buxtorf, 20 May 1613, Basel, UB, MS G I 60, 317; 21 September 1613, Basel, UB, MS G I 60, 318.

shop, he included no figure comparable to the modern editor. Correctors, by his account, simply corrected: evidently the printers themselves chose what they published.[28] In fact, though, trusted correctors often served in what would now be editorial capacities. Buxtorf, as we have seen, scouted for texts that might deserve printing and advised Waldkirch about works submitted. In some cases, he seems to have had the power to make editorial decisions. Buxtorf had studied Hebrew at, among other places, the Herborn Academy, where he worked with Johann Piscator, professor of Old Testament theology. By 1605, Piscator was writing to express admiration for Buxtorf's *Epitome* of Hebrew grammar, and to offer thanks for some information about the Jewish year. Evidently, he had submitted work of his own, since he asked his younger friend earnestly "that if you find other things that need correction in my work, please let me know."[29] Buxtorf did exactly that. In the fall of 1607, he explained that Waldkirch had planned to begin printing Piscator's elementary textbook of Hebrew. Buxtorf, like other correctors, knew a great deal about fonts and other material aspects of printing. When he finished his major study of the Targumim, the *Babylonia*, he set out firm instructions for the potential printers at the start of his manuscript:

> To publishers and correctors:
> 1. In printing, following their custom observed in other works, let them maintain the distinction between what we call in Germany running [italic] and ancient [Roman] characters, and adapt the copy to that before it is sent to press. I have given a specimen of this in the first seven leaves, on chapter 15 of Genesis. Passages translated into Latin from Hebrew or Aramaic could be printed in italics, and underlined to that end.[30]

Accordingly, the compositor had consulted with Buxtorf, the corrector, about proper layout before beginning to set Piscator's text. Buxtorf had noticed that

28. Theodor Zwinger, *Methodus apodemica, in eorum gratiam, qui cum fructu in quocunq[ue] tandem vitae genere peregrinari cupiunt* (Basel, 1577), 398–400, esp. 399: "Corrector vulgo, nec male θεωρός vel etiam ἐπίσκοπος: quandoquidem Exemplum Typis expressum cum Exemplari confert."

29. Piscator to Buxtorf, 9 September 1605, Basel, UB, MS G I 60, 315: "Pro Epitome tua Grammaticae Hebraeae gratias tibi ago... Rogatum autem te velim ut et alia si quae correctione digna in illo meo opere observasti, mecum communices, idemque ab aliis, qui isthic sunt, viris doctis impetres."

30. Buxtorf, *Babylonia* MS, Basel, UB, MS F IX 41, Dv: "Monita. Pro Editoribus et Correctoribus.

I Ut in edendo, pro suo more et in reliquis observata consuetudine, diversitatem characterum, Currentium et Antiquarum, ut nos in Germania vocamus, observent, et ad eam exemplar, antequam typographiae committatur, adaptent. Cujus specimen feci in prioribus septem foliis, ad XV. cap. Genes. Quae sc. ex Hebraeo vel Chaldaeo latine sunt versa, possent charactere currenti edi, et eo fine lineolis subduci."

one verb form was missing. "Hence," he explained, "I thought the work would be imperfect, if that time were missing. And hence we agreed that the work should be sent back to you."[31] Perhaps feeling that this explanation was somewhat brusque, Buxtorf then offered a second one, appealing to the state of Waldkirch's workforce.

> Even if it had been complete, it could not have been printed this summer. Before now Waldkirch had four skilled compositors for Hebrew texts. Two were busy with my Lexicon. The third went off to Geneva, where a Hebrew Bible is being printed, with an interlinear translation, after the form of the Antwerp edition. The fourth, Jacob, whom you know, and who was not the least competent of [Christopher] Corvinus's workmen [at Herborn], quarreled with Waldkirch and left him. So I can attest that the book could not have been printed.[32]

If these explanations can both be trusted, they suggest that Buxtorf played a significant role at the press. Apparently, he had oversight over the details of format and mise-en-page for Jewish books, decided that certain books should not be added to the list – and knew the actual condition and abilities of the workers who were on hand at a given time better than Waldkirch himself did.[33]

He was also a diplomat. After returning Piscator's work for revision in what seem fairly brusque terms, Buxtorf promised that he himself would act as corrector if the missing portion could be added to the MS.[34] This mixture of precise criticism and sympathetic interest had its intended effect. Far from taking offense, Piscator thanked Buxtorf for intervening and preventing him from publishing an

31. Buxtorf to Piscator, 10 September 1607, 154: "Interea D. Waldkirchius volebat incipere impressionem tuorum Rudimentorum linguae Hebraicae. Literarum positor mecum de formae constitutione confert. Ego librum perlustrans video deesse Participium Praeteritum Pahul, eius tabulam et notas in tabulam. Vnde putavi opus imperfectum futurum, si id Tempus desideretur. Vnde consultum fuit, ut ad te remitteretur."

32. Ibid.: "Vtut perfectum fuisset, tamen hac aestate excudi non potuisset. Waldkirchius antehac quatuor hebraicae literaturae peritos Positores habuit; duo in meo Lexico fuerunt occupati: tertius Genevam discessit, ubi excuduntur Biblia Hebraea cum interlineari versione, ad formam Antvverpianae editionis: quartus, Iacobus, quem nosti, qui et Corvino inter operas non postremus fuit, cum D. Waldkirchio lites habuit, et ab eo discessit. Hinc testari possum, excudi non potuisse."

33. Basel, UB, FA VI 109, preserves examples of the way Buxtorf prepared a *Vorlage* for the printing shop. This is a Bomberg *Humash* (Venice, 1525), in which the book of Genesis has been cast off in red, presumably in preparation for the Waldkirch edition that appeared in 1611–12 (Basel, UB, F G X 1). Buxtorf has entered notes in German to the compositor. These give instructions for inserting or removing spaces and for replacing individual letters with characters "aus einer kleineren schrift."

34. Buxtorf to Piscator, 10 September 1607, 154: "Quod reliquum, si remittatur, ego correctoris operam libenter praestabo. De his satis."

imperfect work. In fact, he confessed that he had no time either to look for the missing material or to draw up a replacement and would simply use Buxtorf's work in place of his own for his courses.[35]

In the Jewish world as well as the Christian, Buxtorf claimed, he enjoyed prominence as Waldkirch's representative. As early as 1599, he wrote to Waser: "I receive frequent letters from Jews in many places, even from Poland. The reason is Waldkirch's printing shop, to which they often send writings to be printed."[36] Buxtorf knew the Jewish physician Isaac Eckendorf, who was involved in the publication of Nathan ben Yehiel's *Arukh,* a monumental lexicon of Talmudic and Midrashic vocabulary, in 1599. At some point later he wrote to Buxtorf that he hoped to publish a book without vowels in Rashi script, seventy sections (or pages) long, and named a price of 15 gulden. He hoped to print 1,500 copies.[37]

Buxtorf's detailed and unguarded letters reveal the life of a printing house in the handpress age in all its glorious messiness. In August 1608, Waser sent Buxtorf a Hebrew poem in honor of the antiquarian Johann Wilhelm Stucki, author of an innovative treatise on ancient ways of feasting, who had died in 1607. "I was asked as well," Buxtorf informed him, "to add something of my own in the empty page and hand it in to the press."[38] By chance, one of the compositors had just finished another job and could set the pages in question immediately. In this case, speed

35. Piscator to Buxtorf, 12 March 1608, Basel, UB, MS G I 60, 316: "S. Mi domine Buxtorffi. Rudimenta mea linguae Hebraicae, superioribus nundinis Francofurtensibus a te mihi remissa, paulo post nundinas una cum literis tuis mihi reddita sunt. Et gratias tibi ago, tum quod defectu animadverso, interetssisse [intervenisse?] ne praelo committerentur: tum quod tuam mihi operam offers in illis corrigendis, si defectu oppleto ad excudendum D. Waldkirchio fuerint remissa. Sed de defectu illo explendo solicitus interea non fui: neque otium vel ad inquirendum, si inter chartas meas latet, vel ad adornandum, si invenire non potuissem, otium fuit. Atque adeo sententiam de Rudimentis hisce denuo in lucem edendis [sententiam is repeated] mutavi: Volebam ea excudi in usum nostrae scholae: sed nunc magis placet, ut tua Grammatica discipulis nostris proponatur."

36. Buxtorf to Waser, early 1599, Zurich, ZB, MS S-154, 40: "A judaeis crebras literas accipio ex varijs locis, etiam ex ipsa Polonia; typographia Waldkirchij causam illis praebet, ad quam subinde scripta excudenda mittunt." For the date, see Stephen Burnett, *From Christian Hebraism to Jewish Studies: Johannes Buxtorf (1564–1629) and Hebrew Learning in the Seventeenth Century* (Leiden, 1996), 43 n39. A Hebrew letter from an anonymous Jewish scholar from Nuremberg to Buxtorf, dated 1599, is reprinted in Prijs, *Die Basler Hebräischen Drucke*, 528.

37. Isaac Eckendorf to Buxtorf, n.d., in Moscow, Russian State Library, MS Guenzburg 1213/2, letter xiii (copy). A note, probably by the younger Buxtorf, precedes the letter: "Isaac Ekendorf, medicus Iudaeus, pristinam amicitiam quaeret renovare, et inquirit in statum typographiae Hebraicae Basileensi[s]."

38. Buxtorf to Waser, 29 August 1608, Zurich, ZB, MS S-160, 76: S.P. "Cum circa principium mensis hujus Augusti, amice Wasere, ex acidulis redirem, inveni literas tuas cum carmine Hebraeo in obitum D. Stuckij: rogatus etiam, ut paginae vacanti meum aliquod adjicerem, praeloque committerem. Operam ilico dedi, ut utrumque fieret."

did not bring efficiency. Buxtorf confessed that though he gave the compositor the texts, he had not paid close attention, having just returned from a trip.[39] Nonetheless, he had made some changes in Waser's verses, "partly relying on the permission granted me, partly on a more elaborate permission that remained to be requested."[40] But time pressed and no negotiations with the author took place. The printer "asked for the pages in question and snatched them from my hand. Careful reflection was not permitted."[41] Buxtorf claimed to be eager to know what Waser thought of the result. It included at least one typographical error, which a corrector – evidently not Buxtorf – had marked but the printer's worker had nonetheless "corrupted." "It can be corrected in ink," Buxtorf remarked, evoking a standard method for making corrections once it was too late to stop the press.[42] In this case too, many parallels could be cited from the everyday life and work of printers working in Latin or their vernacular languages.[43]

But printing Hebrew posed special problems and imposed special needs. Buxtorf and his employer believed that he was fully competent to correct complicated texts in Hebrew and Aramaic. In 1596, he wrote to Waser that "printing work" was keeping him busy, "for Waldkirch has set out to print another tractate of the Talmud, the one called *Baba batra*. I am at work correcting this."[44] Years

39. Ibid.: "Forte ea septimana quidam compositorum suum opus absolverat, unde factum, ut has paginas statim susciperet. Adjeci festinanter quale quale potui, mente adhuc ex itinere distracta." In another case, Buxtorf complained about the incompetence of the Jew who had prepared a collection of Hebrew letters for Waldkirch's press: Johann Buxtorf, "Decas epistolarum Hebraicarum Latine reddita & notis illustrata," in *Sylvula epistolarum Hebraicarum familiarium*, ed. Johann Buxtorf (Basel, 1603), 47: "ERRATA. Iudaeus qui libellum hunc typographo excudendum obtulerat, indoctus, & corrigendo libro parum aptus, quaedam incorrecta reliquit, ad quae minus exercitati facile impingere possent. Ea restituere hic volui, ne quis illis turbaretur." Buxtorf states explicitly that this Jew was not competent to work as a corrector. Prijs wondered if Jacob Mocherseforim was the object of Buxtorf's criticism (*Die Basler Hebräischen Drucke*, 195), but though this conjecture has turned into a fact in later scholarship, no evidence has turned up to confirm it.

40. Buxtorf to Waser, 29 August 1608, Zurich, ZB, MS S-160, 76: "Paucula in tuo mutavi, venia partim mihi concessa, partim plenius roganda."

41. Ibid.: "Typographus interea chartas petijt, ac e manu surripuit. Solidior meditatio negata fuit."

42. Ibid.: "Si tibi arridet, est quod gaudeam.... Paginae itaque excusae sunt, ac ante octiduum ad te... missae. An acceperis, scire aveo. Auriga fidem promisit, accepto pretio mensurae vini. In tuo mansit erratum ך pro ף, a correctore quidem notatum, sed per operam typographicam depravatum. Penna corrigi poterit."

43. On stop-press correction, see the case studies in *Printing and Misprinting: A Companion to Mistakes and In-House Corrections in Renaissance Europe, 1450–1650*, ed. Geri Della Rocca de Candal, Anthony Grafton, and Paolo Sachet (Oxford, 2022).

44. Buxtorf to Waser, 15 December 1596, Zurich, ZB, MS S-151, 64: "Addere debui superioribus causis labores typographicos quib. detineor rursus. Alterum enim tractatum Talmudicum excudendum suscepit Waldkirchius cui בבא בתרא nomen. huic corrigendo

later, after Ludwig König took over control of the Hebrew press, they began to discuss a large-scale project: an edition of the Hebrew Bible. This massive project involved setting the biblical text, with all its vowel points and punctuation, and the major commentaries, in a different, smaller script. As early as 1608, a specimen was printed, which Buxtorf sent to his learned friends.[45] At this point, however, practical obstacles, including the lack of Hebrew type, prevented further progress.[46] Ten years later, as the project became a reality, Buxtorf's friend Waser provided him with specimens of Bragadin's 1617–19 edition of the Rabbinic Bible. Buxtorf's unfettered comments in a letter to his friend make clear how seriously he took typography, paper, and other technical matters of book production:

> Yesterday part of the Venice Bible, which I had eagerly awaited for a long time, was brought to me. I haven't yet had time to examine the work thoroughly. The physical appearance of some of the pages made me sick to my stomach, almost to the point of vomiting; the points were very faint and detached from their positions, the letters were blotted and full of gaps, and in a word, the filth of the Jews pretended to be the splendor of God. It seemed that no craftsman's hand had printed that holy work, but Jewish filth had vomited it onto the pages with one stinking belch. Certainly they now have a blemish (דופי) [*dofi*] in place of beauty (יופי) [*yofi*], as the prophet says.[47] I wish you had opened it up and looked at it. It's not that I think my frog (*rana*) is a god (*Diana*), but the matter is clear. They outdo us with their better paper. The rest of the work is unworthy of examination by a Christian scholar.[48]

---

operam do. Sed nunc vale et me ut hactenus amato." No copy of this edition – if it appeared – seems to survive (Burnett, *From Christian Hebraism to Jewish Studies*, 40). But Buxtorf could have been preparing a text for publication – or all copies of a small edition could have disappeared.

45. Buxtorf to Waser, 7 February 1608, Zurich, ZB, MS S-160, 45: "In Waldkirchii potestate nihil est. Lud. König quasi principalis typographiae gubernator est, ac omnes sumptus suppeditat. Cum hoc agendum erit. Editionem Bibliorum hebraicorum molimur, et habes hic exemplar, ex quo de forma utcunque judicare poteris."

46. Ibid.: "At ante autumnum non fiet initium editionis, cum interea fundendi sint characteres."

47. Cf. Isaiah 3:24.

48. Buxtorf to Waser, 15 April 1618, Zurich, ZB, MS A 52, 57: "S.P. Pars biblica veneta heri mihi allata, quam diu expectaveram cum desiderio. Nondum fuit otium penitus inspiciendi opus. Externa quarundam paginarum facies mihi nauseam movit, pene ad vomitum, adeo puncta evanida, luxata, literae maculatae, hiantes, et uno verbo nihil nisi sordities judaica splendorem dei ementitur. sic certe, ut non artificis manus opus illud sanctum expresserit, sed judaica sordities foetenti ructu in chartas evomuisse videatur. Sic utique illis nunc est דופי loco יופי, juxta prophetae dictum. Utinam aperuisses et vidisses. Non dico, quod mea rana mihi sit Diana, sed res ipsa est. charta meliori nos superant: reliquum opus docti christiani aspectu indignum est."

Keuchen, to whom Buxtorf sent a specimen of his own project, commented in a way that reflected the corrector's expertise he shared with Buxtorf.[49] Though enthusiastic, he detected flaws in the font and the spacing of letters, as well as the paper:

> The specimen of the Bible that you are now working on is elegant, and it must certainly win the approval of Christians, for many reasons, and especially because, as I see, you have shown great industry in reconfiguring Targum Onkelos to fit the laws of grammar. Still, it's the case that a little something is missing here and there in the proportions of the points, and the equal spacing of the letters, which can easily be corrected when the work becomes continuous. But this precious project calls for stronger and brighter paper.[50]

Producing a Bible with medieval Jewish commentaries must have required many consultations among printer, correctors, and artisans. Through conversations like these, correctors like Buxtorf and Keuchen developed an informed interest in the material side of Hebrew printing.

But producing a Rabbinic Bible required more than careful planning. It demanded the collaboration of Jews, and this posed special problems. As early as 1608, when plans for the project were at a very early stage, Buxtorf explained to Waser that "some learned Jew will take charge of the correction with me," for pragmatic rather than philological reasons: "that will enable us to do much better at selling copies to them [the Jews]."[51] As Waser would have understood, this cynical-sounding comment reflected the material nature of the *Biblia Hebraica*. These elaborate editions, produced at first by Christian printers though edited by Jewish scholars or converts, were Jewish in content but Christian in format. Soon after the invention of printing, elaborate editions of the Vulgate Bible, the Roman *corpus iuris*, and standard authors like Vergil had made canonical Latin texts available in a distinctive form, one that crystallized major parts of their interpretive traditions in single books. Endless waves of commentary printed in small type – one standard version of Vergil came with seven commentaries,

49. Buxtorf and Keuchen were not wrong in their assessment of the Bragadin Bible, which was greatly inferior to the previous Venetian exemplars in terms of layout and quality of printing.

50. Keuchen to Buxtorf, 19 January 1618, Basel, UB, MS G I 60, 320: "S.P. Specimen Bibliorum, quae prae manibus habetis, ut elegans est, sic non potest non probari praesertim Christianis, tum propter alia multa, tum vero, quia video Targum Onkelos ad leges Grammaticas a te magna industria revocari. Verum tamen est, hinc inde in punctorum proportione, ac literarum aequali interjacientia exiguum quidquam deesse; quod facile continuata opera corrigetur. Chartam autem solidiorem & nitidiorem preciosum opus requireret."

51. Buxtorf to Waser, 7 February 1608: "Correcturae mecum praeerit aliquis doctus judaeus, ut et inter ipsos tanto melius distrahi possint."

by interpreters from Servius (ca. 400 CE) to Jodocus Badius Ascensius, scholar and printer – broke on all sides of islands of canonical text, printed in a large and more legible form.[52]

In 1517 and 1524–25, Felix Pratensis and Jacob ben Hayyim respectively adapted this model to the Hebrew Bible. Later versions appeared in 1546–48 and 1568–69. Some Jews bought and used them. But as the Reformation spawned a vast literature of controversial biblical commentary, they also became more and more popular among Christian scholars and theologians, many of whom owned and annotated copies. Buxtorf and König set out to serve both markets. As in the past, Jewish scholars would help to ensure the quality of a very complex piece of printing. But they would also provide front matter in Hebrew, to present the work to potential Jewish readers. Meanwhile Buxtorf's name – and his Latin prefatory material – would make the same text appealing to learned Christians. These practices were not new. The first Rabbinic Bible appeared with a dedication to Leo X in the copies intended for Christian readers, but without that dedication in those aimed at Jews.

On 5 September 1617, Buxtorf and König asked the city council of Basel for permission to produce a "Biblisch Truck," which they described as an edition of the Venice Rabbinic Bible. They emphasized that used copies of older editions had become prohibitively expensive. Jews had long been forbidden to live in Basel, though families or small groups often managed to do so.[53] For this project, however, they would be needed. The council accepted their petition, and allowed König to employ two Jewish correctors, on condition that they work on no other project.[54] As it turned out, he imported three Jews. They were Abraham Braunschweig, who eventually presided over the edition; Mordechai Gumplin or Gumpel, another corrector; and Elijah ben Judah Ulma, a compositor who had worked before, and would work again, for the Hebrew press in Hanau.[55] Gumplin had

52. For the development of printed commentary on classical texts, see Paul White, *Jodocus Badius Ascensius: Commentary, Commerce, and Print in the Renaissance* (Oxford, 2013).

53. See Andreas Berger, *Stadt ohne Juden? Präsenz und Absenz in der Frühen Neuzeit* (Basel, 2024), 52–59.

54. Stephen Burnett, "Johannes Buxtorf I and the Circumcision Incident of 1618," *Basler Zeitschrift für Geschichte und Altertumskunde* 89 (1989): 135–44, at 136–37.

55. Ibid. Braunschweig was actively engaged for a long time after the project reached completion in obtaining books for Buxtorf. See Braunschweig to Buxtorf, n.d., in Moscow, Russian State Library, MS Guenzburg 1213/2, letters vi, x and xvii (copies). At least one other potential corrector was recruited, though he seems not to have joined the team. See Simhah bar Eliakim of Speyer, the scribe, to Buxtorf, n.d., in Moscow, Russian State Library MS Guenzburg 1213/2, letter vii (copy): "First, my lord should do what he told me yesterday on account of the letter that he wishes to write to the lord [name illegible] in Geneva and to send it via me. Furthermore, with regard to the 24 (Bible) that you wish to print, and as

worked in Basel before, on an edition of *Seder ha-tefillot rosh hodesh* that appeared in 1611.[56] They would join Buxtorf and König's Christian employees to produce the edition – an immense job, which the team carried out with astonishing speed. In June 1618, Buxtorf had finished preparing the copy for the Pentateuch and was at work on Judges. He was also preparing the Targum and Masorah.[57] At some point in July or August 1619, hundreds of Rabbinic Bibles, so immense as to be unwieldy, reached completion. All of them shared a handsome title page in the best Jewish style, modeled on the façade of a temple. Those aimed at Christians had a Latin description of the book in the center of the title page and a long preface by Buxtorf, very much in the vein of the *Jüden Schul,* on its other side (see Fig. 4.1). Those intended for Jews had a Hebrew description of the book by Braunschweig and a short address to the reader by him on the verso (see Fig. 4.2).

Collaboration of this kind was proverbially difficult. In theory, it should have produced efficiency in the use of time, since Jewish correctors and compositors would not work on Saturday or Christian ones on Sunday. In 1618, Keuchen explained his practices: "Since I have no public task, I make it my custom to read through everything that needs correction on Saturday. On Sundays the Jewish compositor corrects this, and on Monday mornings my reader [a junior worker who read the copy aloud while Keuchen followed in the proofs] and I finally read it over again."[58] As Keuchen remarked, the compositors working on Sunday would have been Jewish – and, probably, overseen by a Jewish corrector. This way of organizing work seems very practical. In practice – as Abraham Braunschweig lamented – the difficulties caused by collaborative Hebrew printing made the labor involved "heavier and more protracted than the earth and more expansive than the sea" (Job 11:9).[59] Many of the difficulties emerged directly from the production process.

---

my lord said, that he would take me as corrector. It is my request and prayer to my lord that I [should become corrector?] My lord knows what he said to me. For with the help of the blessed Lord whose name is great, I assured him that I would bring him signatures of about 300 people who want to buy the said Bible. I will bring you great profit with the help of God for your publication because I have many relatives and friends in Germany."

56. Prijs, *Die Basler Hebräischen Drucke,* 327.

57. Burnett, *From Christian Hebraism to Jewish Studies,* 173. He was presumably engaged in cutting up copies of the 1546–48 *Biblia Hebraica,* correcting them by hand and adding other materials where necessary.

58. Keuchen to Buxtorf, 19 January 1618; printed by Burnett, "Hebrew Censorship," 217 n44: "Und weil ich noch liber a publico munere, pflege ich den Sambstag, was zu corrigieren, allein zu überlesen, welches Sontags vom Judischen setzer corrigiert, und von mir und meinem lectore Montags morgens letzlich überlesen word."

59. Braunschweig, in *Biblia sacra Hebraica & Chaldaica,* ed. Johann Buxtorf (Basel, 1619), vol. 4: sig. K, fol. 1r (Joseph Derenbourg, "L'édition de la Bible rabbinique de Jean

Figure 4.1. Latin title page from *Biblia Hebraica,* ed. Buxtorf (1618–19). Princeton, Princeton University Library 2019–0001F.

Figure 4.2. Hebrew title page from *Biblia Hebraica,* ed. Buxtorf (1618–19). Princeton, Princeton University Library 2019–0001F.

A printing house in the handpress era set out to produce completed sheets of each product that was actively being printed every day. To achieve this, ideally, the compositors set a full sheet and a proof was pulled. The reader then read the copy text aloud to the corrector, who emended the proofs and gave them back to the compositor. After correction of the errors, the pressmen pulled a second proof, which the reader and corrector examined in the same way. Another set of corrections was entered, and a third proof pulled. This the reader and corrector compared with their own second proof. Once they had given final approval, the pressmen printed hundreds of copies of the sheet. They lifted these to racks near the ceiling of the shop, where they would dry more quickly in the warm air. Then they repeated the process.[60] A shop with multiple presses could produce more than one sheet each day.

So much for ideals. In his preface to the errata list of the Bible edition, to which we will now turn, Braunschweig described the version of this neat plan that correctors and printers actually enacted in König's shop (see Fig. 4.3).[61] Some of it will already sound familiar. As Pavel Sládek has shown, Jewish correctors created "a peculiar subgenre of paratext called 'apology' or 'defence,'" in which they complained about the impossibility of avoiding errors and "appealed to the readers' acquaintance with printing technology to elicit their solidarity."[62] Buxtorf knew at least one example of this genre: the letter that Isaac Eckendorf prefixed to his edition of the *Arukh* in 1599, in which he described Christian compositors as groping like blind men in the midday sun as they tried to see the forms of the letters.[63]

Buxtorf," *Revue des Études Juives* 30 [1895]: 70–78, provides a fairly loose French translation of the text). A complete translation appears in Appendix 3, below.

60. This is the process prescribed by Zwinger, *Methodus apodemica*, 398–400, and described in operation, in Basel, by Viglius ab Ayatta Zuichemus; see Johann Gerritzen, "Printing at Froben's: An Eye-Witness Account," *Studies in Bibliography* 44 (1991): 144–63, at 162, where the readings of three sets of proofs is described as standard practice in well-organized printing houses: "Correctori autem subservit is quem Lectorem vocant. Collatio enim primæ formæ impressæ cum exemplari, omnino necessaria est: & ut rectè fiat, duorum operam requirit. Solentque in bene institutis officinis tres confici formæ, ordineque singulæ relegi, quo omni ex partè mendæ vitiaque expurgentur."

61. For a translation of the entire text, see Appendix 3, below.

62. Pavel Sládek, "'Before the Law': Jewish Correctors of Early Printed Books," in *Printing and Misprinting*, ed. Della Rocca de Candal, Grafton, and Sachet, 261–77, at 264–67 (quotations from 264). See also the classic study of Abraham Yaari, "Correctors' Complaints about Printing on Shabbat by Non-Jews" (in Hebrew), *Kiryat Sefer* 14 (1937): 524–28; addenda, *Kiryat Sefer* 16 (1939): 500–501.

63. Nathan ben Yehiel, *Arukh*, ed. Isaac Eckendorf (Basel, 1599), verso of title page. Buxtorf's copy, with his signature, is Basel, UB, FA III 24.

Most of these apologies followed similar lines. The corrector was supposed to do two jobs at once: check that previous errors had been corrected and hunt

התנצלות להקורא משכיל עבור הטעיות·
אשר בכל זה הרפוס נמצאות·

ויקרא כל דברי המגילה זאת·
ויוכל לתקן הטעמי׳ והנקודו׳ וכל אות ואות;

להיות שהשכל האנושי לעת מלהשיג או לפעול איזה דבר בתמור ובשלימות בלתי מעות׳ והוא בחסרון זיק האנושי מצד הפתע שקורה דעתו ושכלו מהכיל ומלהעמיד דבר על עומק בורין ואמיתתו׳ או מצד הפעולה שהיא נפלאת ברוב גדלה ועומקה ועצמותה שבלתי אפשרי

ביציאות השתים שהן ארבע אבות נזיקן׳ והמשל בזה שאם היה איזה טעות באות ונרשם להם לתקנו כדת׳

הוסר הקטן יהודה לאלף
והצעיר אברהם בן לאם החר אליעזר
זצל ברונשווין

K I א ו

Figure 4.3. Abraham Braunschweig, apology. From *Biblia Hebraica*, ed. Buxtorf (1618–19). Princeton, Princeton University Library, 2019–0001F.

for new ones that had been introduced in setting the type. But, as Braunschweig sadly reflected, "It is impossible to perform two tasks simultaneously under one supervision." Every copy of a text, even those produced with a pen, letter by letter, by experienced Jewish scribes, had its share of mistakes:

> It must have come to the attention of a wise reader that sometimes in a Torah scroll that has been written out and gone through and read with great scrutiny as far as human powers permit, and read out a hundred times in the synagogue, suddenly there is a discovery of either a defective or redundant letter or one letter exchanged for another.[64]

The new art of printing, however, multiplied every possibility of error: "it is as impossible to print anything without errors as it is for the world to be without air."[65] Collaboration between Jewish correctors and Christian compositors generated even more mistakes than ordinary typesetting. The language barrier was high and steep:

> In particular, this came about on account of the non-Jewish workers who had not performed God's service from childhood and who were completely incapable of reading or recognizing one letter of our holy tongue. Thus, when they embarked on the work, a holy undertaking, and set out to accomplish and bring it to completion, they were like aliens in a foreign land; and though they were professionals in their own language they seemed beginners in the art of printing. As a consequence, the designated work could not be brought to completion as originally conceived without enormous effort and trouble.[66]

Errors swarmed: "When they [the workers] brought a page to the corrector for editing, refining, and purging of errors he would sit and read it and become overcome with despair until he stopped counting [the errors] – there were so many that he was unable to examine them at one sitting." Impatient to finish their work, the compositors pulled letters and vowel points at random from their type cases, introducing more errors – a practice that Braunschweig attributed to their ignorance of the holy tongue.[67]

Time pressures made matters worse. On the eve of the Sabbath and of a feast, and on the Sabbath and the festive days, no Jews were present. On Saturdays, moreover, the Christian workmen hurried to finish work and prepare for their own weekly rest.[68] In those cases there was no opportunity to impose the checks that

64. Braunschweig, in Buxtorf, *Biblia sacra,* vol. 4: sig. K, fol. 1r.
65. Ibid.
66. Ibid.
67. Ibid.
68. Ibid.

were in place during the rest of the week. Even on normal working days, moreover, or so Braunschweig claimed, König ordered his staff to print three full sheets.[69] Correctors had to work continuously: "so that while one was speaking [about one page] another page was produced for him." The corrector could speak harshly to the workers: "Finish your work according to the indications in the corrected text." But they paid no attention: each did what he liked. The corrector could examine the proofs minutely at any point, but once the sheets have been pulled, the "pages were therefore irreparably damaged."[70]

Indeed, at least according to another witness, the course of printing was much bumpier, and the results even worse, than Braunschweig indicated. At some point after the books appeared, Mordechai Gumplin wrote to Buxtorf. He explained that König had originally hired him to correct the proofs, and that he had worked through the summer even though the printer failed to tell his accountant to pay him. Then, however, a rival had appeared and usurped his place – though he was incompetent to fill it. At some point, a fourth Jew, Abraham of Halle or Hallein, whom we shall call Abraham of Halle, arrived at König's shop (we do not know how he gained permission to work in Basel). He undercut Gumplin:

> my master had sent a message to me to Naumburg asking me to be his corrector and that I went off to Basel. I began the correcting and he arranged that I should receive 7 gold pieces every week. But after two or three weeks, a Jew called Abraham of Halle who had never done any print correction and did not know anything about the art of correction came to my master the printer, trespassing [on my territory]. He said that he was prepared to receive half my pay; the master was taken in by him and ejected me from my task as corrector and took the said Jew in my place.[71]

König did not sack Gumplin from the project. Instead, he sent him on the road to sell copies of the Bible to Jews in the Polish commonwealth, promising that he would earn 2 gulden for every copy he sold. Trusting König, Gumplin traveled to the Frankfurt book fair, only to learn that the printer had paid him nothing

69. As Braunschweig is complaining about time pressure, we take the passage in this sense. It might also refer to the practice of pulling three proofs for each sheet, which was standard in good Basel printing houses. But that would not fit the context as well. See the discussion in the notes to the full text in Appendix 3, below.

70. Braunschweig, in Buxtorf, *Biblia sacra*, vol. 4: sig. K, fol. 1r.

71. Gumplin to Buxtorf, n.d., Moscow, Russian State Library, MS Guenzburg 1213/2, letter xiv (copy). Abraham attempted to gain Buxtorf's attention again as is clear from his letter to Buxtorf, 6 August 1619, Moscow, Russian State Library, MS Guenzburg 1213/2, letter xvi (copy). König's decisions were not always balanced, at least in the eyes of Buxtorf and others affected by them. Keuchen to Buxtorf, 21 September 1613: "Quod de Köningii vestri intemperantia conquueraris, cum aliis facis."

but claimed to have sent several hundred copies of the Bible to Poland from the fair, for him to sell.[72] Once Gumplin arrived in Lublin after the fair, however, he found that he and the edition were deep in hot water. He had lost his position as corrector after finishing the Pentateuch and the Former Prophets. The local authorities found that the Latter Prophets, supposedly corrected by Abraham of Halle, swarmed with errors:

> For three rabbis, the heads of the yeshivot of Posen, Cracow, and Lublin, sent for me and they all ganged up against me.[73] They said that they had seen enormous errors in the third part of the printed Bible [twenty-four] that I had brought from Basel. A very strong rebuke was in order for I performed this divine work, a holy task, fraudulently and I had not been careful enough in emending the text as a result of which there were gross errors.[74]

As Gumplin did not need to explain to Buxtorf, these three rabbis were the dominant figures in the Council of the Four Lands, the core institution of Jewish self-government in the Polish-Lithuanian Commonwealth, which often met in Lublin at the time of the spring and autumn fairs. He explained to them that he had not actually corrected this part of the edition. Further, he told them, "It had already occurred to my lord, the printer, to make an errata list and to correct the mistakes as far as possible." They answered that "they did not believe in that Jew anymore since he had made the errors when he was corrector. He would therefore not scrutinize the errors; on the contrary, he would underestimate the errors in order not to expose his shame."[75] But they expressed their faith in Gumplin. If the printer was willing to commission him to list the errors as the basis for an errata sheet, they promised not to criticize the edition harshly.

Gumplin pointed out that the printer might not be willing to pay him in full. At this moment, the critics made a startling proclamation:

> Then they answered that the *Av bet din* [head of the law court, possibly R. Samuel ben Joseph Hildesheim] of Frankfurt would tell the printer that the errata list should be made by R. G[umplin], that a proclamation should be made in all synagogues that no copy of the Basel 24 [Bible] should be bought unless it included the errata list corrected by R. G[umplin] of Poznan.

72. Gumplin to Buxtorf, n.d., Moscow, Russian State Library, MS Guenzburg 1213/2, letter xiv (copy).

73. Nathan Spira (Cracow); in Lublin, Ephraim Zalman Schorr seems to have held the office of *Av bet din*; it is not possible to identify the incumbent for Posen. Our thanks to Ted Fram who helped us on this matter.

74. Gumplin to Buxtorf, n.d.; Moscow, Russian State Library, MS Guenzburg 1213/2, letter xiv (copy).

75. Ibid.

> Therefore, in keeping with their advice I was forced to swear by the Lord, God of Israel to emend the text itself and Targum within an entire year with great precision and intense concentration, word for word, every page, from beginning to end. During this time, the printer could sell [the Bible], but only on the condition that he would give each buyer a copy of the errata list so that each person could correct the errors for himself without any difficulty.[76]

According to Gumplin, these Jewish scholars in effect claimed the right to censor a Christian edition of the Hebrew Bible – unless König met the conditions they set and appointed the corrector whom they approved. This was not the only time that he found himself caught between the Christian and Jewish communities. In another letter to Buxtorf, he asked for a signed document attesting that "the scholars of the Basel academy" gave him permission to have tractate Berakhot of the Talmud printed. He needed this to prove to the heads of the yeshivot of Poland that he had done what he was supposed to do.[77]

It is not surprising that the rabbis objected so strongly to the errors in Buxtorf's edition. The regulation about flawless copies of the biblical text had been laid down in no uncertain terms already in the Talmudic era: "Rabbi Jose said: 'It is not right to be in possession of an uncorrected book [of the Bible] for it is written, Do not let iniquity reside in your home (Job 11:14).'"[78] No wonder, therefore, that the rabbinic authorities of Cracow, Posen and Lublin rejected the defective text of Buxtorf's Bible and refused to approve its publication unless each and every error had been removed. At the time that Buxtorf was encountering rabbinic opposition to the publication of his Bible, the imprecise spelling and vocalization of the Hebrew in printed prayerbooks were coming under similar criticism in Poland. To this end the rabbis of Cracow, Lublin and other cities strongly supported the initiative of Shabbetai Sofer from Przemysl to produce a corrected prayer book based on grammatical principles.[79] The decision to support the use of this "authorised daily prayerbook" is documented in the records of the Council of the Four Lands.[80] In

76. Ibid. On R. Samuel ben Joseph Hildesheimer, see Marcus Horovitz, *Frankfurter Rabbinen: Ein Betrag zur Geschichte der Israelitischen Gemeinde in Frankfurt am M.* (Frankfurt am Main, 1883), 2: 19–24. Samuel was very active in reorganizing the Frankfurt community in the aftermath of the Fettmilch uprising (1612–16), which targeted the city's Jews (see Chapter 5 below).

77. Gumplin to Buxtorf, n.d., Moscow, Russian State Library, MS Guenzburg 1213/2, letter xi (copy). See also Gumplin to Buxtorf, n.d., Moscow, Russian State Library, MS Guenzburg 1213/2, letter xv (copy).

78. BT, Ketubot 19a.

79. On Shabbetai Sofer's prayer book see Stefan Reif, *Shabbetai Sofer and his Prayer-book* (Cambridge, 1979).

80. Israel Halperin, *The Records of the Council of the Four Lands, vol. 1: 1580–1792*, revised and edited by Israel Bartal (Jerusalem: The Bialik Institute, 1945; repr. 1975), 91, 31–33.

1617, the Council gave its approbation to Shabbetai's prayerbook, stressing that efficacy of prayer depended on correct reading of the text. This approval was repeated in various sessions over the next two years. What is striking is that in the session of 2 March 1618 they stressed the spiritual significance of Shabbetai's grammatical achievement. The Council invoked the kabbalistic notion that the letters like the body move the soul, that is, the vowels. In the strongest of terms they warned against misplacing one vowel or letter.[81] Bahya ben Asher, the thirteenth-century kabbalist, had expressed the implications of erroneous reading: "One who adds or detracts one vowel destroys the world and extirpates Torah and its roots."[82] In practical terms this meant that the Council insisted that precentors and their congregation acquire at least one copy of Shabbetai's prayer book for the use of the community. It is quite likely that Gumplin would have known about the proceedings of the Council; in any case, both he and Braunschweig were certainly aware of the implications and repercussions of being "in possession of an uncorrected book."[83]

The Christian printer agreed to the conditions that the Jewish scholars had set. But the process of creating the errata list was far from simple. Gumplin had not had the chance to discuss the list with Buxtorf and was uncertain about how it should be laid out and printed:

> I am very sad that I was unable to speak with my master about all the details – how and what to do with the errata list and as to which script and letter type should be used – whether with or without vowels. For example, Genesis page 9 recto it is written *me'al* (מְעַל) *penei ha-adamah* [from off the earth].[84] It should be *me'al* (מֵעַל). Is it sufficient that the *me* is printed with its vowel or should it be printed *me'al*, – the letter should be written with a *tsere* under the *mem*, i.e., without vowels or accents or *dageshim*?[85]

He managed to consult R. Samuel in Frankfurt about the task. Though Samuel disclaimed competence "in the art of printing and letters," he remarked that a vocalized errata list would be preferable, and Gumplin agreed.[86] But he made clear that he still urgently needed to know Buxtorf's view. He also wanted Buxtorf to consult with König about the compensation he would receive. At this point – as so often in Gumplin's hard life – chaos began to reign. König promised to consult R. Samuel but failed to do so. R. Samuel again insisted that he was not an expert – a point on which Gumplin did not entirely agree, since, as he pointed out, "He did,

81. Ibid., 96, 34–36.

82. Bahya ben Asher, *Perush al ha-Torah* ad Gen. 18:3.

83. See p. 129 and n78 above for the text of this quotation and its source.

84. This is in Genesis 7:4 and indeed the text is incorrectly spelled מְעַל instead of מֵעַל.

85. Gumplin to Buxtorf, n.d., Moscow, Russian State Library, MS Guenzburg 1213/2, letter xiv (copy). In the errata list the letter *mem* is correctly vocalized.

86. Ibid.

however, know a little about the trouble that he experienced when correcting those first five quires in Frankfurt last year."[87]

Gumplin could only express his eagerness to discuss the whole matter with Buxtorf. He claimed that he should receive 12 batzen (silver coins) for each sheet that he went through for errata. In the past he had been paid half a gulden, probably 7 batzen. Now he remarked that "to correct the commentaries I swear that I prefer to correct . . . commentaries over two bogen than one bogen with vowels and Targum." He also seemed to feel that in the past he had not been able to do uniformly precise work: "But in fact, the correction of the commentaries was not at all carried out correctly to my satisfaction. But the correction of the vowels and accents that I was carrying out over this summer when I was at home, and when I went away they went off with a large part of my text with which I had been satisfied."[88]

Braunschweig eventually oversaw the creation of an errata list (see Fig. 4.4). It covered the Hebrew Bible and the Targum, as well as, in a further section, the accents. He also provided it with a detailed foreword in which, as we have seen, he explained the errors as the result of printing house practices and the fast work tempo imposed by his "generous master" König. He did not blame the master printer or Buxtorf.[89] Instead, he took full responsibility for the errors and their correction. He stated that he had asked König to have the text submitted to an expert review, which would yield an errata list "so that the reader can easily correct the errors. Thereby, the Lord's Torah will be made perfect and later generations will realize that we did our utmost to ensure that there would be no possibility of superseding what we have accomplished."[90] König agreed, and generously paid for the revision of all twenty-four books. Gumplin, whom we now know well, actually composed the list.[91] He took a long time and was guided by Braunschweig.[92] Braunschweig's vivid panorama of life in the lower depths of Buxtorf's world of books omitted the fact that Abraham of Halle had replaced Gumplin while the Bible went through the press. If Braunschweig could not rival Gumplin in the textual criticism of the Bible, he was clearly a more adroit and dramatic narrator. He was equally eloquent when he complained to Buxtorf that he had "nearly drowned under all the troubles and difficulties that I encountered from above and below had God not assisted me."[93]

87. Ibid.

88. Ibid.

89. Braunschweig, in Buxtorf, *Biblia sacra*, vol. 4: sig. K, fol. 1r.

90. Ibid.

91. According to Derenbourg ("L'édition de la bible rabbinique, 77 n6), the initials מ"ס that normally represent the word for bookseller in this case could perhaps also refer to "a corrector of books."

92. Braunschweig, in Buxtorf, *Biblia sacra*, vol. 4: sig. K, fol. 1r.

93. Braunschweig to Buxtorf, n.d.; Moscow, Russian State Library, MS Guenzburg 1213/2, letter xiv (copy).

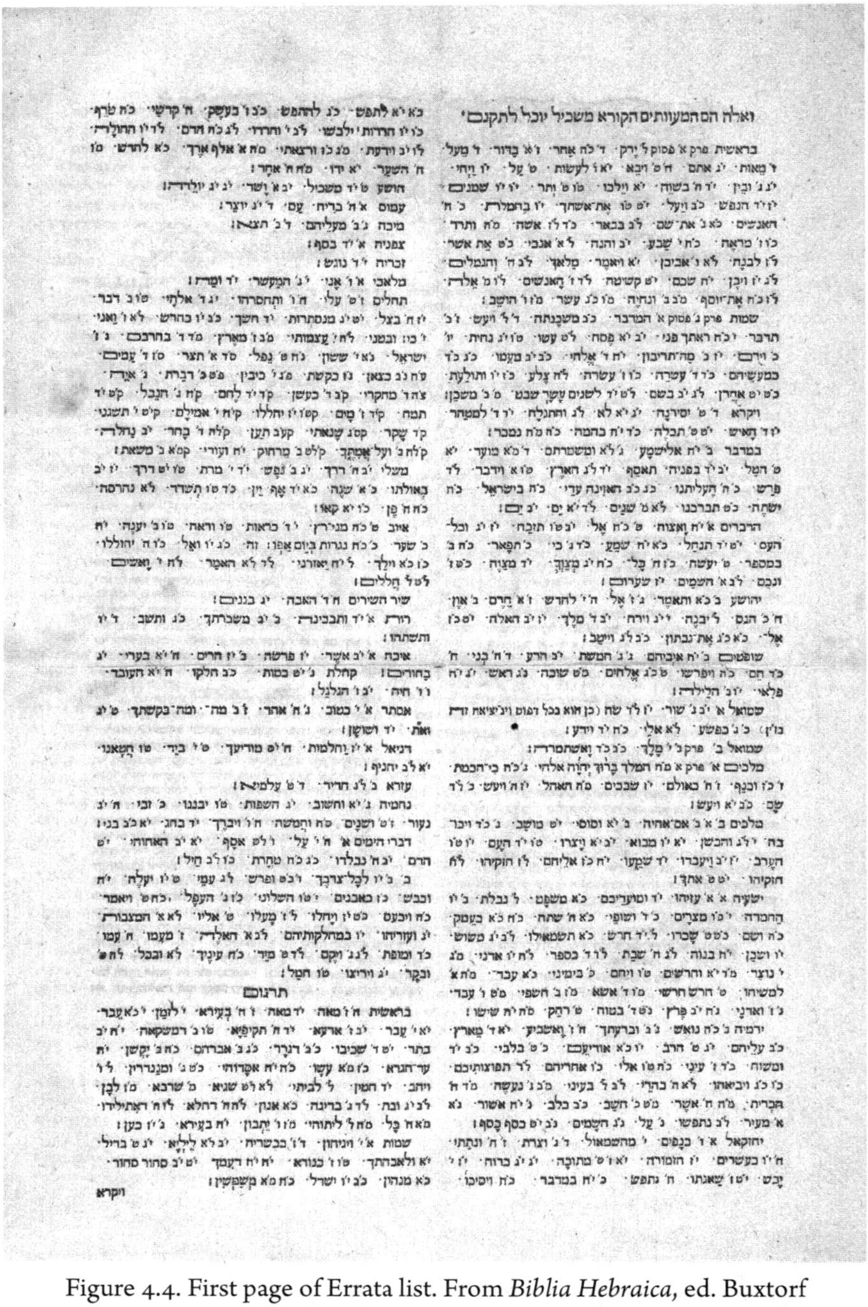

ואלה הם המעותים הקורא משכיל יוכל לתקנם

Figure 4.4. First page of Errata list. From *Biblia Hebraica*, ed. Buxtorf (1618–19). Princeton, Princeton University Library, 2019–0001F.

Another disaster also interfered with the publication. This time it was not friction between Jews and Christians, but their joint participation in a Jewish ceremony, that caused the explosion. In May 1619, printing was almost finished. Braunschweig's wife, who had accompanied him to Basel, gave birth to a son. Georg Martin Gläser was serving as Oberstknecht, a city official responsible for supervising Jews in Basel. Braunschweig asked him for permission to hold a circumcision ceremony, and to invite both Jewish and Christian guests. Gläser agreed. The ceremony took place on 2 June, in the presence of Buxtorf and others, including the younger Ludwig König, the printer's son.

When a minister complained to the city council, the private ceremony became a public scandal. The Jews were imprisoned, and Abraham Braunschweig's wife was expelled from the city with her infant son. Gläser was jailed for exceeding his authority. Buxtorf and the other Christians who had been present received an official rebuke from the council. He and König were fined 100 Reichstaler each, more than Buxtorf's annual salary as professor.[94] As Buxtorf awaited his sentence, he sent Waser a bitter, despairing account of the incident. He took it as a disastrous ending for both his current project and his long-term career: "This is certain: all the Jews (the three printers who had received permission previously from the Council) are to be expelled, and the book, which is almost finished, will lie dead. Some say that I should be cast out with the book. What an honor! What a reward for my boundless efforts!"[95] But the elder König intervened. Though the Council imposed a massive fine on Braunschweig, he and the other Jewish workers were allowed to remain in the city, living with König and Buxtorf, until they completed the edition. Buxtorf, who had considered moving to Heidelberg, also stayed.[96] The Bible appeared, and became a standard – though also a controversial – work for decades to come.

This story poses two complex historical questions: what sort of personal and intellectual relations did Buxtorf himself have with the Jews who worked for him, especially Abraham Braunschweig, and what sort of edition did he and they produce? As we will eventually see, the two inquiries are inextricably connected. Any Christian who presided over the printing of books in Hebrew (or Aramaic or Yiddish) was likely to end up working closely with Jews. Passing remarks and occasional documents record the times when Buxtorf brushed shoulders with

94. The story is well told in Burnett, "Circumcision Incident."

95. Buxtorf to Waser, 15 June 1619, Zurich, ZB, MS F 167, 47, printed by Burnett, "Circumcision Incident," 143: "Hoc certum, judaeos omnes (tres typographiae ministros, quib. antea facultas a Senatu concessa) expulsum iri, et ita liber jacebit qui pene in fine est. Sunt qui dicant, et me cum libro ejici debere. En honorem, en praemium pro immensis meis laboribus. Ita sum animo perturbato ob tam iniqua hominum judicia, ut quo me vertam, nesciam."

96. Burnett, "Circumcision Incident," 137–39.

Jewish print professionals. In January 1602, Jacob Mocherseforim from Meseritz, who had come to Basel to oversee the first edition of a collection of stories and legends in Yiddish, the *Mayse bukh*, lent him a paper manuscript, part of which he copied.[97] Almost twenty years later, Buxtorf told readers of the *Biblia Hebraica* that "Jews had shown him various leaves" of the edition currently under way at the shop of Bragadin in Venice, a potential rival to his own.[98] Here as elsewhere we glimpse the anonymous traffic of Jewish correctors and compositors who moved between Venice and Basel as opportunities determined.[99]

We know a little about these men and their relations with Christian employers and customers. Two decades before Buxtorf's career in Hebrew printing began, the Jewish corrector Israel Zifroni, whose letters Buxtorf studied and translated, had worked for Ambrosius Froben in Basel on a notorious, censored edition of the Talmud, which appeared in 1578–80.[100] Zifroni had many troubles. In the spring of 1585, locked in a bitter struggle with Froben and his son-in-law, he found himself and his wife unable to leave for Italy, as he wished, and confined to an inn in the Basel suburb of Allschwil – which, unlike Basel, still allowed a small Jewish community to remain.[101] A few years later, back in Venice, he still sounded depressed, though he remained eager to supply useful information to his contacts from Basel. He informed Jacob Zwinger that during the day, he could be found at the printing house of Giovanni di Gara. At night, however, he explained, "aloggio in geto come gli altri Hebrei" – "I dwell in the Ghetto like the other Hebrews." He also recorded the death of a bookseller, Pietro Longo, drowned in the Canal Orfano at the orders of the Inquisition.[102] For all his troubles, though, Zifroni

97. Buxtorf, copybook, Basel, UB, A XII 20, 242–48. See Chapter 1, above.

98. The edition was produced from 1617 to 1619.

99. Buxtorf, "Lectori benevolo S.D.," in *Biblia sacra*, vol. 1, verso of title page: "Denique hoc ipso tempore quarta eorundem Bibliorum editio Venetiis sub praelo est, quae sequenti anno nuncia fama prodibit, sed elegantia typorum & nitore impressionis cum praecedentib. editionibus neutiquam comparanda, quantum ex diversis foliis per Judaeos nobis ostensis, manifeste deprehendimus."

100. See Abraham M. Habermann, "The Printer Israel Zifroni and His Son Elishma and Their Books," in his *Studies in the History of Hebrew Printers and Books* (in Hebrew) (Jerusalem, 1978), 215–92; Marvin Heller, "Ambrosius Froben, Israel Zifroni and Hebrew Printing in Freiburg im Breisgau," *Gutenberg-Jahrbuch* 80 (2005): 137–48.

101. Israel Zifroni to Theodor Zwinger, 12 May 1585, Basel, UB, MS Frey-Gryn I 15, fols. 514r–v; 15 May 1585, ibid., fol. 515r; undated note, ibid., fol. 517r.

102. Zifroni to Jacob Zwinger, 29 January 1588, Basel, UB, MS Frey-Gryn I 15, 516r: "et quando V.S. mi uoglia trouare la mia residenza e de giorno nella stamparia de Mr. giouan de garra in cale de pignioli et di notte aloggio in geto come gli altri Hebrei et se V.S. non uuol scriuergli ouero uolete una mia lettera acompagnata con la vostra auuisatione che faro tanto quanto uorete, saprete di nuouo come la notte passata fu anegato in Canal orfano per l'Inquisition pietro longo libraro forse conosciuto da V.S. ma senza dubio conosciuto dal Sr. Theodoro et da tutti gli Basiliensi, et questo gli e auenuto (per quanto si dice) per hauer

had warm relations with some Christians. They seem to have included Buxtorf's friend Waser, who referred to him as "our Hebrew Israel" in a letter of his own to Zwinger.[103] Zifroni responded rapidly and efficiently to Waser's requests for books that were evidently difficult for a Christian to find or buy: multiple copies of a Jewish edition of the Pentateuch, divided into weekly synagogue readings, for teaching purposes, and a *Sefer minhagim* or book of customs.[104] He also corresponded with Buxtorf about the printing of the Talmud.[105] His relations with Christians were clearly mixed, yet he felt able to confess his feelings to some of them and ask for help. Jacob Zwinger, whom he evidently regarded as a well-wisher, certainly knew his way around a printing house.

Did Buxtorf feel affection or gratitude toward Braunschweig? After the scandal of the circumcision, he insisted that he had not. To Waser – who, if anyone, might have responded with sympathy to Buxtorf's attendance at a Jewish ritual – he explained that he had gone not to satisfy curiosity but to denounce the Jews. His pious purpose absolved him from blame: "Cannot a Christian watch a circumcision without violating his conscience, if his purpose is to rebuke them for this abuse, as I did publicly, and bring them back to the circumcision of the heart, as I also did?"[106] Yet Buxtorf perhaps protested too much. Anyone who read the *Jüden Schul* would have known that he had attended a circumcision ceremony years before, taking note of the sermon (of which he made fun). It seems unlikely that he attended this ritual only because he was carried away by a desire to denounce those who carried it out. He gave a Jew credit – anonymous and somewhat grudging – for his essential role in producing the *Biblia Hebraica*.

---

fatto mercantia de libri prohibiti, et per hauere anco opinioni heretiche." On the career, trial, and drowning of Pietro Longo, see Paul Grendler, *The Roman Inquisition and the Venetian Press, 1540–1605* (1977; repr., Princeton, 2015), 186–89. Zifroni's evidence suggests that Longo was executed on 28 January.

103. Waser to Jacob Zwinger, 1 October 1593, Basel, UB, MS Fr Gr I 11, 18: "O quam mihi rem gratam, mi Jacobe, faceres, si Israelem nostrum Hebraeum alloquerere, ut exemplar unum atq. alterum illius Pentateuchi & quinque voluminum, quae hoc anno Hebraice impressa sunt, cum paraphrasib. Onkeli et Ionathae, ad me mitteret per te, sive per aliam commodit. cui ego gratissime satisfaciam."

104. Zifroni to Waser, 24 September 1583, Basel, UB, MS Fr Gr II 9, 199; Zifroni to Waser, 18 November 1583, Basel, UB, Fr Gr II 9, 198.

105. Zifroni to Buxtorf, n.d., Moscow, Russian State Library, MS, Guenzburg 1213/2, letter xviii (copy).

106. Buxtorf to Waser, 15 June 1619, Zurich, ZB, MS F 167, 47, printed by Burnett, "Circumcision Incident," 143–44: "Interim potestne Christianus salva conscientia circumcisionem spectare, eo fine, ut reprehendat eorum abusum, quod feci publice, ut ad circumcisionem cordis eos reducat, quod itidem feci, ut ex ipsorum traditionibus humanis et caecitate horrenda Dei iram in ipsis cognoscat, gratiam Dei in Christianis pro luce verbi dei praedicet, et caet. cogita et aliquid pro me consolando rescribe."

In describing the corrections he had made to the *Masorah*, or textual apparatus, he noted that he had been "aided by the work of a certain Hebrew."[107] But in his own final note, in Hebrew, he took responsibility himself for the correction of the text, and did not mention the Jew – much less make clear whether he had Braunschweig or Gumplin in mind. Instead he claimed that he, "Johann Buxtorf, teacher of the holy tongue in the famous city of Basel," had been "in charge of the job." He himself had "arranged and corrected what was corrupt, day by day, as is the custom of those who do this kind of work."[108]

Braunschweig referred to Buxtorf, as he did to König, with admiration and respect. In his grandiloquent preface to the errata list, he appealed to readers not to blame the Christian scholar for the faults of their edition:

> The intelligent reader may come across a text that is not clear and declare it to be defective. This might be his reaction on seeing, for example, a letter *bet* instead of a *kaf*, and sometimes, he would notice that the vowels or accents are hardly visible such that it was impossible to make out their correct and complete form. Now I will implore you to pay respect to my lord, the great eagle with plush plumage whose influence has spread abroad; his seat is in the city's high place. He specializes in the grammar of our holy tongue and Targum as well as Masorah. I am referring to the man of Torah, the respected JOHANN BUXTORF – may God protect him – who resides in Basel. He was put in charge of the work of correction.[109]

The rhetoric of this passage is not only respectful but floridly so – enough to suggest artificiality to a twenty-first-century reader. By the standards of contemporary Hebrew prose, however, Braunschweig's epithets were conventional, but not excessive, and Buxtorf knew it. In an earlier, private letter to Buxtorf, Braunschweig had written "May you be like Ephraim and Menasseh" – the blessing that a Jew would normally have said over his son.[110] Evidently Braunschweig felt he had to profess both respect and affection for Buxtorf.[111]

107. Buxtorf, "Lectori benevolo": "Nos vero his singulis, opera Hebraei cujusdam adjuti, quotiescunque nostram animadversionem & censuram incurrerent, curatricem manum adhibuimus, labesque & maculas, quascunque deprehendimus, diligenter eluimus." The words "opera Hebraei cujusdam adjuti" are underlined in Buxtorf's copy of the book, Basel, UB, FG II 19. The "Hebrew" was Braunschweig: see p. 242 below.

108. Buxtorf, *Biblia sacra*, vol. 4: fol. 66v.

109. Braunschweig, in Buxtorf, *Biblia sacra*, 4: sig. K, fol. 1r.

110. The letter is cited in Meyer Kayserling, "Richelieu, Buxtorf père et fils, Jacob Roman," *Revue des Études Juives* 8 (1884): 74–95, at 84. Braunschweig states "I will be at your service day and night" before giving Buxtorf his blessing.

111. Note also the beginning of one of his letters to Buxtorf, in which he wrote: "May the One who bestows knowledge and understanding on man endow in great exultation

It seems certain that they shared some philological goals for the *Biblia Hebraica*. Braunschweig praised Buxtorf's mastery of specific fields, all of them relevant to particular aspects of the editing of the Bible: Hebrew grammar, the Aramaic of the Targumim or ancient translations, and the Masorah – the apparatus of vowel points, accents, and marginal notes, compiled by Jewish scholars in Palestine in the last centuries of the first millennium C.E. Keuchen, as we have seen, noticed the care that Buxtorf had expended on the Aramaic of the Targumim. So did the Franeker professor Sixtinus Amama, who wrote that "whenever I compare the Aramaic text in the Venice edition with yours, that seems to be unadulterated medieval claptrap by comparison with yours."[112] He also compared Buxtorf's treatment of the Masorah favorably to that of his predecessors.[113] Buxtorf chose these same points for emphasis in the title page copy and letter to readers of the Bible.[114] His Bible, he admitted, offered essentially "the same biblical work" as its predecessors.[115] But it had four special characteristics of its own: "the Hebrew text, the Aramaic text, the Masorah, and the Commentaries of the Hebrews."[116]

It is not clear what Buxtorf saw as especially important about the Hebrew text of his Bible. In his letter to readers, he remarked only that "I have left the Hebrew text in its most ancient and truest purity and substance, down to the tiniest accent.

---

in every corner of the world my friend, the great scribe. There is no wisdom greater than his – he is the distinguished Johanan Buxtorf, may his rock preserve him. I received his pleasant and kind words and his clear speech, and I was happy to [hear about] his health and his perfected wisdom and I blessed the blessed one who apportioned him some of His glory." Braunschweig to Buxtorf, n.d., Moscow, Russian State Library, MS Guenzburg 1213/2, letter xv (copy).

112. Sixtinus Amama to Buxtorf, 10 March 1620, Basel, UB, MS G I 59, 261: "Quoties textum Chald. Venetum cum tuo confero, videtur is mihi purus putus Gothismus prae tuo."

113. Ibid.: "Vt nihil jam dicam de restauratione Masorae, quae misere in Venetis fuerat accepta."

114. Thus Buxtorf writes on the title page of the *Biblia sacra*: "*In his nunc primum, post quatuor editiones Venetas*, Textus Chaldaicus, qui TARGUM dicitur, a deformitate punctationis & pravitate vocum innumerarum vindicatus; Loca in MASORA transposita, deficientia, pugnantia, numeris depravata, subsidio diversorum exemplarium & Concordantiarum Hebraicarum, quantum fieri potuit, reposita, restituta & conciliata sunt, ut in praefatione amplius declarabitur.

Studio fido & labore indefesso

JOHANNIS BUXTORFII, linguae Sanctae in Academia Basileensi Professoris Ord."

115. Buxtorf, "Lectori benevolo": "Hac editione [the contemporary Venetian one] nihil obstante, consultum judicavit vir honestissimus, & ex plurimorum librorum editione, de re literaria optime meritus, D. Ludovicus König, civis & bibliopola Basileensis, idem Opus biblicum, sed nostro studio correctius elaboratum, praelo committere."

116. Ibid.: "Quatuor autem sunt, quae hoc Opus praecipue continet, nempe *Textum Hebraeum, Textum Chaldaeum, Masoram*, & *Commentaria Hebraeorum*[.]"

For anyone who adds anything to it or takes anything away from it, or changes anything at all in it, is impious."[117] Yet as any Christian who managed to read Braunschweig's errata list would have known, this was empty boasting. Buxtorf entered many corrections into his own working copy of the *Biblia*.[118] His text retained features that had worried him in the past: for example, the unusual form of the letter *ayin*, which was suspended above the line in the word "forest" in Psalm 80:14 ("The boar from the forest ravages it"). Buxtorf had fiercely denounced the letter and Bahya's interpretation of it in a letter of 1606 to Matthias Martinius.[119] Yet he retained the suspended letter in his *Biblia sacra*.

The other three categories apparently received more attention from him. Fifteen years after the book appeared, and a few years after Johann's death, his son Johann Buxtorf wrote an account of the edition for James Ussher. Johann, his son explained, had planned to accompany the work with "a triple commentary, that is, Masoretic, Chaldaic, and Hebrew."[120] He had published the Masoretic commentary, the *Tiberias*, in 1620, taking the opportunity to trace the history of the Masoretes and their work and to argue that it was considerably more ancient than Elijah Levita had argued.[121] He had also compiled, but failed to publish, his "Aramaic commentary, which contains critical notes on the whole Targum."[122]

117. Ibid.: "Textum Hebraeum in antiquissima & verissima sua puritate & substantia, in minimo etiam apice, reliquimus. Impius enim, quisquis ei aliquid vel addiderit vel detraxerit, aut quovis modo in eo quid mutaverit."

118. Buxtorf, *Biblia sacra*, Basel, UB, FG II 19–20; see Prijs, *Die Basler hebräischen Drucke*, 335.

119. Buxtorf to Martinius, 4 September 1606, in *Bibliotheca Historico-philologico-theologica*, 4, no. 3 (1721): 598–603, at 602: "Vidisti aut videre potes Psal. 80 in חזיר מיער literam ע suspensam supra vocem. Quaeritur quare? Quidam Judaeorum mysterium explicant, ut possunt. Ego legi in libro Cad hakkaemach, cujus R. Bechai author rationem diram. *Litera Ajin*, scribit, *suspensa est: ut sint sic omnes credentes in Suspensum illum, sc. Christum*. Vis plura? Quis porcus ille de sylva? Respondent, Christum esse, qui vineam Israelis suffodit. Quorsum ergo typographi nostri hanc scribendi rationem imitantur, פרש על פניהם si studio facerent. Hoc unum ex secretis, quod paucis notum esse putes. Sed alias de his plura." See above, Chapter 1, p. 33.

120. Johann Buxtorf the Younger to James Ussher, 26 August 1633, in *The Correspondence of James Ussher, 1600–1656*, ed. Elisabethanne Boran, trans. David Money (Dublin, 2015), 2: 616, 620: "Quia autem post absolutam editionem aliud Consilium ceperat, de triplici Commentario addendo, Masorethico scilicet, Chaldaico, et Hebraeo, in Commentarium Hebraicum ista reservabat, vel in peculiarem Tractatum quam parabat de Blasphemiis Judaeorum contra Christianos."

121. Johann Buxtorf, *Tiberias, sive commentarius Masorethicus* (Basel, 1620).

122. Buxtorf the Younger to Ussher, 26 August 1633, 2: 616, translation slightly altered (original 620); "Commentarium Chaldaicum, qui notas continet Criticas in universum Targum, confectum penes me habeo edendum, quam primum meliora his studiis affulserint tempora."

Braunschweig himself would have been the first to say that his and Buxtorf's efforts to produce a reliable version of the Masorah had not completely succeeded. He admitted that the existing versions they had worked from contradicted one another, on points that could not be resolved.[123] The absence of similar self-deprecatory remarks about the Targum suggests that he considered this component of the edition more successful.

But the younger Buxtorf emphasized above all a third, Hebrew commentary, in which his father had meant "to deal with the Hebrew language and the purity of the Hebrew codices," and which had remained incomplete.[124] He was writing, after all, because Ussher had criticized the edition on another ground, one that Braunschweig had quite naturally failed to mention: "because it followed copies castrated and expurgated by Roman censors, such as the later Venetian editions." In particular, it omitted passages by Jewish commentators in which they criticized or rejected Christian interpretations.[125] Rather like his father writing to Piscator, Johann the Younger made a series of excuses for these omissions:

> As for this matter, I would like your reverence to know that the exemplar from which the Rabbinic commentaries were printed in our Basel Bible was the third Venetian edition [Bomberg, 1546–48], not because it should be chosen instead of others, but because it happened to be available and was believed to be more correct than the previous editions. Then, the fact that the criticisms which they make against Christians were not restored, took place partly by the order and wishes of our magistrates, who gave permission for that edition on condition that nothing that blasphemed Christ, or that cursed and abused Christians and the Christian religion, should be left in it.

123. Braunschweig, in Buxtorf, *Biblia sacra*, vol. 4: sig. K 1r. For a discussion of the Masorah in Buxtorf's *Biblia*, see Prijs, *Die Basler Hebräischen Drucke*, 334–35, 529–31.

124. Buxtorf the Younger to Ussher, 26 August 1633, 2: 616, translation slightly altered (original 620): "Commentarium Hebraicum, in quo de lingua Hebraica, et Hebraeorum codicum puritate agere volebat, inchoavit, et jam aliquo usque perduxit, sed non absolvit. In hoc quoque cogitabat prolixe et speciatim agere, de Keri uketif, Tikkun Sopherim, de Punctorum vocalium origine, aliisque rebus ad hanc materiam pertinentibus."

125. Ibid.: "Intellexi ex D. Passore et D. Frey R.T. in Bibliis Parentis p.m. id desiderare, quod castrata et a censoribus Romanis expurgata exemplaria, posteriores scilicet editiones venetas, sit secutus, atque id monere in libro suo Anglicano, ex occasione loci ex Aben Esra, Gen. 37.35." Ussher criticized Buxtorf's Rabbinic Bible for following earlier editions that had been censored in his *Answer to a Jesuit* (Cambridge, 1835), 275–76. This criticism was accurate, as the younger Buxtorf admitted. The particular passage at issue in Ussher's polemic, however, was ibn Ezra's remark, in his comment on Genesis 37:35, that the word *Sheol* referred not to Hell but simply to the grave. Buxtorf's edition in fact contained that polemical anti-Christian remark, even though it was exactly the sort of comment that he normally leaped upon and condemned as anti-Christian.

> And anyone who knew the persecutions and very severe hatred that my father brought on himself in that city by producing that Bible, will not be amazed that he was wary of printing those sort of cavils of the Jews.[126]

The elder Buxtorf had happened to choose a censored base text, which omitted numerous passages, and then the Basel authorities would not allow him to restore the missing passages.

This discussion focused on the most difficult technical problem posed for Christians by printing a Rabbinic Bible. Every rabbinical commentator had argued, often forcefully, against Christian readings of verses and passages in the Hebrew Bible. Some had expressed anti-Christian opinions. The book necessarily had a problematic reputation in the Christian world – one not unlike that of the Talmud. Buxtorf told Waser that he was disliked in Basel "not only for viewing the circumcision, but also because of that Jewish book, for whose printing I am responsible."[127] In his letter to readers, Buxtorf claimed that he had done something novel to the rabbinical commentaries, "which the Jews had added to this work in the past, for their own special uses."[128]

In the *Biblia* itself, Buxtorf made clear that he accepted only those aspects of Jewish commentary that did not openly conflict with Christian teachings:

> The fourth part of the work contains the commentaries of the most learned Hebrews, which the Hebrews have added to this work in earlier editions, and set out to be read by both Hebrews and Christians. It was no part of my duty or my plan at this time to separate them or to reject them in their entirety. There is honey in them, there is gall. The honey is the part that has to do with clarifying the nature of the Hebrew language and helps to offer an accurate explication of the literal and grammatical sense of Scripture. We have included that, as all other Christian scholars who have attained an accurate

126. Buxtorf the Younger to Ussher, 26 August 1633, 2: 616, translation slightly altered (original 620): "Hoc quod attinet, sciat velim T.R. exemplar illud, ex quo in nostris Bibliis Basileensibus expressa sunt Commentaria Rabbinorum, fuisse tertiam editionem venetam, non quod illud prae caeteris eligeretur, sed quia ita se forte obtulit et prioribus editionibus emendatius esse credebatur. Deinde quod illa quae contra Christianos faciunt non fuerint restituta, factum esse, partim mandato et voluntate Magistratus nostri qui ea lege permisit editionem illam, ut ne quid vel in Christum blasphemi, vel in Christianos et Christianam Religionem maledici et contumeliosi in iis relinqueretur. Et qui novit persecutionem et odia gravissima, quae occasione Bibliorum istorum in hac ipsa Urbe, contra se excitavit Pater meus, non mirabitur, ipsum sibi ab istius modi Judaeorum cavillis exprimendis cavisse."

127. Buxtorf to Waser, 15 June 1619, Zurich, ZB, MS F 167, 47, printed in Burnett, "Circumcision Affair," 143: "Ego in magna invidia apud omnes, non tantum ob spectatam circumcisionem, sed et propter librum judaicum, cujus impressionis ego sim author."

128. Buxtorf, "Lectori benevolo": "... *Commentaria Hebraeorum*, prout jam antea Hebraei in suos praecipue usus, illa huic Operi adjecerunt."

> knowledge of the Hebrew language have done until now. The gall is the part that comes up in perverse or false treatment of questions, and especially in those passages that deal with the promised Messiah and his person and office. We leave this entirely to them[.][129]

On its own, this longer comment could mislead the reader into thinking that Buxtorf had retained the rabbinical commentaries as a whole. In fact, the special feature of his treatment of them was an enhanced and stringent selectivity. Though he took the rabbinical commentaries from an earlier edition that had already been censored, he did not simply follow this source for guidance. Sometimes he mangled already mangled passages even further beyond recognition.[130] Buxtorf, moreover, did his best to conceal the omissions from casual readers by substitution of words and ingenious typographical expedients.

Johann the Younger suggested that Buxtorf meant to deal with the omitted materials explicitly and openly in another context: "In addition, my father had collected all those passages, and wished to set them forth in the preface, which he wanted to prefix to it." Even after he changed his mind, Johann the Younger attested, "he reserved those passages for a Hebrew commentary, or for a special treatise that he was preparing about the blasphemies of the Jews against the Christians."[131] Johann the Younger's testimony is anything but clear. He suggests that Buxtorf could have planned to discuss anti-Christian remarks in a "Hebrew commentary" that dealt with "the purity of the Hebrew codices" and the nature of Jewish proposals to correct them" – an unlikely pairing. Then, however, he

129. Ibid.: "Quarta pars continet Hebraeorum doctissimorum Commentaria, quae ab Hebraeis in praecedentibus editionibus huic Operi adjuncta fuerunt et tam Hebraeis quam Christianis legenda proposita. Ea vel separare vel plane rejicere nec officii nec consilii nostri hoc tempore fuit. Est in iis mel, est in iis fel. Melleum quod ad linguae Hebraicae proprietatem melius explicandam, quodque ad literalem grammaticumque Scripturae sensum genuine illustrandum facit commodumque est suscipimus. quod & omnes alii viri docti, quotquot accuratiorem et perfectiorem linguae Hebraicae cognitionem inter Christianos assecuti sunt, hactenus fecerunt. Felleum, quod in rerum tractatione perversa aut falsa occurrit et maxime in iis locis qui de promisso Messia ejusque persona et officio agunt, id totum ipsis relinquimus[.]"

130. See Piet van Boxel, "Hebrew Books and Censorship in Sixteenth-Century Italy," in *Jewish Books and Their Readers: Aspects of the Intellectual Life of Christians and Jews in Early Modern Europe*, ed. Scott Mandelbrote and Joanna Weinberg (Leiden: Brill, 2016), 75–99, esp. 89–92.

131. Buxtorf the Younger to Ussher, 26 August 1633, 2: 616, slightly altered (original 620): "Ad haec, collegerat illa loca omnia Pater meus, et in praefatione quam illis praefigere volebat proponere voluit. Quia autem post absolutam editionem aliud Consilium ceperat, de triplici Commentario addendo, Masorethico scilicet, Chaldaico, et Hebraeo, in Commentarium Hebraicum ista reservabat, vel in peculiarem Tractatum quam parabat de Blasphemiis Judaeorum contra Christianos."

writes that his father "also intended to join to these a treatise on the reading of rabbinic writings."[132] Perhaps that would have dealt with the omitted passages. But its relation to the special treatise about Jewish blasphemy remains unclear. Two points, however, seem certain. For all his efforts, Buxtorf did not manage to excise all of the problematic passages from the commentaries. And for all the hard and effective collaborative work that went into Buxtorf's project – and for which Braunschweig offered such warm praise – in the end, he worked at cross-purposes not only with his Jewish helpers but also with the Basel authorities. Expert corrector and censor, and skilled diplomat, that he was, Buxtorf could never have turned out a *Biblia Hebraica* that fully satisfied its Christian stakeholders, to say nothing of its potential Jewish users.

At the end of his letter to readers – which he dated 22 August 1618, probably the date when printing began – Buxtorf promised that a fuller and more precise explanation of what he hoped to accomplish would follow. There would be an introduction, to be written and printed, as usual, at the very end of the project:

> On the further use of these four sections of this biblical work, and on the end and goal [*fine & scopo*] of this edition and everything we have done to illuminate it, we will set forth a specific and adequate set of instructions when the work reaches completion, if God grants it. To this I refer every Christian and fair-minded reader at this time, asking him to think well of these efforts of mine.[133]

Buxtorf's terminology is revealing. As we have seen, from the early sixteenth century onward, humanist and Protestant theologians had given the term *scopus* a central place in the formal discipline of hermeneutics. Theologically, Phillip Melanchthon and Matthias Flacius Illyricus agreed on almost nothing. But both held that every text had a single "scopus," a goal to which it pointed.[134] Once a skillful interpreter had identified that goal, the remaining exegesis of the passage followed readily. In using the term, Buxtorf was claiming that his edition, like the Hebrew Bible it presented and elucidated, had a single overriding point.

132. Ibid., 2: 616 (original 620): "Cogitabat etiam his subnectere Tractatum de Lectione Rabbinorum."

133. Buxtorf, "Lectori benevolo": "Sed de quatuor harum partium, quae in hoc Opere biblico continentur, ampliori usu, hujusque editionis & omnium, quae in ea illustranda praestitimus, fine & scopo, specialem & sufficientem Instructionem una cum completo hoc Opere, Deo annuente, proponemus, ad quam Christianum & benevolum quemque Lectorem hoc tempore remittimus, & nostris hisce studiis & laboribus favere optamus."

134. On the role of the *scopus* in the hermeneutics of Melanchthon and Flacius, see Kathy Eden, *Hermeneutics and the Rhetorical Tradition: Chapters in the Ancient Legacy and Its Humanist Reception* (New Haven, 1997), chaps. 5–6, esp. 84–85 n10, 98–100.

In fact, though, as the circumcision incident showed most graphically, the edition did not and could not have a single point, any more than it had a single shape. For the most part, as Buxtorf himself admitted, he and Braunschweig were making marginal, technical alterations in a corpus that had been fixed before either of them was born. Imagined buyers had an impact of their own on the book's formats. Christians and Jews wanted their book to serve different ends and were encouraged to buy it by sharply different introductory paratexts. The Basel authorities, for their part, doubted that a Rabbinic Bible could be made Christian at all. It is possible that their furious response to Buxtorf's actions dissuaded him from writing the preface he had planned. But it is also possible that Buxtorf realized the impossibility of presenting this extraordinarily complex project in a coherent way. An edition constructed at the point where so many fields of religious, political, and commercial force intersected was bound to have to more than its share of cracks. Buxtorf, moreover, lacked control over many aspects of the process of publication. He worked closely for many years with the elder Ludwig König, and his daughter married König's son, also named Ludwig. Yet the testimonies of Braunschweig and Gumplin reveal that in these years, Buxtorf did not follow the operation of the printing house as closely as he had when he worked with Waldkirch early in the 1600s. His Jewish correspondents clearly assumed that he did not know which responsibilities each of them had undertaken, or when – exactly the sort of information that fills his letters on Waldkirch's operations.

What we do not know is whether Buxtorf ever became aware of the deepest contradictions in the book – those produced by the radically different ways in which he and Braunschweig envisioned their joint project. The point of censoring the rabbinical commentaries – as the Basel authorities said even more clearly than Buxtorf – was to make clear that messianic passages and predictions in the Hebrew Bible referred to Jesus, the Christian Messiah who had already come, rather than to a Jewish Messiah whose coming was still in the future. Abraham Braunschweig, as we have seen, referred to Buxtorf both in letters to him and in texts about him with unflagging, highly rhetorical respect. But he was a Jew. Braunschweig complained that Jewish correctors found it hard to control Christian compositors. In 1615, Keuchen warned the authorities in Hanau that it was also hard for Christian authorities to control Jewish correctors: "Censure is of no use, if the corrector can add or remove everything that he wishes."[135] In the Basel *Biblia Hebraica* – as in other large projects of its kind – the Jews had the last word. At the bottom of the last page appears a colophon in Hebrew, signed by the compositor, Elijah ben Judah Ulma. The rest of the page is taken up by a long text in verse by Abraham Braunschweig. In the very last line, he invites the reader to await the Messiah "in

135. Keuchen, quoted in Burnett, "Hebrew Censorship," 220 n66.

our exile."[136] While this Bible began with a letter to the reader, dated "from the birth of Christ our Savior," it ended with an invocation of another savior entirely. The supposedly humble corrector, the interloper who complained that no one listened to him, had enough agency to give the whole text a conventional Jewish identity. No preface that Buxtorf could have written would have done justice to Braunschweig's aspirations – expressed, just as indelibly as Buxtorf's, in their collaborative print creation.

136. Braunschweig, in Buxtorf, *Biblia sacra*: "May the Lord strengthen our weak hands in this exile of ours; may he speedily comfort us with our righteous Messiah, as we were promised by his prophets." Buxtorf did not annotate this passage in his working copy, Basel, UB, FG II 20.

CHAPTER FIVE

Buxtorf and Censorship

The printers of Basel had more room to experiment than most of their colleagues. But when one of their projects pushed at the boundaries of the allowable, it was easy to pass beyond them and attract the hostile attention of the authorities. On the other hand, it was also possible, sometimes, to circumvent those boundaries.[1] In 1542, Johannes Oporinus set out to print an edition of the Qur'an in Latin, along with older refutations of it. He did not ask for permission. When professors began to agitate against his project, the authorities confiscated his materials and jailed him. After news came from Zurich, Wittenberg, and elsewhere that the good and the great of the Lutheran and Reformed churches approved of the plan, the publisher was released, the texts were returned, and Theodore Bibliander, the scholar charged with preparing the text, went back to work. The censors forbade Oporinus to mention his own name or that of Basel in the book. But they missed the marginal notes, in which Bibliander remarked that Muslims were morally superior to Christians and insisted that the passages in which Mohammed ordained that all believers in God should be tolerated and protected expressed his true sentiments – though at times he seemed oblivious of that point.[2]

1. For the broad background, see Christian Scheidegger, "Die Bücherzensur in der Schweiz von der Reformation bis 1600," *Zwingliana* 45 (2018): 365–435; on censorship in Basel in the late Renaissance, see Alban Norbert Lüber, "*Die Basler Zensurpolitik* in der zweiten Hälfte des 16. Jahrhunderts," *Basler Zeitschrift für Geschichte und Altertumskunde* 97 (1997): 77–141.

2. Martin Steinmann, *Johannes Oporinus: Ein Basler Buchdrucker um die Mitte des 16. Jahrhunderts* (Basel, 1967), 24–35; Carlos Gilly, *Spanien und der Basler Buchdruck bis 1600: Ein Querschnitt durch die spanische Geistesgeschichte aus der Sicht einer europäischen Buchdruckerstadt* (Basel, 1985), 16–20. For the wider context, see Thomas Burman, *Reading the Qur'an in Latin Christendom, 1140–1560* (Philadelphia, 2007), esp. 178–98.

Censorship in Basel

Basel censors included such cosmopolitan scholars as Bonifacius Amerbach, who knew the great world of scholarship at first hand. They did their jobs carefully. If versions of a given controversial passage had to go before the city council for critical scrutiny, they were translated into German. The council sometimes defended its action in a published text.[3] "Mice-eyed decipherers," in the words of Thomas Nashe, scrutinized many books that could be dangerous and reported their results.[4] When a respected Basel professor – for example, the irenic Hellenist Sebastian Castellio – disagreed with Calvinist orthodoxy on salvation by faith, the offending passage was removed: not from the first edition of his commentary on Romans, but from later ones.[5] When the Latin Machiavelli published by Pietro Perna in 1580 gave offense because its dedication to the Catholic bishop of Basel was too warm and submissive, Perna set out to replace it. Its author, a professor named Nicolaus Stupanus, sat down in the workroom at Perna's shop and wrote a new, less irritating dedication. Unfortunately, he retained phrases as dangerous as those he cut. Perna refused to pay Stupanus the fee he had promised, which in turn provoked Stupanus to pummel Perna in the Academy building. Happily, the printer was not harmed, but decorum had broken down, and the name of Basel was besmirched. In the end, Stupanus was deprived of several years' salary, while Perna deftly dodged the censors' demand that all copies of the original dedication be destroyed.[6]

Jewish books – and their censorship – could also cause scandals. In the 1570s, Jews began pressing to have an edition of the Talmud printed, to replace some of the copies publicly burned in Italy in 1553 and after. In 1574, the Spanish scholar and bookdealer Casiodoro de Reyna was trying to set up a new printing house in Basel. One of his projects for it seems to have been an edition of the Talmud. The subject was touchy. When he visited the men who would be his professional colleagues, to ask them to support his bid for citizenship, he took the greatest

3. Frank Hieronymus, "Gewissen und Staatskirchenraison: Basler Theologie und Zensur um 1578," *Archiv für Reformationsgeschichte* 82 (1991): 209–38.

4. Thomas Nashe, "The Prayse of the Red Herring," in his *Works,* ed. Ronald McKerrow, repr. with corrections, ed. F.P. Wilson (Oxford, 1957; repr. 1968), 3: 218.

5. Sébastien Castellio, *Annotationes in Pauli Epistulam ad Romanos, ex cap. ix,* ed. and trans. Michiel Op de Coul and Mirjam van Veen (Geneva, 2020); cf. Carlos Gilly, "Die Censur von Castellios *Dialogi quattuor* durch die Basler Theologen (1578)," in *Freiheitsstufen der Literaturverbreitung – Zensurfragen, verbotene und verfolgte Bücher,* ed. József Jankovics and S. Katalin Németh (Wiesbaden, 1998), 147–76.

6. Werner Kaegi, "Machiavelli in Basel," *Basler Zeitschrift für Geschichte und Altertumskunde* 39 (1940): 5–51; Gábor Almási, "*Experientia* and the Machiavellian Turn in Religio-Political and Scientific Thinking: Basel in 1580," *History of European Ideas* 42 (2016): 857–81.

possible care not to mention this to them. As he reported to Theodor Zwinger, "I did not discuss the Talmud with Heinrich Petri, not a single word.... I shared my plan with Episcopius (not a word about the Talmud)[.]"[7]

A few years later, Ambrosius Froben actually set out to print a Talmud, with financial support from Simon Günzburg, a wealthy Jewish merchant and Talmudic scholar. Froben and his patron disagreed on how thoroughly – if at all – the text should be purged of anti-Christian elements. His correctors – one of them a converted Jew from Ferrara, Immanuel Tremellius; another a Catholic censor from Brescia, Marco Marini; the third a Calvinist from Geneva, Pierre Chevalier – all believed that the Talmud contained rich information about Jews and early Christians that should be preserved intact. But they were also willing to remove what Tremellius called "the passages in it that are against those to whom only the land was given," in order "to ensure that those would not breathe fire and fury."[8] Meanwhile furious letters came from the capitals of the Reformation, Zurich, Geneva, and Wittenberg, demanding that Froben cease and desist lest Jewish lies corrupt Christian truths.[9] Bonaventura Vulcanius, who was working for Froben as a corrector, needed fast footwork to convince Theodore Beza and others that he was doing his best to persuade the printer to abandon the project even as he arranged for Marini's corrections to be incorporated in the published texts.[10] In many quarters, resentment was slow to fade – especially as the title pages of the individual volumes noted that Marini had corrected and expurgated the volumes "in accordance with the views of the holy Council of Trent."[11]

Less prominent Jewish texts also caught the eye of the censors. In 1573, a Jewish convert named Marcus Lombardus proposed to publish a short work that he titled "A Thorough Report and Explanation of the Actions and Ceremonies of

7. Casiodoro de Rayna to Zwinger, 24 September 1574, in *Bibliotheca Wiffeniana: Spanish Reformers of Two Centuries from 1520*, ed. Edward Boehmer (Strasbourg and London, 1883), 2: 225.

8. Tremellius, note on the flyleaf of a copy of Samson ben Isaac of Chinon, *Sefer Keritot* (Cremona, 1558), London, British Library 1956.b.13; first published by A.M. Habermann in his *Studies in the History of Hebrew Printers and Books* (in Hebrew) (Jerusalem, 1978), 290–91; translated by Michela Andreatta.

9. Ernst Staehelin, "Des Basler Buchdruckers Ambrosius Froben Talmudausgabe und Handel mit Rom," *Basler Zeitschrift für Geschichte und Altertumskunde* 30 (1931): 7–37; Joseph Prijs, *Die Basler hebräischen Drucke, 1492–1866*, ed. Bernhard Prijs (Olten, 1964).

10. Anthony Grafton, *The Culture of Correction in Renaissance Europe* (London, 2011), 132–35.

11. On the Basel Talmud edition see Prijs, *Die Basler hebräischen Drucke*, 175–210. In fact, the censors carried their work of expurgation farther than the Council of Trent had demanded (ibid., 179).

the Jews."[12] Despite its title, this was an extremely polemical pamphlet that drew more heavily on Luther's anti-Jewish tracts than on the author's inside knowledge of Jewish life and customs.[13] Lombardus had other plans for publications. He came to know at least one Basel publisher, Perna, with whom he, like many others, seems to have had problems. A note to his patron in Basel, Basilius Amerbach, reads, in total, "Most reverend and excellent Lord Rector, I ask your excellency, that your excellency may please show me what I should do with that printer Petrus Bernat. Your excellency, Most obedient Marcus Lombardus."[14] Though Amerbach supported Lombardus, he was serving as university rector, with responsibility for press censorship, in 1573, when the *Gründlicher Bericht* was submitted for publication. By chance, surviving notes reveal that he wondered whether the book needed examination: "On 20 [?] June I sent the text of M. Lombardus on the Jews, written in German, via the beadle to [Ulrich] Cochius [Coccius], so that he might examine it, or, if he thought it a question for the Magistrates, send it to them, since they were about to meet then after lunch."[15] Amerbach's note makes clear how small a world the Basel authorities inhabited. Evidently they thought it possible that even a modest work on the Jews, at least if highly polemical, might need modification before it could be published. No records survive to indicate whether the *Jüden Schul* inspired similar thoughts.

Jewish texts of many kinds needed censorship of a less visible, but still indispensable, kind. Many used terms (such as "goi") and included statements (such as references to the future coming of the Messiah) that could give offense to Christian readers. Catholic and Jewish censors made massive, though necessarily sporadic, efforts to remove these from texts already in print.[16] More important, both official censors and correctors and compositors did their best to shape

12. Marcus Lombardus, *Gründlicher Bericht und Erklärung von der Juden Handlungen und Ceremonien* (Basel, 1573); a Latin translation by Conrad Huser, *Tractatus de imposturis et ceremoniis Iudaeorum nostri temporis*, was printed by Perna in 1575.

13. Gaby Knoch-Mund, *Disputationsliteratur als Instrument antijüdischer Polemik: Leben und Werk des Marcus Lombardus, eines Grenzgängers zwischen Judentum und Christentum im Zeitalter des deutschen Humanismus* (Tübingen, 1997), 348–50; Andreas Berger, *Stadt ohne Juden? Präsenz und Absenz in der Frühen Neuzeit* (Basel, 2024), 238–48.

14. Knoch-Mund, *Disputationsliteratur*, 407.

15. Basilius Amerbach, notes, in Basel, UB, MS C Via 31, 49, printed in Lüber, "*Die Basler Zensurpolitik*," 139: "20 (?) Junij misi scriptum M. Lombardi de Judaeis Germanice scriptum per pedellum ad Cochium, ut is censeret, aut, si putaret ad Magistros [or Magistratus] pertinere ad eos mitteret, quì tum post prandium conventum habituri erant." Coccius was professor of the New Testament.

16. William Popper, *The Censorship of Hebrew Books* (New York, 1899); Isaiah Sonne, "Expurgation of Hebrew Books: The Work of Jewish Scholars: A Contribution to the History of the Censorship of Hebrew Books in Italy in the Sixteenth Century," *Bulletin of the New York Public Library* 46 (1942): 975–1015.

and polish Jewish texts before publication. These efforts were more systematic in Italy than elsewhere, but they took place in printing shops in Protestant as well as Catholic cities. Joseph Skloot has recently shown that *Sefer hasidim,* a compendium of Jewish rituals and stories passed down by the Pietists of medieval Ashkenaz, was reshaped twice for its appearances in print: the first time in Bologna in 1518, the second in Froben's shop in Basel in 1580, less than a decade before Buxtorf arrived in the city.[17] Much of the work that Buxtorf did as a censor, as we will see, took similar forms: meticulous, granular efforts to remove problematic elements from texts, often without leaving any verbal indication that this deletion had occurred. Yet his interventions in the Jewish texts he censored were, as we will see, sometimes quite radical, in his distinctive manner.

A First Approach: Walter Keuchen, Censor in Hanau

Buxtorf made his disapproval of Jewish texts and writers clear in many ways. His views mattered, for he spent long hours, year after year, examining Hebrew and Yiddish texts submitted for publication at Basel, demanding changes in at least some of them. Yet it is not easy to reconstruct exactly what he thought about working as a censor, for he left no explicit discussion of what he saw as the duties of a censor of Jewish texts or the skills needed to carry them out. A colleague and former student of his, however, provides some insight into the way a censor addressed these issues. We have already met one of Buxtorf's pupils, Walter Keuchen, who helped his master obtain Jewish books or at least information about them. Keuchen, whose career Stephen Burnett has reconstructed, served as rector of the gymnasium in Hanau from 1612 to 1622. He also censored Hebrew books on behalf of the magistrates of Hanau, one of the few towns in Germany that allowed a Jewish press to operate – so successfully that it would replace Waldkirch and König as the major producer of Jewish texts in the German world.[18] Twenty-two of his memoranda to the magistrates survive, along with one from a local cleric. These provide a rich context for Buxtorf's efforts and help us assemble the more fragmentary evidence for them into something like a whole.

17. Joseph Skloot, *First Impressions: "Sefer Hasidim" and Early Modern Hebrew Printing* (Waltham, MA, 2023).

18. Stephen Burnett, "Hebrew Censorship in Hanau: A Mirror of Jewish-Christian Coexistence in Seventeenth-Century Germany," in *The Expulsion of the Jews: 1492 and After,* ed. Raymond Waddington and Arthur Williamson (New York, 1994), 199–222; Burnett, "German Jewish Printing in the Reformation Era (1530–1633)" in *Jews, Judaism, and the Reformation in Sixteenth-Century Germany,* ed. Dean Phillip Bell and Stephen G. Burnett (Leiden, 2006), 503–27.

In 1609, when the magistrates permitted the establishment of a Jewish press, they turned first to the superintendent of the local church, Heinrich Heidfeld. In a memorandum dated 23 May 1609, he made clear that he did not feel capable of acting as censor: "I do not have enough knowledge of the holy language that I can undertake to make any judgment about Jewish books." The ones that he had been shown seemed eminently respectable: Bibles printed by Christophe Plantin in Antwerp, "which are considered perfectly correct," and prayer books printed by Conrad Waldkirch at Basel, "doubtless with the approval of the theologians." True, he pointed out, all Jewish prayers were blasphemous by definition, since they were not directed "at the true God, who is the Father of our Lord Jesus Christ." He thought it more praiseworthy to try to convert the Jews than to confirm them in the exercise of their "detestable religion."[19] In the end, though, Heidfeld seems to have accepted that a censor must be able to scrutinize Jewish books and detect actual, intended blasphemies.

Keuchen wrote at much greater length to the magistrates. They had asked him to censor an edition of the *Arba'ah turim* or *Tur*, a legal code compiled by Jacob ben Asher in the fourteenth century. Keuchen did the job, on his account, in less than two months. He made clear from the start that he knew a fair amount about both printing and Jewish texts. Founding a press, he argued, could spread the fame of Christian rulers (such as the counts of Hanau-Münzenberg, who had admitted the Jews to their territory, or the city magistrates): and a Hebrew press would do so more effectively than a Greek or Latin one.[20] Hebrew books were unattractive, to be sure. But under their alien covers gleamed "the jewels whose

19. Heidfeld to the Hanau Magistrates, 23 May 1609, Marburg, Staatsarchiv Marburg, SA 81 B 1 Rubr 81 Nr 21, fol. 18r: "Clarissimi consultissimique viri, Tantam ego linguae sanctae cognitionem non habeo, ut de libris Hebraicis judicium aliquod mihi sumere possim. Hi vero libri, qui censurae Consist. sunt subiecti, partim Biblia sunt, ex editione Plantini, quae habentur castigatissima, partim duo precationum libelli, qui judaeis sunt in quotidiano usu, Basileae a Waldkirchio excusi, cum Theologorum sine dubio approbatione. Quanquam omnes Iudaeorum preces, quamcumque etiam pietatis habeant speciem, blasphemae sunt, quia directae non sunt ad verum illum Deum, qui est pater Dn. Nostri Iesu Christi. Mallem nos id potius laborare, quomodo Iudaeos Christo possemus lucrifacere, quam ut in sua, quam detestamur, religione, per nos confirmentur, fiantque tanto obstinatiores." For the length of time Keuchen spent on his task see ibid., fol. 20r, where he states that the book had been sent to him "ante septimanas forte 7. vel octo ab Amplitudine vestra[.]"

20. Keuchen to the Hanau Magistrates, n.d.; Marburg, Staatsarchiv Marburg, SA 81 B 1 Rubr 81 Nr 21, fol. 20r: "Quibus rebus Princeps ac civitas ad perfectam laudem ascendere solent, eae, cum a vobis, viri amplissimi, in urbe Hanoviae, summa cura & studio fundantur, est profecto quod exteri gratulentur & gaudeant. Inter illas vero cum Typographia q. praecipuum est instrumentum, quo fama boni Principis, & Reipubl. existimatio quam latissime extendatur & augeatur; tum vero summae laudis fuit consilium vestrum, de hebraea typographia erigenda, per quam melius efficiatur, id, quod intenditur, quam Graeca aut Latina unquam fieri possit."

brilliance adorned Plato and Aristotle, the best of philosophers, and the holy Fathers. There is no stream of true wisdom that does not flow and ebb from the Hebrew spring."[21] Christians had a duty to save the Jews, the people of the Messiah, who had preserved the Scriptures and to whom he had promised salvation, from their own obstinacy.[22] Though most Christians failed to realize it, they needed knowledge of Hebrew in order to dispute with and convert the Jews. Lack of Hebrew books had prevented them from learning the language in the past. The Hanau Press would be a major aid in bringing the Jews into the Christian fold.

Keuchen, however, also made clear that he was wary of both Jews and Jewish texts. The press would carry out its task, he explained, only "if it is overseen carefully and without tricks."[23] Some Christians might fear that "in the books of the Jews, blasphemies against the Christian religion are spat out in all directions," and Christian magistrates ridiculed.[24] And here Keuchen became more modest, professing himself unable to give a firm judgment. As he admitted, he had read few of the rabbis in their entirety and could neither affirm nor deny that their books contained such things. Yet he knew – and pointed out, as Heidfeld had – that many Jewish books had been already printed once or twice before at Venice or Cremona. So long as "another reading shows that these had already been censored in a careful way," he declared, "I would not think there is a reason why they should be held back from the press."[25] On one point, then, he and Heidfeld seem to have agreed: to be safely publishable, the Hebrew texts submitted should already have

21. Ibid.: "Nam si mercantes Belgae gloriantur, urbes suas utrique Indo celebratas esse, vestrae urbi major gloria surget, si tam nobiles merces ad remotiss. oras transportet: quae quamvis saepe sordidis involucris tectae sint, tamen gemmae latent, quarum splendore Plato & Aristoteles summi philosophi & Patres S. ornati erant. Neque vero usquam verae sapientiae rivulus est, qui ex fonte hebraeo non fluat ac refluat."

22. Ibid.: "Deinde vero maxime, quod ex sola hac Officina procedente aptissima Christianis arma adversus ipsorum Judaeorum obstinaciam: Quorum saluti certe invigilare plus consentaneum est Legibus Christi, quam ullius alterius Gentis. Ipsi enim sunt gens Messiae nostri secundum Genealogiam, ipsi foedus Dei scriptum ad nos inviolatum traduxerunt, ipsi praeclaram promissionem reinsitionis in corpus mysticum, in N.T. consecuti sunt, atque ipsi illi sunt, quibus Christus quodam Testamenti loco in luctu S. mortis suae reliquit, ut Deus ipsorum ignorantiae ignoscat. Quae omnia quamvis plaerosque Christianos hactenus parum novisse videatur [MS videantur], ut studio hebraeo diligentiorem operam navarent, sine cujus cognitione cum Judaeo agi omnino impossibile est, tamen id praecipue librorum hebraeorum penuriae adjudicari debet."

23. Ibid.: "Quod itaque Typographiam consultissime a vobis, viri prudentissimi, Judaeis instituere permissum est, si caute ac sine dolo, ut equidem confido, administretur, talia opera proditura sunt, de quibus gratulaturi sibi sunt viri boni & christiani."

24. Ibid.: "Nam quod fortassis timeri posset, in libris Judaeorum blasphemias in Religionem Christianam hinc inde evomi, & rideri aut quoquo modo contemni Magistratum..."

25. Ibid.: "Attamen si certo exploratum habeatur, quos libros excusuri sunt Judaei vestri, eos jam antea Venetiis, Cremonae, semel bisve impressos: postea vero etiam, si rursus

undergone censorship by Christians. The exact confession such censors belonged to evidently mattered less than their shared Christian identity, as is clear from the variety of churches to which the correctors of Ambrosius Froben's Talmud belonged.

Despite Keuchen's admission of his inadequacy, he apparently scrutinized the whole text of the *Arba'ah turim* for blasphemies, taking other editions into account as well as the one in question

> Perhaps seven or eight weeks ago your excellency sent me, unqualified as I am, the Book *Arba Turim*, to be examined. It has been printed not only at Cremona and Trent [Riva di Trento], but also at various places in the Holy Empire.[26] Once I had read it in its entirety from end to end, I found nothing explicitly said against religion or against the magistrate.[27]

If the text had contained subtler statements of blasphemy or sedition to which a Christian reader might object, he explained, either the earlier censor had deleted them or he himself, as he read, "so far as my slender experience allowed, corrected, changed, or deleted them."[28]

Keuchen also made clear that he had not hesitated to rewrite passages that could have offended a Christian reader. He provided the magistrates with a specimen of the sort of work he had done, "so that from one example your Prudence may judge them all."[29] In book 2, the *Tur* declared:

> This is the foreskin, by which the nations are polluted, as it has been said: because all the nations have foreskins. And everyone who breaks the covenant of Abraham our father, because he is not circumcised or has his foreskin,[30]

relegantur, clare appareat, non oscitanter dudum censuratos fuisse, non putarem causam esse, quare a prelo arcerentur."

26. There were many editions of *Arba'ah turim*, including two incunabula. Keuchen refers explicitly to those printed in Cremona (1558) and Riva di Trento (1561).

27. Keuchen to the Hanau Magistrates, fol. 20r: "Veluti Liber Arbaturim, qui mihi, licet impari, relegendus ante septimanas forte 7. vel octo ab Amplitudine vestra transmissus fuit, non tantum Cremonae, & Tridenti, sed in ipso etiam S. Imperio diversis locis impressus est. Quam cum ego a capite ad calcem totum perlegerim, nihil contra Religionem, vel adversus Magistratum expresse dictum reperi."

28. Ibid.: "Sed siquid forte hinc inde obscurius scriptum erat, unde sensus aut consequentia a lectore Christiano non probanda elici posse videbatur, id omne, vel á Censore seu revisore ad id officium a Magistratu ordinato deletum est, vel a me nunc hac postrema relectione, juxta experientiae tenuitatem, directum, mutatum, vel deletum est."

29. Ibid.: "Quae qualia sint, ut ab uno omnia Iudicet Prudentia vestra, exemplum apponam."

30. Keuchen mistranslated the phrase "moshekh orlato" – which means "one who disguises his circumcision," not "has his foreskin."

> even if the divine law and good works are in his hand, he has no part in the world to come.[31]

Keuchen began by pointing out that the rabbis had condemned other peoples for rejecting circumcision long ago, before the ceremonial law was abrogated. Because the *Tur* repeated these ancient utterances, however, someone might think that these words now referred also to the people of the baptism. The other necessary corrections, he explained, were more or less similar to this one: not radical changes but adjustments of wording designed to clarify the meaning of the text.[32]

Keuchen justified the mildness of his practice by arguing that the *Tur* was not an anti-Christian text: "the author's intention was not deliberately to write against the nations, but he mentioned them, whenever the dealings of the Jews with other peoples required it."[33] He went on to provide the magistrates with a detailed summary of the book, which he characterized as an effort to rein in the excesses of the Talmud, so that a Jew could more easily determine his actual duties.[34] From the information Keuchen had provided, he explained, the magistrates could

31. Keuchen to the Hanau Magistrates, n.d., fol. 20r: "Libro secundo de circumcisione sic habetur: De. et despreta, hoc praeputium est, quo foedatae sunt gentes, ut dictum est: quia omnes gentes praeputiatae sunt. Et omnis dissolvens foedus Abrahamae patris nostri, quia circumcisus non est, aut retinet praeputium suum, quamvis sit in manu ejus Lex divina, & opera bona, non habet partem seculi futuri." He is referring to *Yoreh De'ah*, Hilkhot millah, par. 260.

32. Ibid.: "Reliqua omnia non absimilia sunt."

33. Ibid., fols. 20r–v: "Praeterea non est propositum Authori huic, quicquam ex instituto contra Gentes agere, sed illarum meminit, quoties ad id mutua Judaeorum cum reliquis populis commercia coegerunt."

34. Ibid., fol. 20v: "Totus enim liber, in quatuor sectiones dividitur: quare prima dicitur, Orah hajim, id est semita vitae, qua continentur Preces, quae, quales sint, ubi, quando, & quomodo dici debeant, quibus adjungitur, de frontalibus, penniculamentis, etc. aliaeque paene infinitae observationes sacri cultus, quas ne mente concipere Judaeum posse, nedum observare crediderim. Secunda appellatur Joreh dea, docens scientiam, qua comprehendunt omnia, ut Authoris verbis utar, quibus quisque opus habet, ex iis, quae hodie in usu vigent, ut Mactatio, & quae circa illam, in sanguine, ruptura membrorum, adipe etc. It. mixtura, vinum consecratum, Idololatria, usura etc. & reliqua, quae omnia recensere, nihil attinet. Tertia vocatur, Ebhan haezer, id est lapis adjutorum, quae tota vertitur in rebus ad Matrimonium pertinentibus; quis, quae, ducere vel nubere teneatur, sponsalia, dos, nuptiae, repudium, connubium cum uxore fratris demortui, & quae circa haec etc. Quarta Hoschan hammischpat, pectorale Iudicii, complectitur omnia Judiciorum genera: quae, quales, quid judicent hoc tempore Iudices, de Testimonio & testibus, de mutuato & commodato, de Actione, de Juramento, de contractibus, quaeque circa illos, de acquisitione, possessione, haereditate, furto, rapina, damno etc. Et ut, quod res est, dicatur, nihil hic libet aliud esse, quam restrictio quaedam Talmudici effraenis excursus. Quae enim illis in omnes partes sine modo late diffunduntur, hic quidem intra limites arcte coguntur & finiuntur, ut sciatur, quid sequi, quod Legi ac Rationi proximum est."

easily decide whether there was any reason to prevent the Jews from printing the text.[35] He did not know for certain whether the Jews who were responsible for the printing house were honorable or deceitful. Still, he had a somewhat favorable opinion of Seligman Ulma, son of Moses Simeon Levi Günzberg, who edited several of the Hanau editions, "he does not absolutely refuse to listen, like the others."[36] As we will see, there was more than a little irony in this judgment. Though Keuchen claimed to be embarrassed to make such an illiberal request, he asked that he be paid an appropriate fee, both "so that the Jews may not despise the efforts of those who shall be assigned to work as censors in the future," and because he knew the authorities wanted those who did this job to be decently compensated.[37] Later letters stated his charges without hemming and hawing.

Most of Keuchen's later judgments on Jewish books resembled this one in tone and content. He evaluated each text briefly. The *Touchstone* (*Even bohan*) of the Provençal Jewish writer Kalonymus ben Kalonymus (d. after 1328), he wrote in December 1610, was an elegant satire of the corrupt ways of the Jews.[38] About the *Nishmat Adam* written by Aaron Samuel ben Moses Shalom of Kremnitz, he commented in 1611 that it "consists of the sort of argument that its title suggests:

35. Ibid.: "Vos igitur, viri consultissimi, sitne causa, cur Judaei inhibeantur, ab hoc Opere excudendo, nec ne, facile perindere potestis."

36. Ibid.: "De fide autem an dolo illorum, quibus cura typographiae librorumque correctio commissa est, mihi nihil certi constat; nisi quod me de Judaeo Seligman opinionem haut sinistram concipere, facit, quod, ut caeteri, ferreas aures non gerat." Seligman Ulma worked on many of the Hebrew books printed in Hanau. As we will see, he seems to have imposed his own will with regard to the corrections of the books and retained the original Hebrew passages that Keuchen had condemned. Apparently, Seligman, by then called Seligman zur Sonne, according to the Konsistorium Protocols of Hanau (1613), converted to Christianity. His conversion is not registered in the documents and no trace of the alleged conversion is detectable in the works that he corrected. Thus, e.g., in the *Ginnat egoz* of 1614, he refers to "our Torah" and adds the conventional plea for the coming of the Messiah. The problematic sentence in the protocols "Seine discipuli weren gewessen Drusius und Buxtorfius" should not be taken as a reference to Seligman's teaching of Hebrew to Buxtorf and Drusius; rather it suggests that Seligman had discussed the foreshadowing of Christ in the messianic prophecies in the Old Testament with these two scholars. See Eckhard Meise, *Konversion und Assimilation: Taufen von Menschen fremder Konfessionen in Hanau bis ins 19 Jahrhundert* (Hanau: Selbstverlag des Hanauer Geschichtsvereins, 1999), 32–35.

37. Keuchen to the Hanau Magistrates, n.d., fol. 20v: "Laboris denique mei, in hoc libro relegendo indepti, quamvis repensam ullam illibenter posco, ne tum aliorum, quo forte imposterum ad censuram deputabuntur, apud Judaeos laboris vilescat aestimatio, tum vero etiam quod vos ita velle sciam, si iniquum non est, pendant, quibus hoc incumbit, 9 vel decem R. Ipsius vero libri exemplum petivissem, nisi jam ante ab aliis mihi promissum esset."

38. Keuchen to the Hanau Magistrates, 14 December 1610, Marburg, Staatsarchiv Marburg, SA 81 B 1 Rubr 81 Nr 21, fol. 23r: "Atque hic satyrica et curta sermonis elegantia in corruptos gentis suae mores, omnesque caeremoniarum abusus mordacius invehitur, adeo ut Subsannatoris nomen apud populares suos sibi invenerit."

the soul of man."[39] The *Ginnat egoz* (*Nut Garden*) of the Spanish kabbalist Joseph Gikatilla, he explained in 1613, instructed the reader in the secrets of the higher worlds, by listing and enumerating the meanings of Hebrew names, letters, and points.[40] Keuchen also summarized the contents of each work. Authorial intent mattered to him. Of Aaron ben Moses Shalom's book, for example, he noted that "the author never crosses the bounds of his subject, so as to harm the inviolate parts of religion. But here and there, things that seemed to be said in a problematic way have been laid out more clearly or removed."[41] For the most part, he found no objection to their publication, even when he found it necessary to censor certain passages.

At times, he expressed more serious disapproval. Discussing certain texts that the Jews hoped to add to a prayerbook, he praised the "elegant brevity" of one of them.[42] Still, he made clear that his reservations were serious: "But some matters are mingled with these, which could offend a piously reared mind: for example, certain gestures of those praying, the secret use of the names of the angels and God, and some other abuses of words or letters." Keuchen showed the uncertainty appropriate to this hard case by asking the authorities to determine "which of such works can be sent to the press."[43] In other cases he was more critical and more decisive. Sending back a Jewish prayerbook, he noted that the text had been printed before at Hanau, "but from a purer copy." The one in question had to be "castrated in many passages: it was printed at Prague, where the Jews are given too much license."[44] Yet he tolerated polemical works by Jews, as Stephen

39. Keuchen to the Magistrates, 13 March 1611, Marburg, Staatsarchiv Marburg, SA 81 B 1 Rubr 81 Nr 21, fol. 25r: "Libellus hic ab Amplitud. vestra Censurae traditus, tali argumento constat, quale titulus refert: Nischmath adam id est Anima hominis." Seligman Ulma appended a postscript in praise of the work (fol. 43a) and also added a paginated index of all the novel topics included in the book.

40. Keuchen to the Magistrates, 20 March 1613, Marburg, Staatsarchiv Marburg, SA 81 B 1 Rubr 81 Nr 21, fol. 44r: "Intendit autem Author libri, occultiora mundi Angelici, Sphaerici & Purin tradere, non disquisitione Philosophica, sed enumeratione & interpretatione Nominum, Literarum, & Punctorum hebraeorum."

41. Ibid., fol. 25r: "Metas vero propositi hujus sui, nusquam Author ita transcendit, ut intemerata Religionis laedat: sed & hinc inde, quae suspectius dici videbantur, apertius digesta aut amputata sunt."

42. Keuchen to the Magistrates, 29 January 1611, Marburg, Staatsarchiv Marburg, SA 81 B 1 Rubr 81 Nr 21, fol. 24r: "Et huic additum est aliud Canticum Unitatis, idem eleganti brevitate complectens." The *Canticum Unitatis* is to be identified as the mystical anthropomorphic hymn *Shir ha-yihud*, published in the majority of Ashkenazic and Polish prayer books.

43. Ibid.: "Sed quia hic quaedam miscentur, quae pie institutam mentem offendere possent; ut sunt singulares quidam orantium gestus, Angelorum Deique nominum secretus usus, & siqui sunt alii vocum aut literarum abusus: Quid talium prelo mitti possit, statuet Prudent. & Amplitudo vestra; cui officia mea debite commendo."

44. Keuchen to the Magistrates, 5 May 1615, Marburg, Staatsarchiv Marburg, SA 81 B 1 Rubr 81 Nr 21, fol. 48r: "Remitto librum Precum Iudaicarum, antea hic Hanoviae quoque

Burnett has shown, so long as they were not directed against Christianity. Thus he saw no reason to prevent Salman Zvi Hirsch from printing his *Yudischer Theriak*, a reply to an attack on Judaism written by a Jewish convert to Christianity, Samuel Friedrich Brenz. After all, Hirsch's book was a defense of Judaism, not a critique of Christianity.[45] In any event, the magistrates allowed it to appear only in Yiddish, which made it unlikely to circulate among Christians.[46]

The one book that Keuchen thought the Hanau press should simply reject was the *Vinzenzlied* of Elhanan Heln. This poem, in alternating stanzas in Hebrew and Yiddish, described the 1616 punishment of Vinzenz Fettmilch and his associates, who had expelled the Jews from Frankfurt with considerable violence in 1614.[47] The Jews celebrated Fettmilch's execution as a second Purim. Shown the text, Keuchen commented:

> The pages of this pamphlet tell the story of the Frankfurt riots in Hebrew and German, and in particular mourn the slaughter of the Jews. Yet later they make this less painful with the music of flutes and drums, by the music of which they were brought back into their neighborhood. I recall that a work like this was printed here before. But at that time, some respectable citizens of Frankfurt seemed surprised that such contemptuous things about them could be distributed here with the formal permission of the authorities. For my part, accordingly, I would prefer that the Jews take this stuff to be printed at Basel or elsewhere. But what you gentlemen think best will be done.[48]

The magistrates accepted his view and rejected the book.[49] Political risk evidently weighed more heavily in the censor's balance than even a large number of passages

---

excusum, sed ex puriori Exemplari; quare multis in locis castrandus fuit hic Pragae impressus, ubi Iudaeis nimia datur licentia."

45. Salman Zvi Hirsch, *Yudisher Theriak: An Early Modern Jewish Defense of Judaism*, ed. and trans. Morris Faierstein (Detroit, 2018).

46. Burnett, "Hebrew Censorship," 208–9.

47. Ibid., 209.

48. Keuchen to the Magistrates, 22 July 1616, Marburg, Staatsachiv Marburg, SA 81 B 1 Rubr 81 Nr 21, fol. 64r: "Tumultus frankifortensis Hebraice & Germanice his pagellis recontatur [MS recuntatur], ac praecipue clades Judaeorum deploratur: quam tamen postea tibiis & tympanis, quorum praecentu in vicum suum sunt reducti, solantur. Antehac simile hic impressum memini. sed quia tum temporis non infimi quidam civium Frankfortensium mirari videbantur, talia hic in sui despectum superiorum licentia in vulgus spargi; ego pro me mallem, Basileam haec, aut alio imprimenda judaeos perferre. Sed id fiat, quod vobis, viri Ampliss. melius videbitur."

49. Burnett, "Hebrew Censorship," 209–10. Buxtorf owned a copy of the Hanau 1615 edition of the *Yudischer Theriak* (Basel, UB, FA VI 26: 3) which he read with great intensity, marking all passages that contained vituperative expressions about Jesus.

that had to be "castrated" for religious reasons.[50] The original text has not survived, and the first edition appeared in Amsterdam in 1648 and a later edition in Frankurt in 1696.[51] Johann Wagenseil produced the Yiddish text with German translation in 1699.[52]

Two points are most striking about Keuchen's practices: in a sense, they work against one another. He did his best, as we have already seen, not to pull individual passages out of the context of the whole texts they came from, or to judge their authors on the basis of such fragments. As a good Protestant humanist, Keuchen started his examination of texts by finding their *scopus*. When he assessed the fables of the thirteenth-century Spanish medical man Isaac ben Solomon Abi Sahula, he noted that "the author embraced the *scopus* of this book in five parts."[53] When he evaluated Gikatilla's *Nut Garden*, he told the magistrates, "As is clear from its subject and *scopus*, there is no reason why it should be kept from printing."[54] In these passages Keuchen made clear that Jewish texts, like Christian ones, should be read as wholes. His reluctance to forbid the printing of books that could be made safe by moderate correction stemmed from a principled theory about Jewish books and how Christian scholars should interpret them.

At the same time, he clearly distrusted the actual Jewish printers in Hanau, more sharply as time went on. And he had reason. Unfortunately for Keuchen, the "Jew" who oversaw the printing of the *Arba'ah turim* did not make the changes he had called for. The passage from the laws of circumcision in the second part of the *Tur* called Yoreh de'ah was not changed at all. The text is a combination of a passage from Nedarim (31b) and Midrash Genesis Rabbah (48). The Hanau

50. Note also Keuchen's comments on Moses Yaqar Ashkenazi's *Petah Einayim*, ibid., fol. 39r: "Haec fere sunt quae hoc libro tractantur, digna profecto plaeraque quae Theologis & Philosophis nostris nota essent. Cum tamen quaedam periculosum sit dicendo aut scribendo tractare, nequid sinistre aut adversum fidem nostram de Deo, Personis, & Attributis, reliquisque, ut & summa cujusque sectionis intelligi poterit, proferatur: ultram Censuram in correctura cauto opus erit, ne si alii ea permissa, errorem aut dolum pariat, quod dicto vel verbo uno facile fieret. Duos Ducatos pro Censura Judaeos solvere aequum existimo." This very short moralistic work, which was very popular, had already been published in Cracow, Salonika, and Prague. There is no evidence that it was published in Hanau.

51. On the complicated printing history, see Chava Turniansky, "The Events in Frankfurt am Main (1612–1616) in *Megillas Vints* and in an Unknown Yiddish 'Historical' Song," in *Schöpferische Momente des europäischen Judentums in der frühen Neuzeit*, ed. Michael Graetz (Heidelberg, 2000), 121–37; see n19.

52. The poem is published in Wagenseil's volume titled *Belehrung der Jüdisch-Teutschen Red- und Schreibart* (Königsberg, 1699), 119–48.

53. Keuchen to the Magistrates, Nr 21 fol. 23r: "Quinque autem partibus scopum libri huius Author complexus est."

54. Keuchen to the Magistrates, 20 March 1613, fol. 44r: "Ut ex argumento & scopo appareat, non esse, ut ab impressione cohibeatur."

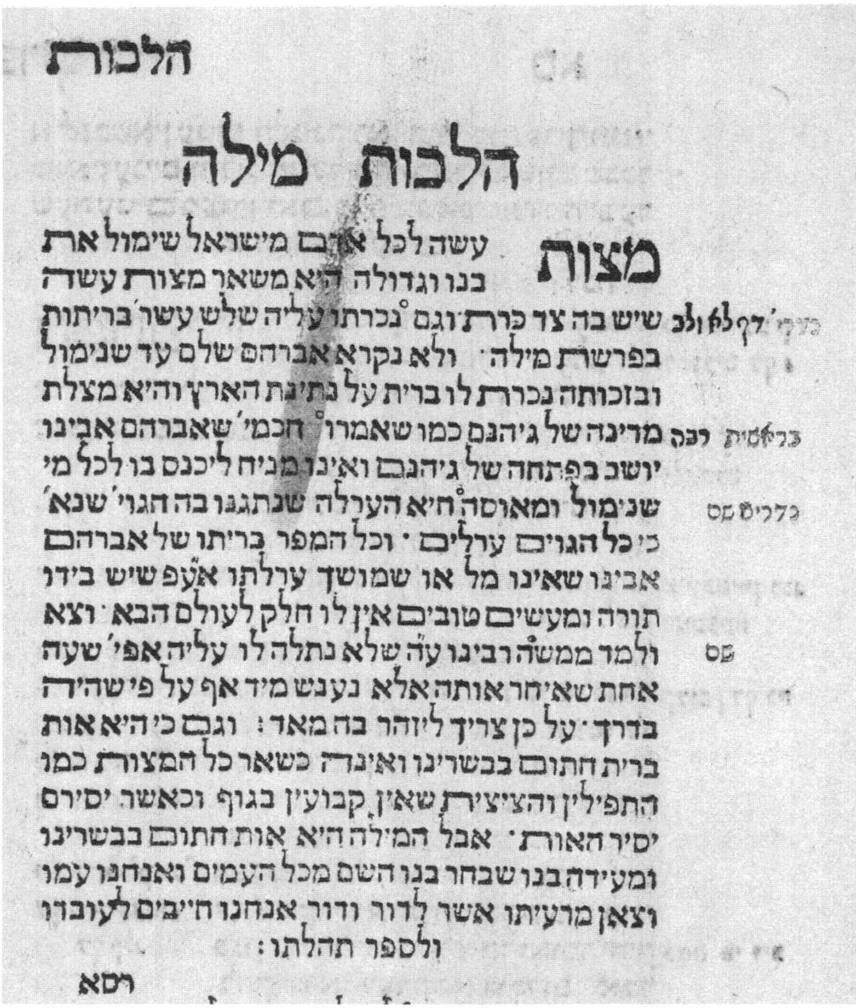

הלכות

הלכות מילה

מצות עשה לכל אדם מישראל שימול את
בנו וגדולה היא משאר מצות עשה
שיש בה צד כרת וגם נכרתו עליה שלש עשר בריתות
בפרשת מילה ׃ ולא נקרא אברהם שלם עד שנימול
ובזכותה נכרת לו ברית על נתינת הארץ והיא מצלת
מדינה של גיהנם כמו שאמרו חכמי׳ שאברהם אבינו
יושב בפתחה של גיהנם ואינו מניח ליכנס בו לכל מי
שנימול ומאוסה היא הערלה שנתגנו בה הגוי׳ שנא׳
כי כל הגוים ערלים ׃ וכל המפר בריתו של אברהם
אבינו שאינו מל או שמושך ערלתו אע״פ שיש בידו
תורה ומעשים טובים אין לו חלק לעולם הבא ׃ וצא
ולמד ממשה רבינו ע״ה שלא נתלה לו עליה אפי׳ שעה
אחת שאיחר אותה אלא נענש מיד אף על פי שהיה
בדרך ׃ על כן צריך ליזהר בה מאד ׃ וגם כי היא אות
ברית חתום בבשרינו ואינה כשאר כל המצות כמו
התפילין והציצית שאין קבועין בגוף וכאשר יסירם
יסיר האות ׃ אבל המילה היא אות חתום בבשרינו
ומעידה בנו שבחר בנו השם מכל העמים ואנחנו עמו
וצאן מרעיתו אשר לדור ודור אנחנו חייבים לעובדו
ולספר תהלתו ׃

Figure 5.1. *Sefer Arba'ah turim*. Hanau, 1610. Yoreh de'ah, hilkhot millah, par. 260, fol. 61b.

edition preserved the text printed at Riva di Trento in its entirety, including the passages that had bothered Keuchen (see Fig. 5.1).[55]

To make matters worse for Keuchen, Seligman Ulma even provided one of the usual apologias for errors at the end of the book.[56] Here he noted that mistakes

55. Jacob ben Asher, comp., *Sefer Arba'ah turim* (Hanau, 1610), Yoreh de'ah, hilkhot millah, par. 260, fol. 61b.

56. Ibid., fol. 140a.

had crept into the text. He referred explicitly to the chaos caused during the nights and on Sabbaths and festival days, when there was no proper supervision at the press. Most strikingly, he testified that he had done his best to eradicate the errors that had occurred, particularly those that had resulted when persons unknown altered the meaning of the text "as they thought fit." It seems likely that he had Keuchen in mind here, and that he deliberately reversed the changes the Christian censor had made.

In Keuchen's March 1613 evaluation of the *Nut Garden*, he noted that he had not had the time to examine the work in full detail.[57] He also made a suggestion to the magistrates:

> I wish that you would require of the printer by your authority that he send me the individual leaves [the proofs], before they were printed, so that I could go over them. In fact, I would consider this necessary in all the books that they either are printing or will print in the future: that the printer be required by your decree not to go to press with anything that the censor, if he were present in the city, had not gone over one last time. What would be the use of censorship if what has been removed, changed, etc. should be entrusted once more to the fancy of the Jews, with no observer present? They could restore things as easily as the censorship had changed them.[58]

Again, it seems likely that he had grounds for his suspicions. Seligman Ulma added another "corrector's address" at the end of the *Nut Garden*. Here he repeated in virtually identical terms his lament about the shoddy work of non-Jewish compositors (and his own efforts to purify the text).[59] It seems highly likely that in this case, too, Seligman discarded Keuchen's interventions.

57. Keuchen to the Magistrates, 20 March 1613, Marburg, Staatsachiv Marburg, SA 81 B 1 Rubr 81 Nr 21, fol. 44r: "Ego tamen tempore exclusus, & quod impressionem maturet necessitas, singula ubique non potui exactissime perpendere."

58. Ibid.: "Proinde vellem, vestrae Amplitudinis etc. authoritate hoc Typographo imponeretur, ut singula folia, antequam imprimerentur, mihi revidenda mitteret. Quamvis idem in omnibus libris, quos vel imprimunt vel impressuri imposterum fuerint, necessarium judicarem; ut singulari Amplit. Vestrae decreto Typographus teneretur, ne quicquam prelo commiteret, quod a censuratore, si praesens esset in loco, ultimo revisum non fuisset. Quid enim censura prodesset, si, quod sublatum, mutatum etc. erat, Judaeorum libidini, nullo praesente observatore, rursus committeretur? Eadem enim facilitate isti restituere possent, quod Censura mutaverat."

59. The "corrector's address" (in Joseph Gikatilla, *Ginnat egoz* [Hanau, 1614], fols. 77r–78v) follows conventional lines. In rhyming prose, Seligman complains that the errors in the book came about through the stubbornness of workers who would not listen to their teachers, worked at fixed times every day and did not have their heart in their work. More destruction was wrought on the eve of Sabbath and festivals when "the demons were generated": they worked in haste in order to get [paid] double and there was nobody to

Keuchen repeated his warning to the magistrates twice in 1615.[60] Assessing an edition of the Talmudic tractate Niddah, which deals with the laws of menstruation, in 1617, he again urged the need to take special care.[61] Though this text had no offensive contents, he still suspected that the Jews who were paying for its publication would restore any and all omissions unless daily inspections prevented them.[62] Keuchen saw most Jewish writers, apparently, as ignorant of the truth but not malevolent in their intentions. He did not extend the same faith and credit to the printers with whom he worked, or their backers. His suspicions, as we have seen, were well founded.

A Second Approach: The Evidence for Buxtorf's Work as Censor

Like Keuchen, Buxtorf was a dedicated censor of Hebrew books. No official document attests to his work in this capacity, but "mute evidence" of his censorial profession is paraded all over the books he owned and in the assembly of extracts strewn throughout his notebook.[63] More revealing was what he did not record:

---

take charge and oversee their work. "Who can correct the errors and negligence which was committed through ignorance?" Seligman then goes on to talk about the problem of the manuscripts, which derived from scribes who lacked knowledge and were not careful in their transcriptions. There is no explicit reference to Keuchen.

60. Keuchen to the Magistrates, 3 May 1615, fol. 48r, referring at the start to an earlier letter on the same theme which is not in the Marburg dossier: "Hoc nunc solum rogo, ut, quod ante septimanas sex septemve scripto monui de Judaeorum nostrorum in excudendis libris periculosa libertate, recordari v. Amplit. velit. Nam quid Censura juverit, si praeterea sit nulla inspectio, si correctori omnia, quae velit, liceat addere vel detrahere, si denique Typographus fide vestrae Amplit. obstrictus non teneatur? Sed vestra prudentia hic facile disponat."

61. This text was published in Hanau in 1618. A Latin approbation is given at the bottom of the title page: "Massechet Nidah in quo de Menstruatis et puerperis agitur. Recognitum, ac iuxta mentem S. Concilii Tridentini approbatum." On this edition and other tractates published in Hanau, see Burnett, "Hebrew Censorship," 206–7, and Marvin Heller, *Printing the Talmud: A History of the Earliest Printed Editions of the Talmud* (Brooklyn, 1992), 354–56.

62. Keuchen to the Magistrates, 9 May 1617, Marburg, Staatsarchiv Marburg, SA 81 B 1 Rubr 81 Nr 21, fol. 70r: "Nec habet haec pars, cujus causa rejiceretur, quicquam apertae impietatis, praeter nudum Talmudi nomen. Sed quia complura hinc inde suspecta loca recisa olim fuerunt, ne illa in nostra hac typographia per dolum rursus irrepant, necessarium videretur, ne Judaei sumptum hic impendentes, omnia suo libitu, ut hactenus factum, agerent. Nihil enim censura prodest, nisi operarij ipsi teneantur non aliter imprimere, quam qua lege & recognitione ipsis libri permittuntur. Nec hoc tantum, sed quotidiana fere inspectione operum opus esset. Sed vestrae prudentiae totum committo."

63. "Mute evidence" is the expression used by Stephen Burnett, who briefly refers to Buxtorf's censorial work in *From Christian Hebraism to Jewish Studies*, 35–38.

for example, his excision of a story from the first edition of the *Mayse bukh*, which we have already examined in Chapter 1. It is not certain whether Keuchen would have seen so extensive an intervention as necessary.

Elsewhere, in formidable detail, Buxtorf scrutinized a wide range of printed prayer books, festival liturgies (Mahzorim), books of Penitential Prayers (Selihot), and Lamentations (Kinot).[64] He filled thirteen pages with his findings, which made clear both the depth of his knowledge of individual texts and the limitations of his research (see Fig. 5.2). Without exception all the prayer books undergoing inspection represent the Ashkenazic rite. Apparently, Buxtorf had no interest in the religious world of Sephardi or Levantine Jews, or else was not required to examine liturgies pertaining to other Jewish communities. Indeed, all the Jewish liturgical texts printed at Basel catered solely for the needs of Ashkenazic Jews. Moreover, his *Jüden Schul*, which purports to represent Jewish life, portrays only local Ashkenazic customs and laws. Oriental, Sephardic, or Italian Jews have no place in this German-centered work. Isaac Casaubon, who did not come close to matching Buxtorf's Jewish learning, pointed out that when Buxtorf described Jewish liturgies, he quoted only Ashkenazic ones.[65] Yet Buxtorf cast his net far and wide, beyond the printed sources easily available on the market, and drew on personal connections to gain access to rare sources. His notes, for example, refer to a reading drawn "from the manuscript of Hortin." This is a reference to a manuscript now in Bern's Burgerbibliothek (MS 409), a *mahzor* with Selihot for Yom Kippur (the Day of Atonement), probably written in the early fifteenth century, in semi cursive Ashkenazic script. In Buxtorf's time it belonged to the pastor and Hebraist Samuel Hortin of Bern, whom he names.[66]

When it came to the practice of censorship, Buxtorf remodeled texts much more forcefully than Keuchen. The entire focus of his close reading centered on the poetic sections of the liturgies, the *piyyutim*. He chose well: these were, in Michael Fishbane's words, "the heart and soul of Jewish tradition." These poetic creations, which were inserted into central parts of the liturgy, were composed over the centuries by consummate poets and scholars of diverse backgrounds. The texts are often complex, laden with biblical and rabbinic allusions and echoes, and sometimes grounded in the particular historical situation in which the poets

64. Buxtorf, copybook, Basel, UB, MS A XII 20, 254–64, 283–85.

65. Anthony Grafton and Joanna Weinberg, with Alastair Hamilton, *"I have always loved the Holy Tongue": Isaac Casaubon, the Jews, and a Forgotten Chapter in Renaissance Scholarship* (Cambridge, MA, 2011), 146.

66. Buxtorf, copybook, 256: "hoc in manus. Hortini." For a description of the manuscript, see Olivia Franz-Klauser, *Die hebräischen Handschriften: Verzeichnis mit Einleitung* (Bern, 1999), 18–19. Buxtorf's hand is noticeable on the manuscript with the words "S. Hortinus me sibi vendicat." He dated it to 1426.

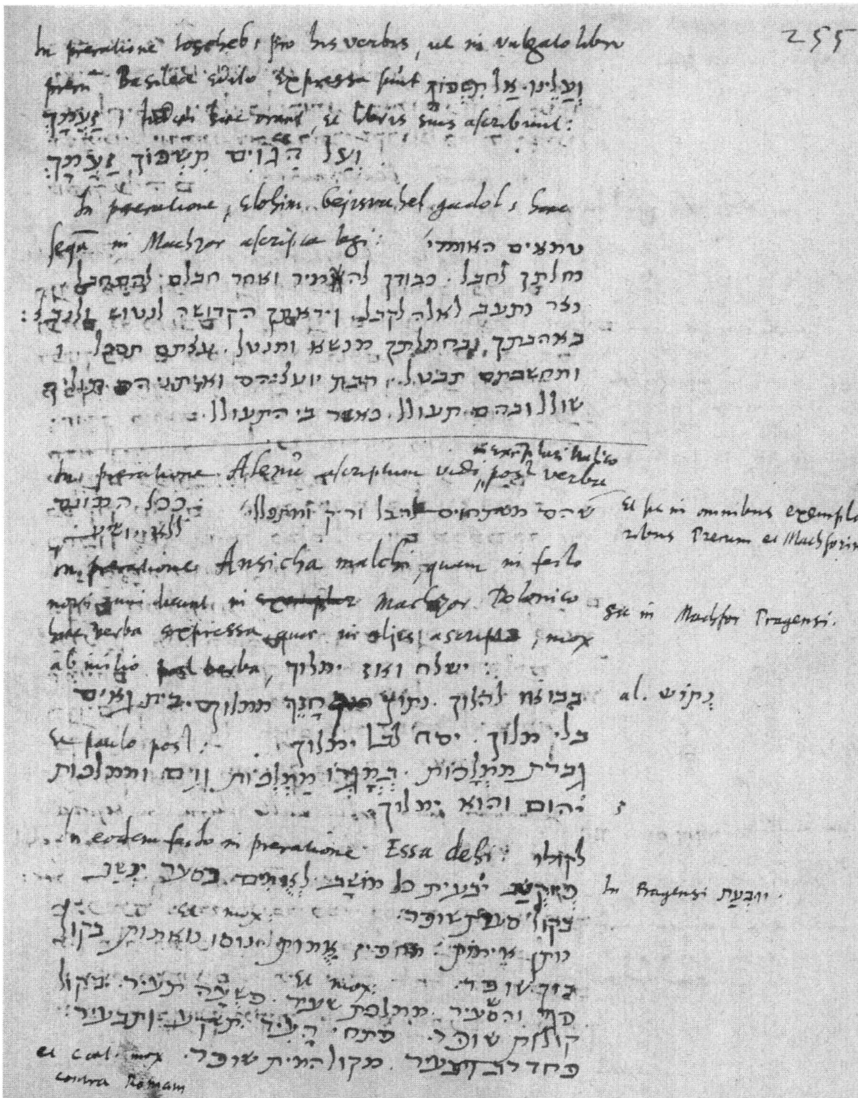

Figure 5.2. Buxtorf collates and censors prayerbooks.
Basel, Universitätsbibliothek, A XII 20, 255.

found themselves. The existential conditions of dispersion and exile are constant themes. Pleas for God's salvation are occasionally intertwined with expressions of hatred for Israel's enemies and of yearning for their downfall.

Buxtorf often excerpted out of context, wrongly assuming that Jews were constantly parading anti-Christian convictions in their writings.[67] He also read the *piyyutim* with this heightened sensitivity. For example, in his copybook he singles out Kinot for Tisha b'av and refers to the poem אהלי אשר תאבת ("My tent which from before creation You desired to join to the Throne of Glory..."). Eleazar Killir, to whom the poem is attributed, speaks of the destruction of the Temple by "robbers," "adversaries," the "uncircumcised," and so on. These designations, according to Buxtorf, refer to Christians; and yet, the whole poem is simply a general outcry against the devastation of the Temple, and neither implicitly or explicitly alludes to Christians.[68] It is indisputable, however, that in the case of many of the liturgical poems under consideration, he was not far off the mark. Like Keuchen, he was aware that prayer books had been transmitted in a variety of forms. Unlike Keuchen, though, he sought to identify or reconstruct the originals from which the texts before him had been copied or printed. As he leafed through the prayer books, he indicated the places where the offensive words had been removed in one version, and then wrote out in full the lines that he had recovered from the manuscript or from a different printed text.[69]

67. See Anthony Grafton and Joanna Weinberg, "Johann Buxtorf Makes a Notebook," in *Canonical Texts and Scholarly Practices: A Global Comparative Approach*, ed. Anthony Grafton and Glenn W. Most (Cambridge, 2016), 275–98, reprinted with revisions as Chapter 1 above.

68. Buxtorf, copybook, 263: "In libello קינות variis titulis nos appellant in precat. אהלי אשר תעבת pag. 11, item in precat. ואתה אמרת pag. 18. Item in precat. הטה אלהי pag. 19, 2. Et in precat. אמרתי vindictam subitaneam contra Christianos precantur pag. 27." The page numbers refer to the edition of the Kinot printed in Basel by Waldkirch in 1599. The last poem that he mentions, written by Kalonymus ben Judah the Younger, does have clear anti-Christian allusions.

69. E.g., Buxtorf, copybook, 255: "In precatione iosheb [יושב בגבהי מרומים] pro his verbis, ut in vulgato libro prec. Basileae edito expressa sunt ועלינו אל תשפוך זעמך (and do not cast Your anger on us)" – this is a reference to *Seder tefillot* (*Order of Prayers for the whole year)* (Basel, 1579), fol. 216v – whereas the original text, as in the 1536 Augsburg edition produced by Hayyim ben David (Shahor), fol. 6r, reads ועל הגוים תשפוך זעמך (And cast Your anger on the nations). Again, on 262 Buxtorf refers to the poem *Arehu ha-yamim*, an alphabetical poem by Simeon bar Isaac Hazak, pointing to the omission of the lines for the letter *kuf* and *resh* that proclaim the downfall of the nations ("In precat. Arechu hajamim quam in [MS in in] festo Rosch haschana dicunt") and noting in the margin that these lines are omitted from other editions, especially the Venetian ones, but appear in the 1536 Augsburg edition ("haec ex alijs Venetis praecipue omissa: in Augustano exemplari expressa"; see fol. 14v of the Augsburg edition). The fluidity of these texts, as a consequence of sensitivities to the requirements of ecclesiastical censorship, means that ultimately it is difficult to come across any "pristine" text; for in two copies of the Augsburg edition (London, British Library 1971, fol. 15; Oxford, Bodleian, Opp. Add. Fol. II.153), the problematic line seems to have been censored, thereby rendering the passage meaningless. On these occurrences, see Piet van

סליחות

מסודר כפי סדר ומנהג בני מדינת אשכנז ולועזים יז״א עם
פירוש שהתחיל לחבר הגאון מהר״ר מרדכי בר יודא ז״ל הנקר׳
ר׳ מארדוש ונתבקש בישיבה של מעלה והתנו הגאון מהר״ר צבי
בן מהר״ר חנוך ז״ל מלא מקומו להשלים אותו וגם התנה על יחו
וגם אנחנו המדפיסים החתומים בסוף הספר הזה העיר יי׳ את
רוחינו לחנו׳ בו וללקוט בתר לקוטי להוסיף ולגרע במקו׳ הצריך
ובפרט באותן הסליחות אשר לא נמצאו בהסליחות שנדפסו
מקדם בארץ פולין ובפראג היינו מוכרח לחפש ולבדוק ולעיין
בספרים ישנים וללקוט משנשנים אחת הנה ואחת הנה ואח׳
לאחת למצוא החשבון עד שהיו לאחדים בידינו וכדי לזכות
בהם הרבים אזרנו כגבר חלצינו להיותינו נמנה בכת הזריזים
מקדימים והמזכה את רבים ובכן נבקש מאל העוזר שיעזרנו
להתחיל ולגמו׳ כחפצנו בלי מכשול ופגע רע אמן כן יהי רצון
סדר סליחות׳ שלבנה צחות׳ וכשמש זורחות׳ לבני עליה׳
מפורשות יפות׳ זכות וחפות׳ בי״ג נפות׳ הליכותיה צופיה׳
מלאכ׳ ההדור׳ החהל׳ בתפאר׳ שנת שמו ליציר׳ לפק מעיה׳

נדפס פה
פראג הבירה

ובבית הקצין והישיש רבי
מורדכי בר גרשם הכהן זל
המחוקק שליט

Figure 5.3. *Selihot*. Prague, 1587. Basel, Universitätsbibliothek, FA IV 3.

The ultimate goal of some of these transcriptions and comparisons comes to the fore when we examine two printed books, one of which was clearly produced under Buxtorf's supervision. Basel University's Library owns a magnificent, censored edition of the Selihot – the penitential prayers recited during the period between the New Year and the Day of Atonement and on other occasions in the liturgical year (see Fig. 5.3). This volume, printed in Prague in 1587 according to the rite of the most important Ashkenazi community of Frankfurt am Main, bears the unmistakable signs of Buxtorf's pen (and possibly also that of his son).[70] A complex enterprise is under way. It includes compilation of manuscript evidence that provides the "original" offensive expressions that Buxtorf pits against the readings of the printed text. Above all, the volume of Selihot is a repository of red marks for collation and Buxtorf's additional censorial intrusions, all in preparation for a new printing of the text.

Manuscript and Print

Examination of one of the poems recited on the first day of the penitential period illustrates the method and scope of Buxtorf's enterprise. In the poem by Benjamin ben Zerah (1050), an invocation to God, imploring compassion for His humiliated people, one line clearly hit home hard for a reader like Buxtorf. The Prague edition leaves traces of the original text when it reads ...פורכי עמך אשר למו שוחחים (the oppressors of your people who bow down to themselves [למו]).[71] In the margin, Buxtorf noted the original reading, as indicated in the manuscript, which had "who prostrate themselves to the dead one" (למת; see Fig. 5.4).[72] It would appear that on reading the poem in the printed text, Buxtorf rightly perceived that the phrase "bow down to themselves" was unintelligible. He consulted his

Boxel, "Hebrew Books and Censorship in Sixteenth-Century Italy," in *Jewish Books and Their Readers: Aspects of the Intellectual Life of Christians and Jews in Early Modern Europe*, ed. Scott Mandelbrote and Joanna Weinberg (Leiden, 2016), 75–99, esp. 89–92.

70. *Selihot* (Prague, 1587), Basel, UB, FA IV 3. The book was printed by Mordechai ben Gershom, also known as Markvart. Its intended readers were those living in Ashkenazi lands and Ashkenazim living in Italy (*lo'azim*). Other printings of this *Selihot* have additional material at the end of the volume. Thanks to Olga Sixtova for clarifying the meaning of *lo'azim* in this context. On the complexity of the rites, see Lucia Raspe, "Tradition, Migration, and the Impact of Print: Local Rites of Seliḥot Recitation in Early Modern Ashkenaz," *Jewish Quarterly Review* 113 (2023): 83–10.

71. *Selihot* (Prague, 1587), Basel, UB, FA IV 3, fol. 3v.

72. The original reading is found in *Selihot*, Leiden, UB, MS Or. 4721 (Scaliger 4), 36a. It is quite likely that Buxtorf had access to this manuscript, which had once belonged to Scaliger. See below, n94.

Figure 5.4. Buxtorf consults a manuscript of the *piyyut*. *Selihot*. Prague, 1587. Basel, Universitätsbibliothek, FA IV 3.

manuscript and was confronted with the anti-Christian sentiment that he had all along suspected was there.[73]

Expurgation

Already in his copybook Buxtorf had singled out another penitential poem, this one ascribed to Rabbenu Gershom (960–1028), known as "Me'or ha-golah" ("Light of the Exile").[74] One of the earliest Ashkenazic authorities known to us by name, Rabbenu Gershom served as head of the academy in Mainz and was renowned not only for his halakhic expertise but also for his *piyyutim*. This poem, composed in strophes of four lines, with a monorhyme at the end of each line, begins with a plea to God:[75]

אליך נקרא איום ונורא

> To You we will call, O awesome and fearful one, Do not hide Your face from us, seeing our travail.

The first line of the fourth strophe consists of an adaptation of two biblical verses. Deuteronomy 32:17 reads: "They sacrificed to demons, not to God; to gods they had never known, new ones who came but lately [whom your fathers did not fear]" – יִזְבְּחוּ, לַשֵּׁדִים לֹא אֱלֹהַ - אֱלֹהִים, לֹא יְדָעוּם חֲדָשִׁים מִקָּרֹב. But the biblical verse is totally transformed here, as the poet writes:

טמא ומת חדש הבא מקרוב

> The impure one dies, a new one who came but lately.

This is followed by a line adapted from Proverbs 17:18:[76]

טבו מה אצלי ערובתו לערב

> What do I have to do with one such as him, to stand surety for him?

73. *Selihot* (1587), fol. 22v. The original reading is also preserved in other editions, among them those of Lublin, 1603, and Prague, 1608.

74. Buxtorf, copybook, 263. Buxtorf wrote out part of the poem "In precatione Elecha nikra quam die sexto penitentiis dicunt" (i.e., for the sixth day of the penitential period between the New Year and Day of Atonement), and then singled out other lines in which the poet asks God to send His wrath on those who afflict them. Buxtorf suggests that Christians are the unidentified persecutors.

75. For a critical edition of the poem, see *Seder ha-selihot ke-minhag Polin ve-rov ha-kehillot be-Erets Yisra'el*, ed. E.D. Goldschmidt (Jerusalem, 1965), 177–80.

76. Proverbs 17:18: "A man void of understanding is he that strikes hands and becomes surety in the presence of his neighbor."

The meaning of these two lines is not obfuscated. And as we see, Buxtorf will not permit such a frontal attack on Jesus to remain embedded in the Jewish prayer. According to Simha Goldin, Rabbenu Gershom wrote this poem not only to support his community in time of persecution, but also to persuade recent converts to Christianity of the emptiness of their adopted religion and to appeal to them to return to the fold.[77] In the 1587 text Buxtorf erased the objectionable words and emended the text, substituting the word *ta'ut* (error) for *tame u-met* (the impure one dies).[78]

This was not a suggestion but a directive. In 1599–1600, Jacob ben Abraham Pollack (Mocherseforim) did Buxtorf's bidding and printed another edition of the *Selihot*, in Basel.[79] The title page (Fig. 5.5) vaunts this work as a totally new production and makes the conventional claim that it had been corrected with due consideration (*betakhlit ha-iyyun*): in this case, censored.[80] Buxtorf's emendation of the unacceptable words is inserted in the appropriate place: "the error that has come of late," with the rest of the lines remaining intact (Fig. 5.6).[81] The removal of the offensive words has rendered the verses meaningless.[82]

The pious congregants who owned a copy of Buxtorf's edition of the *Selihot* would have either silently corrected the text or simply glossed over it. Yet this reading of the text did not necessarily prevail and contaminate other editions.[83] The failure to produce a truly official text is proven from the case of Hanau, where, as we have seen, Keuchen, Buxtorf's former pupil, attempted to control the content of Jewish books printed under his aegis. There, in 1616, a so-called new edition

77. Simha Goldin, *Apostasy and Jewish Identity in High Middle Ages Northern Europe: "Are you still my brother?"* (Manchester, 2014), 7–9.

78. *Selihot* (1587), fol. 22v.

79. This edition of the *Selihot*, which is a replica of the Prague 1587 edition, but lacking the commentary, provides an index to the *Selihot*. It explicitly states that it is arranged according to the rite of Frankfurt and other Ashkenazic communities as well as those of Poland, Bohemia, and Moravia.

80. Prijs, *Die Basler hebräischen Drucke*, 272–73, describes this 1599–1600 edition but is not aware of the changes to the text. He therefore expressed surprise as to why it should be described as a novelty when it simply appears to be a copy of the Prague 1587 *Selihot* on a smaller scale with omission of the commentaries and appended Selihot.

81. *Selihot* (Basel, 1599–1600), fol. 71v.

82. This text is expurgated in the 1536 Augsburg edition, fol. 56v, with spaces and meaningless phrases, and the entire *piyyut* is missing from the 1548 Venice edition. The 1584 Cracow edition, fol. 65r, presents a slightly different version: טפל ולא עקר חדש (new, secondary and unimportant). The brief commentary on the poem makes the meaning of the line explicit: "your idolatry is secondary and unimportant, recent and new."

83. See, e.g., the *Selihot* printed in Amsterdam (1664), which also retains the original words *tame u-met*, and later editions such as that of Metz (1822).

סליחות

ראה זה חדש אשר כבר
לא היה לעולמים סליחות קטון
מסודר כפי סדר ומנהג דקק ורנקבורט
ושאר קהילות ומדינות האשכנזים וגם
מנהג מדינות פולין ופיהם ומרהרין כמוהו
לא נדפס עד היום הזה כאשר עיניכם
במשרים תחזינה ובראותכם יופי
המלאכה פתחו פיכם בשיר
וברנה אמן

וגם הוגה בתכלית העיון

במצות יעקב בר אברהם ז״ל מק״ק מעוריטש ממדינות רייסין
אצל בבריסק דליטא הנקרא יעקב מוכר ספרים :

ותהיה ראשית מלאכתו יום ד׳ ג׳ תשרי בשנה
טובה בשנת הנה לא ינום ולא ישן
שומר ישראל

Figure 5.5. *Selihot*. Basel, 1599–1600. Basel, Universitätsbibliothek, UBH Aleph D IX 9.

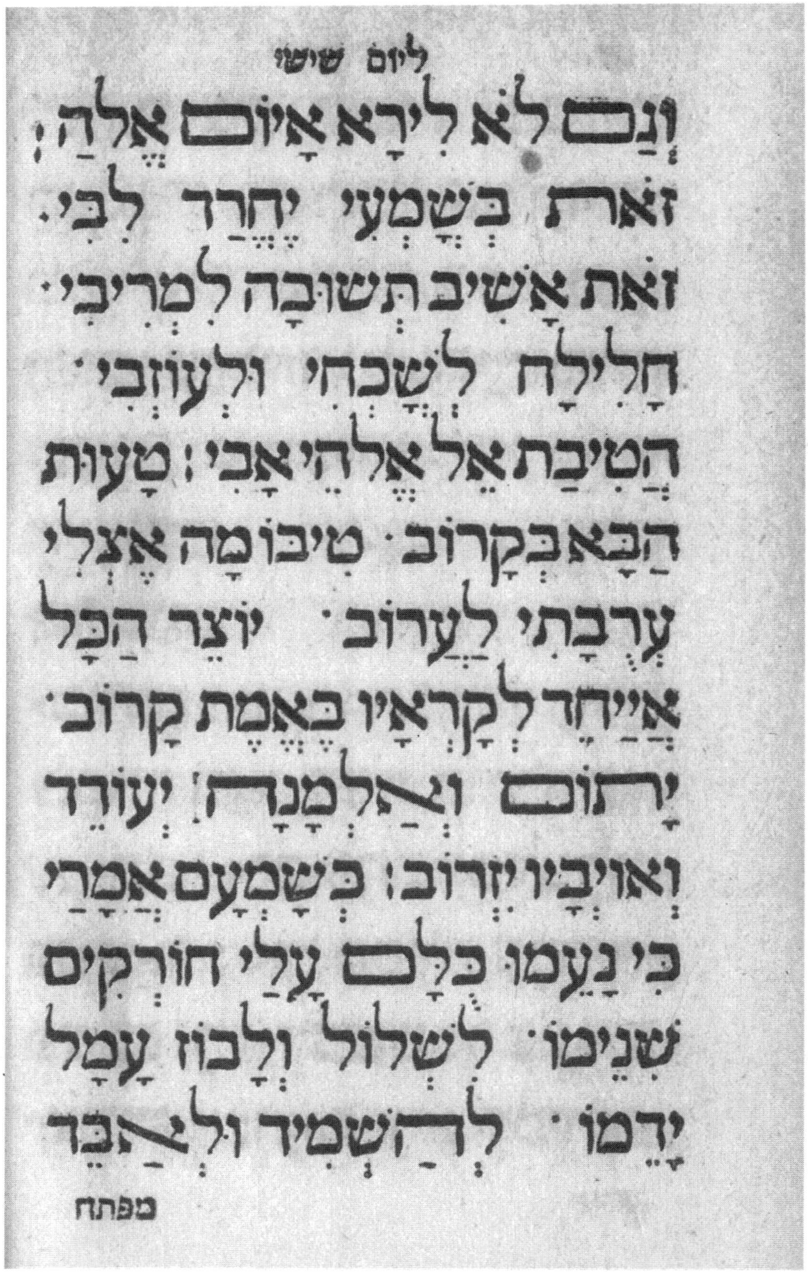

ליום ששי
וגם לא לירא איום אלה;
זאת בשמעי יחרד לבי.
זאת אשיב תשובה למריבי·
חלילה לשכחי ולעוזבי·
הטיבת אל אלהי אבי: טעות
הבא בקרוב· טיבו מה אצלי
ערבתי לערוב· יוצר הכל
איחד לקראיו באמת קרוב·
יתום ואלמנה! יעודד
ואויביו יזרוב: כשמעם אמרי
כי נעמו כלם עלי חורקים
שנימו· לשלול ולבוז עמל
ידמו· להשמיד ולאבד
מפתח

Figure 5.6. Buxtorf's emendation in print. *Selihot*. Basel, 1599–1600. Basel, Universitätsbibliothek, UBH Aleph D IX 9.

of the *Selihot* was printed. Though sanctioned "cum licentia superiorum," the offensive words "The impure one dies, a new one who came but lately" remained intact.[84] Perhaps Keuchen did not mislead the magistrates when he told them he lacked deep expertise in Jewish texts.

Premeditated Censorship

Censorship can take many different forms. It could be suggested that Buxtorf performed a better job in another case. Here his red pen was directed at a poem – a *pizmon* – recited on the third or fourth day of the penitential period between the New Year and Day of Atonement. Composed by Isaac ibn Ghiyyat (1038–1089), a Spanish commentator and liturgical poet, the poem would have been recited in the early hours of the morning. It begins appropriately:[85]

באשמרת הבקר קראתיך אל מהולל
יערב לך חין ערכי יום לבי לך סולל
ותגיה את חשכי וכאור בקר יהולל
מלכי ואלהי כי אליך אתפלל

In the early morning I call upon thee, O praiseworthy God.
Let my prepared prayer be agreeable to you
and lighten my darkness on the day when my heart appeals to You,
and may it be as bright as the light of the morning. O my king, my God – for it is to You that I pray.

The second stanza takes on a rather different tone:

צרי מזעימיך באף חרפו עמך
ונשאתי אמיך ברגש שאון קמיך
שפך עליהם זעמך וחרון אף ממרומיך
וקנא אל לשמך אשר בגוים מחלל

My enemies enrage You (*mazimekha*), in fury they taunt Your people (Zeph. 2:8).
But I keep fear of You even amidst the tumultuous uproar of those who rise against You.
Pour out Your wrath against them and Your fury (Ps. 69:25) from Your heights.

84. *Selihot* (Hanau, 1616), fol. 33v. Though in this instance the Hanau edition retained the original text of the poem, that was not always the case. See below.

85. *Selihot* (1587), fol. 54v.

Be jealous for Your name (Ezek. 39:25) which is profaned among the nations [goyim] (Ezek. 36:23).

Once again, Buxtorf is enraged by what he regards as explicit antagonism toward his co-religionists. Erasing the guilty words, he was propelled to rewrite the lines, preserving rhyme and rhythm even as he reset part of the poem. His emendations (Fig. 5.7) were implemented on the relevant page of the new Basel edition (Fig. 5.8).

The lines now read:[86]

צרי עבדיך באף חרפו עמך
The enemies of Your *servants* taunt Your people in anger

שפך רוח משמיך על ישראל עמך
Pour out wind/spirit from Your heavens over Israel Your people.

וקנא אל לשמך אשר בארץ מחלל
Be zealous for Your name which is profaned *on earth*.

The original sentiment of the poem has simply been reversed. Buxtorf's emended text bears some similarity to the wording used in the censored Venice (1548) and Heddernheim (1546) *Selihot* and the *Mahzor* of Sabbioneta/Cremona (1557–60), which similarly recast the original text into a prayer on behalf of Israel.[87] But in both cases, the text reads *ruah kodshekha* (Your holy spirit), which has a rather more familiar and meaningful ring than Buxtorf's *ruah mishamekha* (wind/spirit from Your heavens).[88]

A Jew, Jacob bar Abraham Pollack of Mezeritch, assisted Buxtorf in producing a bowdlerized form of the *Selihot*. He condoned Buxtorf's censorship or even colluded with the Christian in tampering with the texts of the great *payyetanim* of the past. Yet he also seems to have managed, like the printers in Hanau, to evade his Christian censor's attention. At the end of the volume, Jacob used the same colophon printed by Gershom ben Judah for the Prague edition. Here he expressed the hope that with God's help he would be able to publish the whole

86. *Selihot* (Basel, 1599–1600), fol. 200r.

87. *Selihot mi-kol ha-shanah ke-minhag ha-ashkenazim*... (Venice, 1548), fol. 60r. This reading is retained in the 1600 Venice Bragadin edition. The Sabbioneta Mahzor from which Buxtorf made extracts in his copybook, 254 ("Ex Machzor Savionetae impresso in 4°"), reads "the enemies of your servants" but then reads "pour out Your Holy spirit on your people Israel" and preserves "Be zealous for Your name which is profaned by the gentiles."

88. The 1616 Hanau edition discussed above (fol. 89r–v) preserved the first part of the problematic text, but then gave the same reading: "pour out wind from Your heavens on Israel your people and be jealous for Your name which is profaned on earth."

Figure 5.7. Buxtorf rewrites Isaac ibn Ghiyyat's *piyyut. Selihot* (Prague, 1587). Basel, Universitätsbibliothek, FA IV 3.

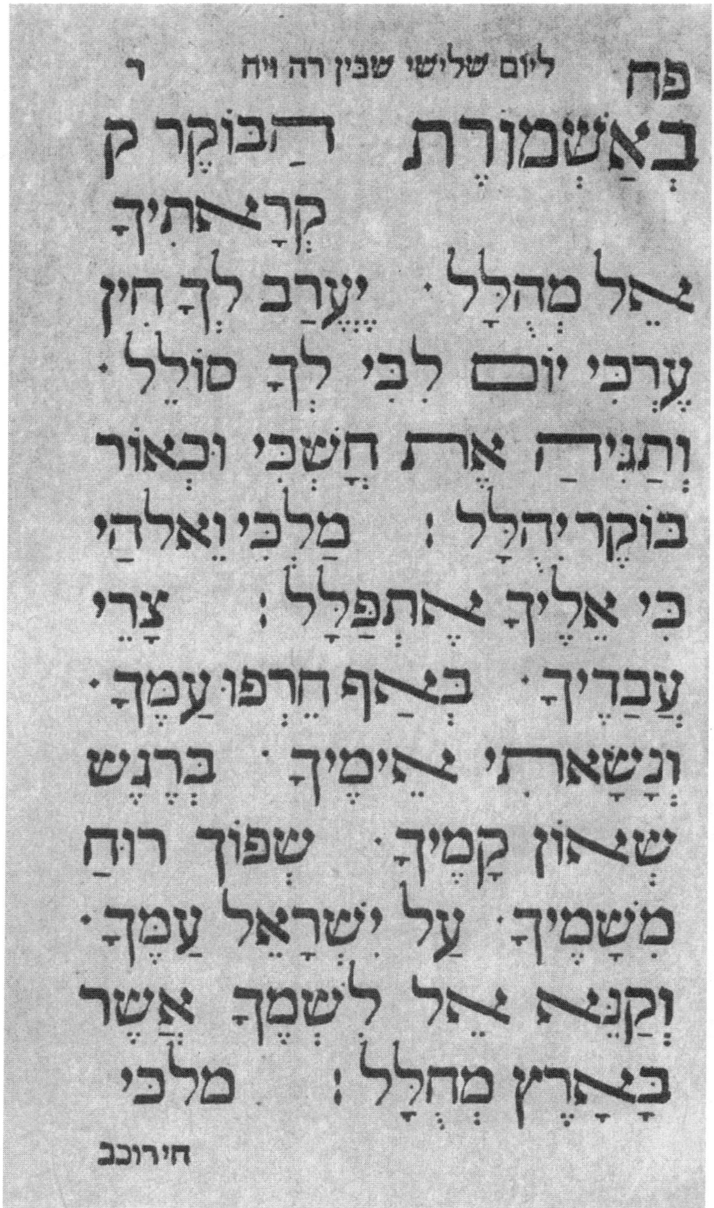

פח ליום שלישי שבין רה ויח ד

באשמורת הבוקר ק
קראתיך
אל מהלל · יערב לך חין
ערכי יום לבי לך סולל ·
ותגידה את חשכי וכאור
בוקר יהלל : מלכי ואלהי
כי אליך אתפלל : צרי
עבדיך · באף חרפו עמך ·
ונשארתי אימיך · ברגש
שאון קמיך · שפוך רוח
משמיך · על ישראל עמך ·
וקנא אל לשמך אשר
בארץ מחלל : מלכי

חירוכב

Figure 5.8. Buxtorf's rewritten *piyyut* appears in his new edition. *Selihot*. Basel, 1599–1600. Basel, Universitätsbibliothek, UBH Aleph D IX 9.

stock of prayers "and by their merit we will go up from exile."[89] Though both Gershom and Jacob appear to be referring to the size of the volume and not to its castrated contents, the "prayer" – that is what they call it – is addressed to their fellow Jews.

They were not the first Jews to tinker with these verses. Long before they had set about their professional task, the Jewish printer Gershom Soncino had produced a text of the *Selihot* in Fano in 1504.[90] Some copies provide the complete unbowdlerized text; others present those telling empty spaces that advertise to all and sundry that something unacceptable to those in power has been omitted. As Daniel Goldschmidt wrote in his authoritative edition of the *Selihot*: "Internal later censorship by Jews of the diaspora attempted to smooth over and blur the expressions of hatred, and sometimes they changed words in the texts (without regard to the structure of the poems). This caused different readings in the printed edition but did not disguise the original meanings of the poets."[91] It did not take a scholar of Buxtorf's caliber to detect these lines, but the production of a new edition based on such changes goes far beyond the mere act of censoring or expurgating. It required a deep knowledge of Hebrew and a sense of poetry.[92] We should bear in mind that Buxtorf was later to devote many pages of his *Thesaurus Grammaticus* to Hebrew prosody. But let us make no mistake. Buxtorf's interventions here do not open up new spaces for independent creative readings.[93] His efforts at imposing a new edition of the *Selihot* on the Jewish community failed – not miserably, because some extant versions of the *Selihot* do omit the offensive words. Yet the original, or rather, more authentic versions of *Selihot* survived Buxtorf's shears and pen, as we can see not only from the texts produced by the great scholars of liturgy, Daniel Goldschmidt and Abraham

89. *Selihot* (Basel, 1599–1600), last page.

90. E.g., for the poem attributed to Gershom ben Judah discussed above, the words "the impure one who died," "tame u-met," are missing in the Fano edition, fol. 55v.

91. *Seder ha-selihot*, ed. Goldschmidt, 13.

92. There are other examples of Buxtorf's metrical changes of the text. For example, on fol. 56v (for the piyyut "eikh ukhal"), the verse "How has ... the kingdom of iniquitous kings been prolonged" (וארכה מלוכה למלכי בני עולה) been changed to "their kingdom has been prolonged and does not end" (וארכה מלכותם ולא כלה). This change, which is implemented in the Basel edition (fol. 207r), is not found in other versions. See also fol. 62v (*piyyut afas hod* for the fourth day between Rosh Hashanah and Yom Kippur), where Buxtorf has reduced the vehemence of the sentiment, and this change is then implemented in the 1600 Basel version (fol. 232r). Once again, we have not found this small change documented in any other version.

93. This means that it is not possible to interpret these interpolations and tampering with the texts in a positive light, as Amnon Raz-Krakotzkin argued in his widely cited *The Censor, the Editor and the Text: The Catholic Church and the Shaping of the Jewish Canon in the Sixteenth Century* (Philadelphia, 2007).

Habermann, but also, for example, from the version published in Lublin in 1603 and from a semi-censored version printed in Venice in 1600.[94]

But as we have mentioned, Buxtorf was an omnivorous reader of Hebrew books. As he grew older, his interest in these books was put to other purposes. In the *Bibliotheca Rabbinica*, the useful but eclectic companion to Jewish literature and learning that Buxtorf published in 1613, he identified and described versions of dozens of Jewish texts. He included entries on the titles *Mahzor* and *Selihot*:

> *Selichos*: a collection of certain prayers that they use for special feasts, has this title. It has this title because they were established only for mercy and the remission of sins. The book is printed in all formats, because it is so widely used, and is published with short comments. Scaliger left a manuscript in parchment.[95] It was printed in folio at Prague, and again in Yiddish there.[96]

This comment is nearly dispassionate, and it certainly does not reveal the former zealot, the Buxtorf who tried to bring the impious Jews to order. Buxtorf the censor has become Buxtorf the Hebrew bibliographer.

The similarities and differences between Buxtorf and Keuchen are clear: in particular, neither of them – as Keuchen clearly understood – could always keep his Jewish collaborators under strict control. Both rewrote Jewish texts. But the range and ambition of Buxtorf's corrections were far greater, even if they fell short of offering plausible substitutes for the passages he altered or excised. Buxtorf's practices as a censor reveal the same unremitting distaste for Jewish texts that accompanied his deep knowledge of them. Yet the two men worked from very different hermeneutical principles. Curiously, it seems that Buxtorf, for all his prejudices, was better than Keuchen at the collaboration with Jews that the process of printing Hebrew texts required. Sometimes – as in the case of the poem that

94. *Selihot* (Lublin, 1603); *Selihot* (Venice, 1600). Abraham Meir Habermann, ed., *Selihot u-fizmonim* (Jerusalem, 1943/4) (an edition of poems of Gershom ben Judah).

95. He is referring to the Hebrew MS Leiden, UB, Or 4721 (Scaliger 4) in Leiden University Library, dated 1347 by the scribe Judah ben Moses. We are grateful to Arnoud Vrolijk, who sent us a digitized version of the manuscript.

96. Buxtorf, *Bibliotheca Rabbinica*, in his *De abbreviaturis hebraicis liber novus & copiosus* (Basel, 1613), 313: "סליחות *Selichos*. Liber precum quarundam sic appellatur, quibus utuntur in festis principalibus, & appellantur סליחות, quia pro misericordia et remissione peccatorum tantum sunt institutae. In omnibus formis liber excusus est, quia eius magnus usus. Est & cum brevibus Glossis editus. Scaliger manuscriptum in membranis reliquit. In fol. Pragae: ibidem etiam Hebraeo-germanice." The Yiddish translation was printed in 1602. On the history of the Yiddish translation of the *Selihot*, see Lucia Raspe, "Three Sixteenth-Century Translations of Penitential Prayers into Yiddish," *Jewish Studies Quarterly* 26 (2019): 224–42.

appears at the end of his edition of the rabbinical Bible – he missed a trick. But it appears that when he thought it necessary to change a Jewish text substantially, he usually succeeded. Keuchen knew that he could not exert that level of control. Censorship of Jewish writing in the German-speaking Christian world was no more uniform, in theory or practice, than censorship of texts by Protestants in the Catholic world.[97]

97. For contrasting views of the practice and impact of censorship in Italy, both founded on exhaustive research, see Gigliola Fragnito, *Rinascimento perduto: la letteratura italiana sotto gli occhi dei censori (secoli XV–XVII)* (Bologna, 2019), and Hannah Marcus, *Forbidden Knowledge: Medicine, Science and Censorship in Early Modern Italy* (Chicago, 2020).

CHAPTER SIX

Tiberias: Buxtorf on the Masoretic Vowel Points and Accents

In 1772, Johann Salomo Semler, professor of theology at the University of Halle, dedicated a German translation of Elijah Levita's *Massoret ha-massoret* of 1538 to Moses Mendelssohn – a provocative act. The Jewish philosopher for whom Hebrew's accents and vowels constituted an intrinsic ingredient of the holy tongue could hardly have taken this dedication as a welcome conferral of collegial esteem. Semler insisted that scholars should reject the prejudices of their predecessors and accept Levita's theory that the Masoretes invented the graphic notation of the Hebrew vowels and accents after the completion of the Talmud. His Jewish reader must have found this view unacceptable. An implicit response to Semler may be detected in Mendelssohn's "*Or li-netivah*," the introduction to his German translation of the Bible.[1] There, he formulated and propagated his concept of a divinely vocalized and accented text, which had been providentially transmitted to humanity.[2]

Semler's dedication is significant in one particular respect. So are Mendelssohn's rejection of Levita and his espousal of the views of one of Levita's most vocal Jewish opponents, Azariah de' Rossi (ca. 1511–1577), in the course of a discourse on the philosophy of language. The two men's opposed positions draw attention to the many ideological misreadings and misinterpretations of sources that the Masoretic *nekudot* (vowel points) and *te'amim* (accents) have provoked

1. On Mendelssohn and Semler, see Avi Lipschitz, "A Natural Yet Providential Tongue: Moses Mendelssohn on Hebrew as a Language of Action," in *Language as Bridge and Border: Linguistic, Cultural and Political Constellations in Eighteenth to Twentieth Century German-Jewish Thought*, ed. Sabine Sander (Berlin, 2015), 1–15. See, too, David Sorkin, *The Religious Enlightenment: Protestants, Jews, and Catholics from London to Vienna* (Princeton, 2008), 186.

2. Moses Mendelssohn, *Or li-netivah* (Berlin, 1783), section 1, sig. 1–2, fol. 6v. The section ends with the statement: "Thus, it is inconceivable that the Torah was edited in any way even in relation to the knot of the letter *yod*; for as long as heaven covers earth, it was preserved and set out in our possession and will never be forgotten by our offspring."

and sustained over the centuries. As so often happened when early modern scholars examined the history of Judaism and Christianity, all parties remained steadfastly attached to their own ideological positions.[3]

But ideology sometimes fosters scholarship and opens avenues hitherto closed. *Tiberias,* Buxtorf's book-length treatise on the Masorah, offered the neophyte an array of tools for penetrating the dense thickets of Masoretic terms, cryptic sigla, and other obscure data that adorned Hebrew Bibles, both printed and manuscript.[4] It also contained a spirited defense of the antiquity of the vowel points and accents. Buxtorf marketed his book as a response to Levita's *Massoret ha-Massoret.* In his view, that work had misled theologians as well as grammarians, once they were able to read the bestseller in Sebastian Münster's partial Latin translation (1539).[5] From the start, as we will see, Buxtorf set out to refute Levita. Over time, however, as he met criticism from other Christian experts, Buxtorf refined his arguments even as he tried to defend them. By the time he drew up the *Tiberias,* he was prepared to reenter the vowel point fray with a different armory from that of any of his predecessors. He would confront Levita's work in detail, as well as the Masorah itself. He would also examine what it meant that Tiberias had been the seat of the Masoretes, the first time that anyone raised this historical question.[6] For this inquiry as well as for the

3. See, e.g., Matthias Pohlig, *Zwischen Gelehrsamkeit und konfessioneller Identitätsstiftung: lutherische Kirchen- und Universalgeschichtsschreibung, 1546–1617* (Tübingen, 2007); Jean-Louis Quantin, *The Church of England and Christian Antiquity: The Construction of a Confessional Identity in the 17th Century* (Oxford, 2009); *Sacred History: Uses of the Christian Past in the Renaissance World,* ed. Katherine van Liere, Simon Ditchfield, and Howard Louthan (Oxford, 2012); Katrina Olds, *Forging the Past: Invented Histories in Counter-Reformation Spain* (New Haven, 2015); Nicholas Hardy, *Criticism and Confession: The Bible in the Seventeenth-Century Republic of Letters* (Oxford, 2017); and *Confessionalism and Erudition in Early Modern Europe: An Episode in the History of the Humanities,* ed. Nicholas Hardy and Dmitri Levitin (Oxford, 2019).

4. The clearly pedagogic purpose of the work is on show in the last section of Buxtorf's work: Johann Buxtorf, *Tiberias, sive commentarius Masorethicus* (Basel, 1620), "Commentarius Masorae didacticus sive מפתח למסורת clavis Masorae" (213–322), which provides essential information about how to read the Masorah, with a sample chapter on how to read the Masora magna on the first chapter of Genesis.

5. Levita's *Massoret ha-Massoret* was first published in Venice (Bomberg), in 1538. In 1539 a bilingual edition was published in Basel by Henricus Petri that contained Sebastian Münster's partial Latin translation. All references will be to the 1538 Venice edition of the Hebrew, unless otherwise specified, and to Christian D. Ginsburg's bilingual Hebrew/English edition: *The Massoreth ha-massoreth of Elias Levita,* ed. and trans. Christian Ginsburg (London, 1867; 2nd ed. New York, 1968).

6. A preliminary discussion about Tiberias appears in Levita's Hebrew letter to Sebastian Münster in his edition of David Kimhi's *Commentary on Amos* (Basel, 1531), fols. 4r–v. Having been asked by Münster to explain who the Tiberians were, Levita referred to the

other arguments he marshaled throughout the book, Buxtorf availed himself of a mixed bag of Jewish sources. Though his inspiration remained confessional, he produced a historical reconstruction of the ancient Jewish academies that reads surprisingly like some of the histories of the Jews written in much more recent times, pioneer works on which we still draw, though we no longer subscribe to their larger theses (see Fig. 6.1).[7]

Buxtorf was no friend of the Jews, as we have seen. But Buxtorf also unashamedly portrayed himself as a serious lover of Hebrew books. True, in this as in other cases Buxtorf began by working with standard sources in predictable ways and proclaiming that he was certain about his conclusions. Over time, however, he mustered a dazzling array of Hebrew sources to promulgate his thesis. And the more material he brought to bear, the more willing he became to entertain, and even to admit, certain misgivings and doubts.[8] As we will see, adopting new methods and applying them to new sources went along with a newly self-critical attitude on his part.

Buxtorf was by no means the first scholar to study the Masorah at Basel. Long before his time, the baptized Jew Matthaeus Adriani, who worked for Johann Froben as a corrector and taught Hebrew at Tübingen, Louvain, and Wittenberg, drew up a Hebrew-Latin Psalter, with some aids to learning, in collaboration with Bruno Amerbach. Adriani wrote the Hebrew and translated it; Amerbach took down the Latin.[9] At 22:17 the Vulgate reads *foderunt manus meas et pedes meos* (they pierced my hands and my feet), a passage traditionally taken by Christians

---

Roman settlement of Tiberias and cited R. Jonah, who had stated that the Tiberian Jews were experts in the Holy Tongue and spoke it more fluently than any other Jews.

7. For a general description and overview of the *Tiberias*, see Stephen Burnett, *From Christian Hebraism to Jewish Studies: Johannes Buxtorf (1564–1629) and Hebrew Learning in the Seventeenth Century* (Leiden, 1996), 203–39. For a detailed and nuanced discussion of the debate and its repercussions, see Timothy Twining, "The Early Modern Debate over the Age of the Hebrew Vowel Points: Biblical Criticism and Hebrew Scholarship in the Confessional Republic of Letters," *Journal of the History of Ideas* 81 (2020): 337–58, and his forthcoming book, *The Limits of Erudition: The Old Testament in Post-Reformation Europe* (Cambridge).

8. In a letter of 1 January 1623 to Louis Cappel (Zürich, ZB, MS F 45, fols. 246v–247v), Buxtorf explicitly conceded that his position was not completely proven and that the serious implication and danger of supporting Cappel's view meant that he was simply not prepared to air such a view in public.

9. Hebrew-Latin Psalter, ed. and trans. Matthaeus Adriani, Basel, UB, MS A VII 2. See the description in Joseph Prijs, *Die Handschriften der Universitätsbibliothek Basel: Die Hebräischen Handschriften: Katalog auf Grund der Beschreibungen von Joseph Prijs*, ed. Bernhard Prijs and David Prijs, with Stephen G. Burnett and Thomas Willi (Basel, 1994), 19–20. Adriani described his method, in a colophon written in two columns, Latin and Hebrew: "Literam e litera verbum e verbo librum spalterium de lingua sancta ad linguam latinam . . . transtuli" (ibid., 19, from Psalter, ed. Adriani, fol. 273r).

as predicting or foreshadowing the Crucifixion. Adriani knew well the Masoretic device that retained a given word in the text, as the *ketiv* (that which is written), and placed another word in the margin, as the *kere* (that which is to be read). In

JOHANNIS BUXTORFI

TIBERIAS

Sive

COMMENTARIUS

MASORETHICUS

TRIPLEX

Historicus

Didacticus

Criticus

Ad illustrationem Operis Biblici Basileensis conscriptus

Quo

PRIMÙM HISTORIA MASORETHARUM Tiberiensium, sive à quibus Masora conscripta sit, excutitur: quid item tractet, quæ ejus methodus, & quo fine inventa sit, amplè explicatur

SECUNDÒ CLAVIS MASORAE TRADITUR, per quam ad maxima & minima Masoræ penitùs cognoscenda, facillimus datur aditus.

TERTIÒ CASTIGATIONES INNUMERORUM mendorum per universam Masoram proponuntur.

BASILEÆ

Typis LUDOVICI KÖNIG.

M. DC XX.

Figure 6.1. Buxtorf, *Tiberias*. Basel: König. 1620. Title page. Princeton, Princeton University Library, 2019–0001F.

this case, the Masoretic text read not *karu*, "they dug or pierced," but *ka'ari*, "like a lion," a reading that could be taken as undermining the Christian interpretation of the text. In the alphabetical list at the end of the second volume of the Rabbinic Bible of 1525, under the letters אר, Jacob ben Hayyim ibn Adonijah had written in a marginal note:

> In some correct texts (*meduyyakim*) I found the *ketiv* spelt *karu* and the *keri*, *ka'ari*. However, I looked at the list of those words which have *vav* at the end of the word – i.e., *karu* – but whose *keri* is with a *yod*, i.e. *ka'ari*, but I did not find any instance of it in the list, nor was it to be found in the list of variants between Western and Eastern codices.[10]

Adriani had Amerbach note down both readings, as given by Jacob ben Hayyim: "aliter in margine scriptum caari quod erat sicut leo" – "this was written differently in the margin, *caari*, because he was like a lion."[11] Virtually all of the scholars who took an interest in the Masorah between Adriani and the end of the sixteenth century examined this passage.[12] At the beginning of his career, Buxtorf pursued the study of the Hebrew text in an equally conventional way, starting from issues and cruxes that had been widely discussed by others. In his copybook he entered a note on the terms *ketiv* and *kere*. He insisted that they represented not variant readings, one or both of which might have been generated by scribal errors, but complementary possibilities, which belonged to the original sacred text:

> Variant readings in the Old Testament did not arise from some doubt, as many have thought, but were transmitted for particular reasons. Moses himself pointed out to Joshua the passages in his five books that had divergent readings and explained the reasons for them. He in turn explained them to his descendants, down to the prophets. The reading of the whole biblical text is absolutely uniform, but its written record has variants. The word to be read is always placed in the margin; the written one, in the text. The Jews call the latter Keri, and the former Kethibh. Various authors have explained why they wrote the word differently in the text: that is, this is done on account

10. *Torah Nevi'im Ketuvim*, ed. Jacob ben Hayyim ibn Adonijah (Venice, 1525), 2: sig. 113 iiv. On this edition, see Jordan S. Penkower, "Jacob ben Ḥayyim and the Rise of the *Biblia Rabbinica*" (PhD diss., Hebrew University of Jerusalem, 1982 [in Hebrew with English summary]). For a more recent discussion of ibn Adonjah's working practices, see Yakov Z. Mayer, *Editio Princeps: The 1525 Venice Edition of the Palestinian Talmud and the Beginning of Hebrew Printing* (in Hebrew) (Jerusalem, 2022), 64–81.

11. Psalter, ed. Adriani, fol. 39r.

12. Anthony Grafton, "Some Early Citizens of the *Respublica Litterarum Sacrarum*: Christian Scholars and the Masorah Before 1550," *Reformation* 23 (2018): 6–28.

> of some remarkable mystery, which was known in antiquity to the learned of that time.[13]

He then explained a number of well-known cases in point. At Ezra 4:1–2, for example, "the adversaries of Judah and Benjamin," hearing that the Jews were building a temple to the God of Israel, "drew near to Zerubbabel, and to the heads of fathers' houses, and said unto them: 'Let us build with you; for we seek your God, as ye do; and we do sacrifice unto Him since the days of Esarhaddon king of Assyria, who brought us up'" – only to receive the dustiest of answers from the Jews. Buxtorf translated the verse into Latin and identified the *ketiv* and *kere*: "it is written ולא [and not] and it is read וְלוֹ [and to him]." "It was not without reason," he explained,

> that Ezra inserted a *Ketibh* that differed from the *Keri*. Through the *Ketibh* [and not] he wanted to show that they had not sacrificed to the God of the Jews but to the gods of the Gentiles, as can be seen at 2 Kings 17:27 ff., and through the *Keri* he reported the words of the enemies, who said that they sacrificed "to him," that is, the God of the Jews, which was false.[14]

Even at this early stage, Buxtorf, relying on Ibn Ezra's explanation of the double significance of the *ketiv* and *kere,* set out to prove the stability of the biblical text.[15] Neither the *ketiv* nor the *kere* was a variant, for both belonged to the original text. They brought out complementary aspects of its meaning. He must have agreed

13. Buxtorf, copybook, Basel, UB, MS A XII 20, 294: "De קרי et קתיב [corrected to כתיב] seu varia lectione quae in Biblijs hebraeis occurrit. Varietas lectionis in V.T. hebraeo non orta est ex aliquo dubio, ut multi existimant, sed ob certas rationes tradita fuit. Moses ipse, quae in suis quinque libris diversam habent lectionem, indicavit Iehosuae, earumque rationem explicavit: is posteris suis usque ad prophetas. Lectio totius Codicis biblici est tantum unica, scriptio autem diversa. Vox legenda semper ponitur in margine; scripta, in textu. Illa, Keri; haec Kethibh a Iudaeis appellatur. Quare autem aliter in textu vocem scribserunt [*sic*], hoc singuli autores explicarunt, hoc nimirum fieri ob singulare aliquod mysterium, quod antiquitus notum fuit illius temporis Doctoribus."

14. Ibid., 296: "Ezdrae 4.2. scribitur לא et legitur לו. Aedificabimus (dicebant hostes Iudaeorum) vobiscum, nos enim sicut vos inquirimus Deum vestrum, eique (scribitur ולא, legitur וְלוֹ) [in margin: 'id est, et non; id est, et illi'] sacrificavimus a diebus Esarchaddon, qui nos hic transtulit. Non absque ratione posuit Esdras Ketibh aliter quam Keri. Per Ketibh enim indicare voluit eos non sacrificasse Deo Iudaeorum sed dijs gentilibus, ut videre est 2 Reg. 17. v. 27 et seqq. et per Keri retulit verba hostium dicentium, se illi, sc. Deo Iudaeorum, sacrificasse, quod falsum fuit." Levita had also referred to this example in the third introduction to *Massoret*, 15; ed. Ginsburg, 108.

15. See Ibn Ezra ad loc.: "The *kere* is with a *vav*, but it is written with an *alef*, meaning 'We do not sacrifice to any God besides Him.'"

with Johann Drusius, who wrote to him in August 1596 that "if a given passage has variant readings, it is not therefore corrupt."[16]

It did not take Buxtorf long to discover Levita. As early as 23 December 1593, he informed his friend Caspar Waser that "through work and sweat I have managed, up to now, to translate Elijah's books *Masoreth Hammasoreth* and *Tuv ta'am* into Latin."[17] His copy of the Hebrew text of Levita's book, in the 1539 Basel edition edited by Münster, contains remarks that show that he used Münster's partial Latin version and collated it against the 1538 Venice edition.[18]

It was no simple task to translate the *Massoret* into Latin.[19] Yet Buxtorf managed to render the entire text, including all but the last of Levita's artful poems.[20] This was clearly an immense learning exercise which, as we will see, culminated in his Masoretic masterpiece, the *Tiberias*. As he translates, Buxtorf also annotates on occasion, adding biblical references and expanding on bibliographical details.[21] Though his rendition is generally precise, he does make at least one error that he would certainly not have made in later life, when his antenna was ever sensitive to anti-Christian sentiment.[22] Levita had expressed great contempt for Jacob ibn Adonijah's edition of the Masorah in the 1525 Venetian Bible printed by Bomberg.

16. Drusius to Buxtorf, 4 August 1596, Basel, UB, MS G I 59, 256: "Non enim si locus aliquis varie legitur propterea corruptus est."

17. Buxtorf to Waser, 23 December 1593, Zurich, ZB, MS S 149, 124, 1: "Studium hebraeum, quantum licet, excolo: labore et sudore mihi comparavi hactenus tantum, ut libellos Eliae Masoreth hammasoreth, et De Accentibus, latinos fecerim."

18. Buxtorf's copy of Levita's *Massoret* (Basel, 1539) is Basel, UB, FA VIII 34. He refers to Münster on sig. [6 6v]: "periculum, Munst. An aliter." On the 1538 Venice edition, see sig. [1 6v]: "In exemplari Vene . . . est ר."

19. The translation of the *Massoret ha-Massoret* (but not the *Tuv ta'am*) is extant in Copenhagen, Det Kongelike Bibliotek, MS E Don. Var. 13. 8°. It seems to have been prepared for publication in 1629 by a former student, the chronologer Hans Jensen Svaning (Johannes Svaningius). See Stephen Burnett, *From Christian Hebraism to Jewish Studies*, 26 n96; on Svaning's private studies with Buxtorf see Daniel Tossanus, *Johannis Buxtorfii senioris . . . Vita et mors* (Basel, 1630), 14. Prijs did not know of this manuscript and speculated that the text had been printed in about 1603. See Joseph Prijs, *Die Basler Hebräischen Drucke, 1492–1866*, ed. Bernhard Prijs (Olten, 1964), n188, 297–98.

20. Sebastian Münster's rendering of *Massoret ha-massoret* and *Tuv ta'am* (Basel, 1539) was both partial and imprecise.

21. E.g. where Levita mentions that his writings had been translated "into the languages of the nations" Buxtorf (MS E Don. Var. 13. 8°, 9) translates 'Et in linguam Latinam translati' and then adds a marginal note: "a Sebastiano Munstero." He thus showed due respect to his predecessor, who was indeed Levita's greatest promoter. See also Chapter 3 above.

22. Buxtorf did not use this passage in the *Tiberias* and though he was aware of Levita's criticism of Jacob ibn Adonijah, he himself always referred to R. Jacob, citing Jacob's famous introduction to the second Rabbinic Bible which was produced before he converted to Christianity.

He articulated his opinion by means of a play on the conventional expression used for the dead, "May he or she be bound up in the bag of life," and the expression "perforated bag" in Haggai 1:6. Levita apparently knew that Jacob had eventually converted to Christianity. As he castigated Jacob for his imprecise printing of the Masorah he made a sharp reference to his conversion: "It [i.e. the Massoretic apparatus] was arranged by one of the learned whose name among Jews [literally 'in Israel'] was formerly Jacob, may his soul be bound up in a perforated bag."[23] Buxtorf, unaware of the meaning of Levita's barbed comment, thought that the text had been arranged in this form for the sake of the rhyme.[24]

When it came to the technical part of Levita's work Buxtorf read with special interest, convinced of his own expertise in Masorah. Where Levita provided explanations of some of the mnemonic devices in the apparatus in the last section of the work, Buxtorf seemed to think that these particular "simanim" did not exist.[25] Yet, clearly Buxtorf was also making every effort to penetrate the obscure wiring of Massoretic technology. In a particularly complicated passage about one of the mnemonic signs linked to the daily reading of Scripture Buxtorf, like Christian Ginzburg, who translated the text many centuries later, explained the allusion to the reading of Scripture in the synagogue on three days in the week.[26]

It is not clear how deeply Buxtorf studied Levita or the Masorah at this point. To render Levita's *Massoret ha-Massoret* into Latin must have been a formative experience. Yet when Buxtorf finally produced his own Massoretic exposition, the *Tiberias,* he translated sections of the text anew. In the letter to Waser he also mentioned that he was at work on translating a third book, which he did not name, and hinted to Waser that he hoped to buy from him some of the Hebrew books that his friend had acquired in Italy.[27] But he did not comment on what he had learned from any of the texts he rendered into Latin. It is possible that they proved too complex and difficult for Buxtorf to exploit at this point.

He certainly took his time before returning to this project. In April 1606, he sent Joseph Scaliger a long letter on his own and others' Hebrew studies. Here he explained that he had just come upon the *Massoret ha-massoret,* "which I translated into Latin ten years ago as a private effort, and I read his discourse on the origin and authors of the Hebrew points and accents thoroughly." By now, certainly, the

23. Levita (Ginzburg, 94).

24. Buxtorf (MS E Don. Var. 13. 8°, 13), marginal note: "propter rhytmum. sic loquitur ex Hag. 1.6. alias enim dicunt in fasciculo vitae ex 1. Sam. 25.29. Isa.28.7."

25. Ibid., 156, 159–60.

26. Ibid., 157 (Ginzburg, 263–64).

27. Buxtorf to Waser, 23 December 1593: "Nunc quoque alium quendam librum hebraeum ad privatum meum usum converto cuius lectione vix dicere possum quantum delecter. Non dubito quin et tu eundem et alios fortasse ejus generis tibi in Italia comparaveris: nolo tamen eum nominare ante quam tu mihi tuorum librorum Indicem miseris."

status of the biblical text interested him deeply. In his copybook, Buxtorf had made clear that he did not want to accept the existence of textual variants in the Hebrew Bible. In his letter, he noted that ancient Jewish sources contradicted Elijah's theory about the age of the vowel points. He quoted passages from the classic mystical texts, the Zohar, on Song of Songs, "than which there is hardly an older book," and the Bahir, "which is likewise quite old."[28] Jews traditionally ascribed the Zohar to Simeon bar Yohai, a sage of the second century CE, the age of the Mishnah. Perhaps uncertain of how much weight to assign to this evidence – as we will see, Buxtorf had not yet seen the Zohar itself – he wound up his short exposition by admitting that he needed to give the whole matter more thought.[29]

Scaliger agreed. In his reply, written in June 1606, he praised Buxtorf's erudition and urged him to carry on with publications that would help young Christian scholars learn to read rabbinic, as well as biblical, Hebrew.[30] But he also made clear that in his view Buxtorf had not understood the nature of the Masoretic project or the value of the sources he consulted. For all Buxtorf's interest in Aramaic and Yiddish, here as elsewhere he treated Hebrew in isolation. Scaliger, by contrast, adopted a comparative approach, examining Hebrew in the light of Arabic, Persian, and other languages, all of which had originally lacked vowels in their written texts. Grammarians, Scaliger argued, devised the vowels for these languages, centuries after they came into existence. Their original lack of vowels had presented no obstacle to readers. Elijah Levita had described "Chaldeans" who told him that their manuscripts were unvocalized, and that they read them as they had learned to in their childhood.

Scaliger generalized Elijah's point. The Jews too had read their books without vowels until the late Alexandrian age, when the grammarians materialized to legislate about what had once been an unruly, accessible natural language. They and the Samaritans still read from unpointed scrolls in their synagogues.

> On the vowel points of the Hebrews, I am equally sure that this is a recent matter and that those who think that they came into being at the same time as the language – than which nothing more stupid could be thought or said – are wrong. For the grammatical rules of a language are created not when it is widely spoken but after it has ceased to be in common use. Who can deny this? The vowels in Arabic were invented recently, that is, many years

28. Buxtorf to Scaliger, 3 April 1606, in *The Correspondence of Joseph Scaliger*, ed. Paul Botley and Dirk van Miert (Geneva, 2012), 6: 381–82. The Bahir is one of the oldest of the kabbalistic writings. It is first mentioned at the end of the twelfth century, but its origins are probably much earlier. See Daniel Abrams, *The Bahir: According to the Earliest Manuscripts* (in Hebrew) (Los Angeles, 1994).

29. Buxtorf to Scaliger, 3 April 1606, 382.

30. Scaliger to Buxtorf, 11 June 1606, in *Correspondence of Scaliger*, 6: 438.

> after the death of Mohammed the impostor. But nowadays, Turkish, Arabic, Persian, and all the other Muslim boys learn to read without points. In the same way, Samaritans and Jews read the Law in their scrolls in Synagogues without points.[31]

Scaliger flatly rejected Buxtorf's evidence from the Zohar, on chronological grounds: "that book is more recent than the Talmud."[32] He claimed that he could draw much more evidence from the Septuagint to prove that its translators had used a text without vowel points than the Jews could draw from the Targumim, though he also admitted that he did not have time to elaborate on his views.[33] Buxtorf, who eloquently expressed his delight at being praised by Scaliger, did not argue with his new friend. Instead, he asked if it would be useful for him to publish annotated editions of his translations of Elijah's grammatical works.[34] Scaliger responded enthusiastically. He noted that he owed a great deal to Elijah's "most erudite offspring," but expected to find it even more attractive once Buxtorf had reworked it.[35]

In one sense, Buxtorf – like many other younger scholars – followed Scaliger's suggestions.[36] In 1609, he published his *Thesaurus Grammaticus linguae sanctae Hebraeae* – a work that, he claimed, would give students a new level of guidance into the mysteries of rabbinical prose and Hebrew verse.[37] Not surprisingly, he seems to have been very much in sympathy with Scaliger's views about the proper way to teach Hebrew and promote Hebrew scholarship.

By the time the book went through the press, however, Scaliger was dead. Buxtorf mourned the great scholar's death in his preface, but he used his finished *Thesaurus* to put forward exactly the views about the Hebrew vowels and the Zohar that Scaliger had refuted. In a long passage that he titled "Addition," he expounded Levita's theories at length, noting that the Huguenot grammarian Pierre Martini, with whom he corresponded, accepted them.[38] Then he rebutted them on multiple grounds. Rejecting Scaliger's views on the Zohar, Buxtorf followed other Christian scholars who believed that the Kabbalah and its written sources were very ancient

31. Ibid., 438–49.
32. Ibid., 439.
33. Ibid.
34. Buxtorf to Scaliger, 4 September 1606, in *Correspondence of Scaliger*, 6: 546.
35. Scaliger to Buxtorf, 23 October 1606, in *Correspondence of Scaliger*, 6: 589–90.
36. See the brilliant review essay by Kristine Haugen, "Joseph Scaliger's Letters: Collaborator, Teacher, Impresario," *History of Universities* 28, no. 1 (2014): 105–47.
37. Buxtorf, *Thesaurus Grammaticus linguae sanctae Hebraeae* (Basel: Waldkirch, 1609), ep. ded.
38. Ibid., "Additio" to chap. 5, on accents, 48–69, at 50–59. A preparatory breakdown of Levita's arguments appears in Buxtorf's copy of the *Massoret* (Basel, 1539), sigs. 5 3r–5, fol. 4r; on sig. 5, fol. 3r he writes: "Eliae sententia de punctis."

indeed – the belief that Pico della Mirandola had put forward when he introduced Christian scholars to the kabbalistic tradition in the 1480s. Pierre Chevalier, a Genevan expert on rabbinic Hebrew, knew passages from the mystical tradition that compared the consonants and vowel points to the body and soul: the points gave life to the letters. He cited these, emphasizing their antiquity, in 1590 in his notes on Antoine Rodolphe Chevalier's Hebrew grammar – a point Buxtorf had not seen fit to mention when he quoted the same passages to Scaliger.[39] Buxtorf's close colleague in Basel, the theologian Amandus Polanus, also knew about the frequent references to the vowel points in the mystical tradition: for example, the oft-cited saying from the Bahir, attributed to the Mishnaic sage Nehuniah ben ha-Kanah, that the relation of the vowels to the letters is like that of the soul to the human body:

> For it is impossible for him to have any life whatsoever unless the soul exists within him, and it is impossible for him to produce a word, small or great, without the soul. The same is true of the vowel point, for the letters are given life by the vowel points, and it is impossible to say a word, great or small, without the vowel point.[40]

Polanus provided further ammunition as well. He cited Jesus's declaration in Matthew 5:16: "Till heaven and earth pass, one jot or one tittle shall in no wise pass from the law, till all be fulfilled." These words, he argued, proved that the Hebrew Bible had been written with vowels as well as consonants – and proved it with absolute certainty for any Christian. True, Polanus admitted, some private copies had lacked points. But "it is certain," he insisted, "that the public and authentic copies did not lack points."[41]

Buxtorf cited Maimonides' description of a famous manuscript, the tenth-century Aleppo Codex, as a case in point: "The book on which we relied in these matters is the book well known in Egypt, which contains all twenty-four books of the Bible. It was in Jerusalem for many years so that other books could be corrected from it."[42] Naturally, Buxtorf did not extend this quotation to include Maimonides' description of how the Masorete Aaron ben Asher had corrected the text. At this point, in other words, Buxtorf's knowledge of the debate – and of the evidence about what the Masoretes actually did – was relatively narrow.

39. Pierre Chevalier, "In Caput IIII Annot.," in Antoine Rodolphe Chevalier, *Rudimenta hebraicae linguae* (Geneva, 1590), 27. He insisted on the antiquity of both the Bahir ("ante compositas Deuteroses seu contextus totius Talmud") and the Zohar ("qui liber est multo antiquior") and argued that they contradicted Levita's claim that no ancient mystical or homiletic text mentioned the vowel points. See Buxtorf's discussion, "Additio," 66.

40. Buxtorf, "Additio," 61–62.

41. Ibid., 63, quoting Polanus.

42. Ibid., 65, quoting Maimonides, *Code*, Laws of Torah Scrolls, 8.4.

Three years later, Buxtorf would write a long letter on the crux in Psalm 22 to the Marburg chronologer Christoph Helvicus. In support of his belief that the *ketiv* was *karu*, which the Masorah itself did not support, he cited still more Christian experts, winding up with Adriani, all of whom attested that they had seen it "in ancient copies."[43]

Yet in taking the side of Chevalier and Polanus, as Buxtorf knew, he also took one side in an older and much more elaborate debate between Jewish scholars about the Kabbalah and the Zohar. Many readers of the *Massoret ha-massoret* seem to have become fixated on one particular passage in the tract's third introduction. There Levita threw down the gauntlet to his readers. In his original text, he challenged them to find any evidence that the rabbis ever referred to vowel points or accents: "But if anyone should prove to me with clear evidence that my opinion contradicts that of the rabbis of blessed memory then I will defer to his opinion and abrogate my view."[44] This is the reading of the first edition (Venice, 1538). But the version more commonly cited was Sebastian Münster's bilingual text, which reads: "But if anyone should prove to me with clear evidence that my opinion contradicts that of the rabbis of blessed memory or is contrary to the true Kabbalah contained in the *Zohar*, then I will defer to his opinion and abrogate my view."[45]

Why was the Zohar introduced in this Basel printing, and was it Levita's insertion? He was still residing in Venice. To be sure, in the first version of Levita's Aramaic dictionary, the *Meturgeman*, he claimed quite explicitly that vowel points and accents were mentioned by name only in the later kabbalistic tracts, not in the Zohar.[46] Yet he did not insert this statement into the published *Meturgeman* of 1541.[47] If Levita entertained this view about the Zohar in 1531, was this still the case in 1538; and why was it introduced in the 1539 printing? Levita copied out various kabbalistic tracts for his patron Cardinal Egidio da Viterbo, but his interest and knowledge may have extended no further.[48] As far as he knew, there were no

43. Buxtorf to Helvicus, 24 December 1612, Basel, UB, MS Ki Ar 189, item 8 (unnumbered pages): "Ex nostris (si fides ipsorum valet) testes sunt Gerhardus Veltwikus, Iohannes Kaum, Wolfgangus Fabricius Capito, Mattheus Adrianus, qui in antiquis exemplaribus scripturam antiquam כארו se vidisse testantur."

44. Levita, *Massoret*, 20.

45. Levita, *Massoret* (1539), 31r.

46. The MS, dated 1531, is in Rome, Biblioteca Angelica Or. 84. See Emma Abate, "Elias Levita the Lexicographer and the Legacy of *Sefer ha-Shorashim*," *Sefarad* 76 (2016): 289–311.

47. See Jordan Penkower, "A Renewed Inquiry into the *Sefer Masoret ha-Masoret* of Elijah Bahur" (in Hebrew), *Italia* 8 (1989): 7–73.

48. London, British Library MS Add. 27,199 contains Levita's transcription of various mystical works, as does Munich, Bayerische Staatsbibliothek MS Heb. 81. On Levita's work for Egido da Viterbo, see Emma Abate, "Filologia e qabalah: La collezione ebraica di Egidio da Viterbo alla biblioteca Angelica di Roma," *Archivio italiano per la storia della pietà* 26 (2013):

shureks and the like in the Zohar.[49] Or perhaps, he preferred to feign ignorance of these texts. We cannot answer the question of Levita's beliefs definitively. What is clear is that the dating of the Zohar and other mystical literature became one of the biggest bones of contention in the vowel-point controversy.

One later reader, Azariah de' Rossi, took Levita's words at face value, and gave him much credit when he wrote in his monumental *Light of the Eyes* (*Me'or Einayim*): "However, notwithstanding these criticisms, it should be noted that from the outset, Rabbi Elias himself put forward the statement, 'I shall succumb to the will of any person who can disprove my argument against our rabbis.' So we would say to him: 'Who will uncover the dust from your eyes, Bahur, you who are chosen from the people?'"[50] After all, de' Rossi pointed out, Levita could not have known the classics of the Kabbalah, since they had not been published in his lifetime:

> Since the kabbalistic works to which we shall refer were not yet in print in his lifetime, how could he have defended his position? But today, blessed be God, the eye can see and the ear can hear the entire texts of the *Bahir*, the *Zohar*, the *Tikkunim* and the *Ma'arekhet ha-elohut* (whose content is based upon the activities of Rabbi Nehuniah ben ha-Kanah, and Rabbi Simeon bar Yohai before the Mishnah was composed).[51]

In chapter 19, de' Rossi, the critical scholar, had warned readers about the non-authentic parts of the Zohar.[52] Now, in chapter 59, he asserted the intrinsic importance of the Zohar and other mystical texts for the issue at hand. Like Buxtorf, de' Rossi felt the need to defend the traditional biblical text from the repercussions of Levita's influential treatise. A deliberate obfuscation is therefore to be found

413–51. Malachi Beit Arié delivered a paper, "Eliyahu Levita as a Scribe, Author-Scribe and Codicologist," at Between Jews and Christians: The Many Faces of the Jewish Polymath Elijah (Bocher) Levita (1469–1549), a conference held in Oxford in 2007.

49. The Zohar that belonged to Egidio da Viterbo (Rome, Biblioteca Casanatense, MS 2971) is defective and does not contain the sections in which the names of the vowels and accents are mentioned.

50. Azariah de' Rossi, *The Light of the Eyes: An English Translation with Introduction and Notes*, trans. Joanna Weinberg (New Haven, 2001), 702. De' Rossi is playing on Levita's pen-name, Bahur (the title of one of Levita's grammatical works), alluding to Psalm 89:20, where the phrase "bahur me-am" (chosen of the people) is used of David, and "who will uncover the dust from your eyes" from Mishnah Sota 5:5. The choice of the Sota passage is significant in that Yohanan ben Zakkai, the supposed author of the Zohar, was castigated there by Rabbi Joshua for not having realized that Job served God not simply "out of fear" but "out of love."

51. Ibid., 702.

52. Ibid., 329.

in his wielding of the topos of loss and rediscovery. The vowel points, he argued, had been lost or corrupted several times, first at Sinai and then again in later periods:

> In the light of this method of mine, all the sages' statements that we put forward at the beginning of the chapter are all equally justifiable: the vowel-points were given at Sinai, in the time of Ezra, and after the close of the Gemara. In other words, when the vowels were rediscovered at those moments in time, after a period in which they had been forgotten or had fallen into disarray in the hands of the people, it was as though they had been given on that very day. Also correct would be our impression that they were instituted with the letters of the script from the time of creation. There was a continual process of neglect and restoration.[53]

De' Rossi's masterpiece was in Buxtorf's library and he made use of it in the *Tiberias*.[54] If this great Jewish scholar held that the Zohar was ancient, Buxtorf could follow him with some confidence.

Buxtorf, accordingly, displayed his evidence in the most prominent place that the *Thesaurus* afforded him. After his dedicatory letter to the Estates of Holland and Zeeland, room remained in the gathering, which Buxtorf filled with a one-page letter to the reader. At last, Buxtorf declared here, he had gained temporary access to the chief primary source in question: "This winter I borrowed from Jews for some weeks the book Zohar, which, all the Hebrews agree, was written before the Masoretes and the publication of the Talmud." Here Buxtorf exaggerated, as he must have known. But he went on to print a series of texts from the Zohar, which he treated both as his own recent discoveries and as powerful evidence that the vowel points were indeed ancient: "I found passages that not only mentioned the points and accents, but also cited the names of some of them.... I will not refuse to be the first to publish these to support the antiquity of the points."[55]

53. Ibid., 707.

54. Buxtorf's copy of de' Rossi's book is Basel, UB, FA V 4. He summarized chapter 59 in a series of marginal notes.

55. Buxtorf, *Thesaurus Grammaticus*, sig.):(v]: "LECTORI BENEVOLO S. Cum hac hyeme a Judaeis commodato acciperem ad aliquot septimanas librum Zohar (qui omnium Hebraeorum consensu ante Masorethas & editum Talmud conscriptum est) inveni loca, in quibus non tantum punctorum & accentuum mentio fit, sed & nomina quorundam, ut Chirek, Cholem, Schurek, Saegol, Tzere, Scheva, Saegolta, Makkaph, Sarka, Schophar holech, exprimuntur. Haec primus publicare non gravabor ad Punctorum antiquitatem astruendam." In Basel, UB, MS A XII 7, 16, there are three pages of extracts, summary, and comments labeled as "Ex Zohar impresso Cremonae, in fol. 1558," in Buxtorf's hand, along with a final page of excerpts from Postel, Scaliger, and others on related subjects. On the first page, Buxtorf notes after a long quotation: "Pag. seq. col. 1 mysteria cabalistica proponit de

Buxtorf's quotations from the Zohar – and his comment about having to borrow a copy from Jews – attracted attention in the Low Countries, to whose dignitaries he had dedicated the *Thesaurus*. Thomas Erpenius, the distinguished Leiden Arabist, already disagreed with Buxtorf, as he would later make clear in print, about the age of the Hebrew vowel points. Still, he sent his colleague in Basel his copy of the Mantua octavo edition (1558–60), and he made clear that Buxtorf could keep it as long as he wanted.[56] The theologian Franciscus Gomarus went to even greater lengths. Erpenius informed him that Buxtorf needed a Zohar. During a business trip to Frankfurt in spring 1614, he visited the book fair, where he "went through the books of the Jews that had been sent there from elsewhere" and found a copy of the text – in this case, the Cremona edition (1559–60). He bought it and sent it to Buxtorf as a gift, urging him to continue in his good work as a Hebraist.[57] It seems clear that Buxtorf, and his project, now enjoyed high standing in the Reformed world of Hebrew scholarship. Yet his interest in the Zohar itself was limited. He never set out to plumb Jewish secrets or unravel the mysterious working of the Sefirot. Rather, he had an urgent need to discover what the ancient sage Simeon bar Yohai had had to say about the vowel points and accents in this work traditionally attributed to him.

The evidence that Buxtorf published, without translation, consisted of four passages taken from the Zohar printed in Cremona. These extracts certainly contained

---

Cholem, Scheva et Kametz." On the third, he remarks, quite reasonably: "De punctuatione per accentus, mirum nihil notasse Sixtum Senensem." The care with which Buxtorf wrote out the extracts from the Zohar attests to the fact that he was exploring territory new to him. Quite likely lack of space, rather than lack of interest, caused him to abridge this collection in the *Thesaurus*.

56. See Arnoud Vrolijk and Joanna Weinberg, "Thomas Erpenius, Oriental Scholarship and the Art of Persuasion," in *Scholarship between Europe and the Levant: Essays in Honour of Alastair Hamilton*, ed. Jan Loop and Jill Kraye (Leiden, 2020), 34–59, esp. 54–57.

57. Ibid., 54–55. Buxtorf's copy of the Cremona edition of the Zohar is Basel, UB, FA III 13. It has a dedicatory inscription to Buxtorf on the bottom margin of the title page: "Doctissimo D. Johanni Buxtorfio Hebraearum literarum ornamento Franciscus Gomarus Theol. D. et professor in schola Middelburgensi dono misit Francofurto a° CIDIDCXIV Nundinis paschalibus." On the front pastedown a presentation letter from Gomarus has been pasted in: "Doctissime Buxtorfi, / S.P. Quum intellexissem Leydae a M^{ro} Erpenio te desiderare Zoharem cuius usum indulsisset tibi, nolui opportunitatem tua studia excitandi negligere. Nam cum negotii causa Francofurtum venissem, lustratis Iudaeorum libris eo aliunde directis, Zoharem reperi: quem aegre decem florenis francofurtensibus redemi. Eum dono tibi mitto: rogoque ut quemadmodum coepisti, Hebraeas literas illustrare pergas. Ita te Deus servet & tuis laboribus aspiret. Plura alias: nunc cogor abrumpere. Vale & ubi librum hunc acceperis per bibliopolam vestrum me certiorem redde. In aliis ubi erit opportunitas tuis egregiis conatibus non deero. Spero autem te cogniturum. Raptissime Francofurto a° CIDIDCXIV Nundinis vernalibus. / Tuus Franciscus Gomarus theol. In schola / Middelburgens, p."

the names of the vowel points and accents, but they were deeply embedded in a highly technical kabbalistic discourse, replete with imagery that could hardly be understood by the ordinary reader. It is therefore not surprising that Buxtorf himself did not translate the texts into Latin. The first of the four passages from the beginning of the Zohar treated the idea of the Radiance (i.e., Zohar), which contained offspring identified as the engraved letters emerging from that point (i.e., Hokhmah, Wisdom, the first of the Sefirot). Within that palace, "the point sowed the seed of three points, *holem, shuruk, hirek*, merging together, becoming a single mystery. The letters are the body and the vowels spirit. All of them range in motion after the intonation and halt with them."[58] In the second example the "secrets" of the various accents – the Zarka, Makaf, Shofar, Holekh, Segolta – are listed as an antidote to disruption of prayer by evil forces.[59] The third example, also taken from "Bereshit," referred to the positioning of the vowels in their relation to the divine realm as constituted by the Sefirot, the emanations or powers by which God manifests Himself.[60] Even more abstruse was the fourth extract from the Zohar, on Exodus (section "This Is the Gate of the Lord" [Psalm 118:1]):

> The letter *lamed* is included in all the Sefirot and all the letters of the explicit and hidden names. It is the point under every letter because she serves her husband. She is a crown on his head. She is from the side of the accents. She is the dot [that is marked above the letters]. She is like the dot of the *Segol* that is below the knees of the king [below Sefirot *netsah, hod,* and *yesod*] as it says, "and the earth is my footstool" (Isa. 66:1). She is at the center with him for then it is called "half a shekel." And this is in the point of the *Shurek* and it is the crown on his head from the side of the *Segulta, Zarka, Makaf Shofar, Holikh, Segulta.*[61]

Some of these passages reappear in the *Tiberias*, as part of Buxtorf's triumphant assembling of evidence about the ancient origins of the Masoretic vowels and accents.[62] Yet here, too, Buxtorf made no attempt to explain the meaning of these abstruse passages overlaid with mythic imagery.[63]

58. *Sefer ha-Zohar*, ed. Reuven Margaliot, 3 vols., 5th ed. (Jerusalem, 1970), 1:15b. Buxtorf cites this passage as "in vestibulo libri col. b." He is referring to the Cremona edition (Cremona, 1559–60), fol. 3, col. b.

59. Ibid., fol. 27r, col. 106 (ed. Margaliot, 1:24a).

60. Ibid., fol. 38v, col. 152 (ed. Margaliot, 1:24b).

61. This passage is from Zohar, "terumah" (from the section "*raya mehmena*"), ibid., 2:158 (Cremona ed., 71 col.1).

62. Buxtorf, *Tiberias*, 76–80. Buxtorf's lead-in to this passage is revealing: "His [the texts he has already cited] nos alia ex iisdem libris magis illustria adjiciemus, in quibus etiam nomina punctorum & accentuum exprimuntur."

63. Further evidence for Buxtorf's lack of substantive interest in Kabbalah is provided by his copy of a popular introduction to its main concepts and principles, Issahar Moses

Buxtorf's private views, however, were not identical to his public ones. As we have seen, in correspondence with Scaliger, Buxtorf had been forced to confront the possibility that the attribution of the Zohar and Bahir to famous Mishnaic authorities was false. True, Buxtorf now admitted that the Zohar was not the work of a single author: the disciples of Simeon bar Yohai had finished it, and it was not published until much later.[64] But he continued to insist that both the Zohar and the Talmud provided decisive evidence for the antiquity of the points.[65] As should be clear by now, in philological terms Buxtorf had not added a great deal to the body of evidence that he had discussed in his letter to Scaliger, though he interpreted it in different ways.

Buxtorf dated the dedicatory letter of the *Thesaurus* to March 1609. In September, the Marburg Hebraist Johann Molther sent him a deeply ambivalent letter. He agreed with Buxtorf that the Zohar was older than the Talmud. "Meanwhile," he added, "Scaliger is also right to say that it is more recent. For he is thinking of its publication, which took place quite late – so late that neither Nahmanides [1194–1270] nor Rav Asher [d. 1327], who made abstracts of the Talmud, saw it."[66] The Bahir, he thought, was older, but it too was a composite work.[67] More generally, Molther viewed the Jewish literary tradition with some distrust: "I am quite worried that such texts lie about their age, or have not reached us in a pure state."[68] Passages that supported Christian doctrine had been deleted, so that quotations

ben Betsalel Katz, *Imrei binah* (Prague, 1610), Basel, UB, FA V 37. Buxtorf has recorded the date of acquisition ("Johannis Buxtorfij 1611") on the title page. He notes that one passage refers to "Cremonensis editio Zohar" (fol. 4r), but apart from this bibliographical point he shows no interest in the book's contents, though he listed them in his *Bibliotheca Rabbinica* (in *De abbreviaturis hebraicis liber novus & copiosus* [Basel, 1613], 257–335, at 275).

64. Buxtorf, "Additio," 67: "R. Schimeon author libri fertur, non quod ipse totum conscripserit, sed quod totius doctrinae author sit. Nam a filio eius aliisque ipsius discipulis demum completus fuit. At multis seculis latuit, et sero admodum publicatus." See Chapter 3, above, on how Buxtorf gives a short report on this question.

65. Ibid., 67–68.

66. Molther to Buxtorf, 1 September 1609; Basel, UB, MS G I 60, 211: "De libro *Hazohar* et *Habehir* quid compertum habeam indicabo. Uterque liber inter Judaeos bene notus est: ille tamen magis, hic minus. Utrumque non solum habuit Reuchlinus: sed Postellus etiam transtulit. Num vero Postelli translationes lucem viderint, non satis scio. Librum Hasohar quidem coepit R. Simeon Filius Jochaj. Sed discipuli ejus consummaverunt. Recte sentis igitur eum antiquiorem esse ipso Thalmud. Interim tamen etiam recte sentit Scaliger, qui recentiorem dicit. Respicit enim publicationem ejus, quae valde sero facta est, adeo etiam, ut neque Ramban, neque R. Ascher, qui Summas Thalmudicas fecit, eum viderint."

67. Ibid.: "Liber Habbehir multo antiquior est: qui tamen ipse quoque non unum autorem habet. . . . Non igitur miror eum a Cevallerio allegatum esse."

68. Ibid.: "Quanquam interim Ego non parum vereor, ne libri tales vetustatem mentiantur, aut certe integri ad nos non pervenerint. . . . Quaedam enim inde citata, non inveni talia."

of them could not be verified.[69] On the other hand, a reference to Constantinople in Targum Jonathan revealed that the work was either forged or adulterated.[70] Yet he admitted that he had seen evidence that very ancient books in Hebrew existed, but had not yet been printed.[71] In December 1612, Buxtorf, after answering some of Molther's philological questions, asked him a series of them: "You," he began, "what do you think about the 18 passages that are *tikkun*[*ei*] *soferim* ['corrections of the scribes' – a technical term in rabbinical literature for changes of wording in the text of the Hebrew Bible].... Who were these *sopherim,* or when was this done?"[72]

At almost the same time, another Marburg scholar, the chronologer Christoph Helvicus, took Buxtorf back to an earlier stage in his discussion of the Masorah. As we have seen, Buxtorf replied at length, arguing that both readings at Psalm 22:17, *karu* and *kaari,* were necessary for the passage to have its full meaning. But once the Jews realized that "the passion of Christ was described too explicitly there, and the Christians used the passage to show their error, then the *ketiv,* that is *karu,* began to be exiled from the text, and the *kere,* that is *kaari,* to be substituted for it from the margin."[73] The Masoretes, "who added their notes to the biblical text some centuries after Christ," found only *kaari.*[74] That explained why they did not record the appearance of *karu* in their standard apparatus, and why Jacob ben Chajim attested that he encountered it as the *ketiv* only in certain "precise manuscripts." God had preserved faint but unmistakable traces of the true reading.[75] As late as 1612, in other words, Buxtorf still believed that the Masoretes

69. Ibid.: "Si quid enim in eis fuit, quod Ecclesiae favere potuit, subtractum est, id quod observavi in More Nefuchim. Quaedam enim inde citata, non inveni talia."

70. Ibid.: "Prodiit ante annos nondum triginta Targum Jonathan, super Legem: Sed quia Constantinopoleos in eo mentio fit, aut suppositium aut adulteratum illud est."

71. Ibid.: "Interim tamen negare nolo, extare inter Judaeos adhuc antiquissimos libros, quos Germania nondum impressos vidit: qualis cum primis est liber Sapientiae, quem Chaldaico idiomate elegantissimo perscriptum, adhuc inter Iudaeos latitare, dubitare non potest, qui Praefationem Ramban in Genesin legit."

72. Buxtorf to Johann Molther, 26 December 1612, Basel, UB, MS Ki Ar 189, Item 8: "Tu autem quid sentis de 18 locis istis, cui sunt תקון סופרים et enumerantur in principio Bammidbar. Qui fuerunt hi Sopherim, aut quando hoc factum? Aben Esra eorum studium rejicit, cui et ego assentior. De LXX senum tredecim locis, jam antea Drusius in Epistolis suis judicium explicavit."

73. Buxtorf to Helvicus, 24 December 1612, Basel, UB, MS Ki Ar 189, Item 8: "Post haec tempora varietas incidit. Cum cognoscerent judaei, nimis aperte Christi passionem ibi descriptam, et se a christianis hoc loco convinci, tunc coepit exulare ex textu כתיב nempe כארו, et substitutum ipsi ex margine in textum קרי nempe כארי."

74. Ibid.: "Inde Masorethae, qui aliquot centenis post Christum annis notas suas ad textum biblicum adiecerunt, tantum invenerant כארי[.]"

75. Ibid.: "Utut autem exstirpare hanc vocem judaei conati fuerint, reliquit tamen Deus ejus vestigia quaedam, ut R. Jacob ben Chajim, qui Masoram in ordinem redegit,

had done their work well after the coming of the Messiah – a position that, as we will see, would shift radically in due course.

The most detailed response to Buxtorf's first publication on the Masorah came from Drusius. He not only agreed with Scaliger about the Zohar but rather thought that he himself had first called the text's recent date to the master's attention.[76] He had learned that the text was late from Abraham Zacuto's *Sefer yuhasin* (Book of genealogies), a chronicle composed in 1504. In the first printing of 1566, the book contained a story that became famous: that the thirteenth-century kabbalist Moses de Leon had been deeply involved in the promulgation of the text of the Zohar. Zacuto also expressed his doubts about the authorship of other works, in a passage that Drusius translated:

> It seems that R. Eleazar Haccaphar was one of the later ones, even though he is cited in the Pirke R. Eliezer. For it is now known that although that book is ascribed to R. Eliezer, it was nevertheless made by later men. So too the Zohar, which is attributed to R. Simeon; and so too the Sefer Jetzira, which is falsely ascribed to Abraham the Patriarch, though it was written by others.

"Here," Drusius commented, "is the testimony of a Jew, which is not to be despised."[77] The antiquity of the vowel points and accents was under threat.[78] Most Christians – like Drusius, apparently – read the 1580–81 printing of Zacuto's *Sefer yuhasin*, which omitted or rather censored the ascription of the Zohar to the activities of the thirteenth-century Moses de Leon.[79] Still, as Drusius showed in his letter, even the censored text suggested that Simeon had not written the Zohar.

---

testetur se in aliquot ספרים מדויקים invenisse scriptum כארו et lectum כארי. Quod autem id in Masoretharum notis non invenerit observatum, causa est, ut dixi, quod tunc כתיב illud communiter erat sublatum."

76. Drusius to Buxtorf, 8 February 1613; Basel, UB, MS G I 59, 260: "Judicium Scaligeri de libro Zohar valde probo. Puto me eum monuisse ea de re."

77. Ibid.: "Ego autem primum didici ex Salmanticensi Judaeo, qui de hac re ita scribit in libro Johasin: Videtur quod R. Eliezer Haccaphar fuerit ex posterioribus, licet citetur in Pirke R. Eliezer. Jam enim cognitum est, quamvis tribuatur liber ille R. Eliezero, tamen a posterioribus eum factum esse, ut liber hazoar, qui tribuitur R. Simeoni: et sic invenitur liber Jetzira, qui falso tribuitur Abrahamo Patriarchae, cum scriptus sit ab aliis. En testimonium Judaei, quod spernendum non est." Buxtorf used this material in his *De abbreviaturis*; see Chapter 3, above.

78. See Anthony Grafton and Joanna Weinberg, with Alastair Hamilton, *"I have always loved the Holy Tongue": Isaac Casaubon, the Jews and a Forgotten Chapter in Renaissance Scholarship* (Cambridge, MA, 2011), 324–25.

79. Abraham Zacuto, *Sefer yuhasin* (Constantinople, 1566), [fols.] 141v–142r. Zacuto's narration used Isaac of Acre's report of Moses Leon's alleged involvement in the promulgation of the text. Apart from the suggestion that Moses de Leon actually transcribed the book from the original manuscript, other claims were aired: there were innuendos that he had

Drusius's opinion of Buxtorf's other arguments was not higher: he professed his surprise that his renowned colleague "preferred to follow the opinion of the unlearned, rather than of those who have won fame for their erudition. Among them Mercier, Veltwyck, Johannes Isaac, and everyone agrees that the points were added by the Masoretes."[80] As to the theological points that Buxtorf drew from Amandus Polanus, they were derivative. The Lutheran Hebraist Flacius Illyricus had argued a quarter century before, soon after what he called the "diabolical hypothesis" of Elijah first appeared, that Protestants could not accept an Old Testament that had been vocalized many centuries after it was first written.[81] Evidently, Drusius saw no reason to allow theological concerns to decide a historical argument – especially when the arguments in question were borrowed from the sinister Gnesio-Lutheran Flacius. Returning to the level of philology, Drusius made clear that the standard sources that Buxtorf worked from in fact provided no historical support for his views: "Anyone who wants to defend this view in a useful way must prove that there was some ancient book with points, or which mentioned points. This cannot be done with the Zohar, since it is a more recent book, as has now been demonstrated."[82]

In his reply, Buxtorf gave little or no ground. He pointed out that he knew the testimony of the *Yuhasin*. He had cited it, though he did not tell Drusius that, in his "Additio" to the 1609 *Thesaurus*.[83] More important, he had drawn from it his current view about the authorship of the Zohar:

> R. Schimeon is said to be the author of the book, not in the sense that he wrote the whole thing, but because he was the author of its doctrine. For it

fabricated the text for filthy lucre, or that he had written the text down through automated writing, "through the power of the Great Name." Even the wife and daughter of Moses of Leon were called upon to bear witness in the case. In his annotation on the passage, Samuel Sullam, the editor of the *Yuhasin*, insisted that the testimonies published by Zacuto were invalid: nothing should impugn the authority of the sealed book.

80. Drusius to Buxtorf, 8 February 1613: "Miror autem te in Grammatica tua, ubi de punctis vocalibus agis, maluisse sequi sententiam indoctorum, quam illorum qui eruditionis nomine clari sunt. in queis Mercerus, Veltuickus, Johannes Jsahac et alii, qui omnes in ea sententia sunt puncta ascripta fuisse a Masoritis."

81. Ibid.: "Quas autem rationes pro contraria sententia producis ex Amando Polano, sunt rationes Illyrici: quae hactenus nihil me moverunt. Missae autem ad me fuerunt ante annos viginti sex aut amplius." Flacius responded to Elijah in a short text that he had shown to his teachers at Wittenberg when he received his degree (Flacius Illyricus, *Regulae et tractatus quidam de sermone sacrarum literarum* [Magdeburg, 1551], 142–60), where he made clear that "incertitudo maxima sacrarum literarum ex hac diabolica hipothesi sequitur (159)."

82. Drusius to Buxtorf, 8 February 1613: "Qui eam opinionem cum fructu tueri vult, debet probare aliquem librum fuisse antiquum punctatum, aut in quo punctorum vocalium mentio fiat. Quod per Zohar fieri non potest cum sit liber recentior, ut iam probatum est."

83. Buxtorf, "Additio," 67: "ut videre est in libro Chronologico Juhasin, pag. 20."

> was finally completed by his son and other disciples of his. But it remained concealed for many centuries, and was published rather late, as can be seen in detail in the *Juchasin*, pag. 42. Hence some doubt that the book is ancient.[84]

As Buxtorf now made clear to Drusius, he had continued to ponder these questions. Perhaps Molther's letter inspired a sharper curiosity. In 1613, in the *De abbreviaturis*, Buxtorf discussed the life of R. Schimeon in detail.[85] He noted how implausible the traditional origin story of the Zohar seemed. Could the work really have come to light only after a thousand years had passed? Could a school have survived so long in Egypt, preserving the work both from men and from worms?[86] "Perceiving this," he explained, "R. Abraham Zachur [*sic*], the author of the Yuchasin, ascribed it not to R. Schimeon but to his son, his son's disciples, and the disciples of his disciples."[87] After all, Buxtorf pointed out, the rabbis were known to name books after men other than their actual authors.[88] He went on to cite a number of examples that Zacuto had previously singled out in the *Yuhasin* – including those to which Molther had called attention, and those to which the text referred in the passage singled out by Drusius.[89] He concluded the passage by making clear the urgent need for an accurate account of the contents of the Zohar: "Faciat qui potest, & cui libri copia" (Let him who can, and who has a copy of the book, do this).[90]

In the *Bibliotheca Rabbinica*, also published in 1613, he offered a more optimistic account:

> Because the author of this book nowhere mentions the Gemara or recalls any book of the Talmud, hence the Jews are convinced that it was composed before the Talmud was composed. As to the fact that it remained unknown for so long, they say that that is not unusual for Jewish books. Even today many still exist, which similarly remained hidden in Egypt and those eastern regions for many centuries. But enough.[91]

84. Ibid.: "R. Schimeon author libri fertur, non quod ipse totum conscripserit, sed quod totius doctrinae author sit. Nam a filio eius aliisque ipsius discipulis demum completus fuit. At multis seculis latuit, & sero admodum publicatus, ut videre est prolixe in libro Juchasin, pag. 42, unde quidam de eius antiquitate dubitant."

85. Buxtorf, *De abbreviaturis*, 164–68.

86. Ibid., 166.

87. Ibid., 166–67: "Hoc animadvertens R. Abraham Zachur, author libri Juchasin, non R. Simeoni, sed filio, filii discipulis, & discipulorum discipulis eum ascribit."

88. Ibid., 167: "Neque hoc Rabbinis novum, appellare libros quosdam ab alieno authore."

89. Ibid.

90. Ibid., 168.

91. Johann Buxtorf, *Bibliotheca Rabbinica*, 287: "Quod author istius libri nullibi Gemarae mentionem faciat, nec ullius unquam libri Talmudici meminerit, inde omnino

No reader of these passages could have believed, as Drusius did, that Buxtorf ignored the scholarly traditions of Christians and Jews.[92]

In his reply to Drusius, Buxtorf reaffirmed his belief that the Zohar was an accurate postmortem compilation based on the author's lectures, and thus both ancient and genuine.[93] More strikingly, he made clear that he was still cogitating about what to make of the complex evidence about the Masorah.[94] Indeed, he wrote as if Elijah's views, which Mercier and others had simply adopted, were as eccentric as they were novel. After all, he pointed out, "all the Jews reject Elijah's opinion. I have very often discussed this with learned rabbis."[95] If the Masoretes really did invent the vowel points, it was extraordinary that no writer mentioned their work.[96]

In the end, he admitted, the question could be settled only by further research into the facts. What he wanted, and could not find, was a connected narrative about the Masoretes and their work: "I certainly have not found a history of the Masoretes up to now in any writer. I think it can be shown that the vowel points existed before the Masoretes. I shall discuss this elsewhere."[97] A few years later, in 1618, Buxtorf corresponded with the brothers Jacques and Louis Cappel, who disagreed with one another about the vowel points. He also worked on the Masoretic apparatus of his edition of the Rabbinic Bible. Then he began to write the *Tiberias*, in which he at last constructed a detailed account of the Masoretes and what they did. On its title page he would claim that he himself had provided

---

judaei putant, ante collectum Talmud librum conscriptum esse. Quod tamdiu delituerit, id aiunt non esse novum inter judaicos libros, quorum multi adhuc hodie extent, qui similiter per multos centenos annos in Aegypto & istis orientalibus locis occultati fuerint. Sed satis." See also the discussion in Chapter 3.

92. For further discussion, see Chapter 3 above.

93. Buxtorf to Drusius, 23 August 1613; Basel, UB, MS Ki Ar 189, Item 8: "Libri Zohar novitas nondum mihi satis probata. Quae in libro juchasin de eo scribuntur, jam pridem legeram, ea adducuntur etiam a me inter Abbreviaturas in רש"בי, cui aliud accedit in Bibliotheca in זוהר ex R. Gedalia. Sed in his attribuitur liber filio et discipulis R. Schimeonis, qui praelectiones praeceptoris a se diligenter exceptas, post mortem ejus in unum corpus contulerunt. Sic adhuc doctrinae istius author manet R. Schimeon, et comportatio et dilvulgatio attribuitur discipulis."

94. Ibid.: "De origine punctorum vocalium nihil adhuc definivi: cogitandum proposui, quod in Grammatica attuli."

95. Ibid.: "Certe omnes Iudaei istam opinionem Eliae explodunt: saepissime cum doctis Rabbinis hac de re contuli."

96. Ibid.: "Si Masoretharum hoc fuit inventum, profecto tam fuit miraculosum, quam quicquam in mundo esse aut fieri potuerit: et quare ejus mille scriptores historici [negative missing] meminerunt?"

97. Ibid.: "Ego certe in nullo authore Masoretharum historiam hactenus reperi. Puto probari posse ante Masorethas puncta vocalia extitisse, de quo videbo alias."

the missing "history."[98] As we will see, in these years Buxtorf's views about who the Masoretes were and what they did changed radically.

To compile the book that Buxtorf saw as necessary, he confronted Elijah's *Massoret ha-massoret* in detail. Before we can explore the Reformed Buxtorf's definitive Masoretic world, accordingly, we must first see how the German Jew Levita embarked on his Masoretic task in the introductions to his work and understand what he actually said about those learned Tiberian Masoretes. Levita's thesis, though generally received as a pioneering and anti-traditional onslaught on the authenticity of the Hebrew Bible, was actually conservative in its content even though provocative in its tone.[99] Take, for example, the following statement: "Indeed, the Masorah is surely a fence around the Torah, and from it are derived essential laws, reasons, and explanations, both literal and nonliteral; and several laws derive from the defective and *plene* readings [that is, the variant spellings of the words as found in the traditional Masoretic text]."[100] Or consider Levita's comment about the accents in his tract *Tuv ta'am*, which was printed after *Massoret ha-massoret* but in the same year: "The rule is that they [the Masoretes] applied the accents as they thought right; and one should not ask why they sometimes put a *zarka* and sometimes a *pazer* or *revia*. Perhaps this was the way they referred to some of the secrets of Torah – for their knowledge was greater than ours – but we do not know why."[101] According to Levita, then, the Masorah was transmitted orally through the generations until the expert Masoretes of Tiberias and later grammarians invented the graphemes and accents and inserted them into the scriptural text. In his view, the work of Masorah continued over the centuries, down to his own time.[102] From the start, in other words, he made clear that the Masoretes had codified profound and valid knowledge, orally transmitted until their time, about the biblical text and its meaning.

Levita wanted his introductions to be read; in fact, he adjured all his Jewish readers to peruse the introductions before delving into the technical materials.

98. The full title of Buxtorf's book was *Tiberias, sive commentarius Masorethicus, quo primum explicatur, quid Masora sit: tum Historia Masoretharum ex Hebraeorum Annalibus excutitur.*

99. Isaac Casaubon was the first to make this point. He could not accept that Levita seemed to argue that the Masorah represented "a restoration of the text's true form rather than a set of innovations." See Grafton and Weinberg, *"I have always loved the Holy Tongue,"* 312.

100. Levita, *Massoret*, 8; ed. Ginsburg, 95.

101. Levita, *Tuv ta'am* (Venice, 1538), 8.

102. This is expressed by Moses Goshen-Gottstein as "a double belief of antiquity of tradition and novelty of graphic notation," which is a "live tradition of endless toil for safeguarding the sacred text." See his "Foundations of Biblical Philology in the Seventeenth Century: Christian and Jewish Dimensions," in *Jewish Thought in the Seventeenth Century*, ed. Isadore Twersky and Bernard Septimus (Cambridge, MA, 1987), 77–94, at 89.

In other words, he wanted to ensure that they knew that the detailed written Masoretic descriptions, rules, and terminology they were about to read were not Mosaic in origin. But his ornate and highly rhetorical mode of writing, in which the three elements of the trivium are all equally on display, is not easy to decode despite his frequent use of printed pointing hands in the margins – a device commonly used by authors, scribes, printers, and readers to draw attention to the significant and innovative points in a given discourse.[103] In his first poem, written in monorhyme, Levita gave a rather fleeting historical survey.[104] He narrated the events that led to Ezra's restoration of Scripture to its "pristine glory" and, subsequently, to the invention of the vowels by the Masoretes, the members of the school who were skilled in this work. Scholars who lived in Tiberias for the most part, they also applied the accents to facilitate the interpretation of the Torah. In this cleverly constructed poem all registers of Hebrew were used (unfortunately they get lost in both Münster's Latin and Ginsburg's English translations). Describing the return from Babylon, Levita wrote: "Ezra who is Malachi, like a ministering angel, went up: a priest and leader, the father of scribes, mother of Scripture and Masorah."[105] This was a choice play on the Talmudic expressions *Em la-mikra* and *Em la-massoret*, literally "mother of Reading [Mikra]" and "mother of Tradition [Massoret]." They occur in debates about words in the biblical text that are read out differently from the way they are actually written. Their presence – like that of the *keri* and *ketiv* – raises a vital question: whether the traditional reading of a word or the written consonantal text takes priority. Thus, for example, when a word read as a plural was spelled defectively, it was regarded as plural in respect to Mikra (the text that is read out), but singular as regards the Massoret (the written consonantal text).[106]

Buxtorf gave surprising prominence to part of this poem, setting the Hebrew and his Latin translation in parallel on the page.[107] He swooped down on Levita's

103. Levita himself explains his use of the pointing hand: "Whenever I have put forward something new in this book of mine or an important principle which no predecessor has ever propounded, a form of a hand in the margin in the apposite place will appear pointing with the finger, as if it were saying, 'See this is something new'" (*Massoret*, 2; ed. Ginsburg, 85–86).

104. The second introduction is composed in rhymed prose, and the third, rhetorical prose (*be-halatsah*).

105. Thus Ginsburg translated these lines: "Ezra then went who is the messenger, like a ministering angel, the priest the prince and the father of scribes, the nursing mother of the scripture and Massorah" (Levita, *Massoret*, ed. Ginsburg, 89). Sebastian Münster translated: "Ezra similis angelo ministeriali, sacerdos, princeps et scribarum pater separavit ab Israelitis Babylonicas mulieres" (thus he did not translate the phrase "em la-mikra u-le-massoret"). See his *Accentuum Hebraicorum... liber: Item liber traditionum* (Basel, 1530), 10.

106. See, e.g., BT, Sukkah 6b or Sanhedrin 4a.

107. Buxtorf, *Tiberias*, 10.

allusion to the rabbinic dictum that identified Ezra, the scribe who, according to the biblical book named after him, reconstituted the Hebrew Bible, with Malachi. This identification had angelic overtones. It insinuated that Ezra's prophetic role was both innovative and salvific.[108] His reconstitution of authentic Scripture and compilation of variant readings signified the birth of Masorah – which took place more than a millennium before, according to Levita, the Masoretes of Tiberias had compiled their apparatus.[109] But Buxtorf also elaborated on Levita's creative composition and adapted its images to suit his purpose. Summing up Ezra's momentous text-critical operation, he wrote: "It seemed that the whole of Scripture had become as it were reborn from a new mother."[110] Levita's imagery had taken on a life of its own. Even though the proficient Talmudist Buxtorf fully understood the implications of the phrases "mother of Scripture" and "mother of Masorah," he chose to retain the maternal image and to represent Levita as saying that Ezra was the *fons et origo* of Scripture.

Later, in his detailed rejoinder to Levita, Buxtorf considered these same expressions in their original context. To support his view that the vowel points were nowhere to be found before the post-Talmudic period, Levita had insisted that the Talmudic rabbis were disputing the correct reading of the scriptural text when they employed such phrases as "This reading of scripture is authoritative" (*Yesh em le-mikra*) or "This traditional pronunciation is authoritative" (*Yesh em le-massoret*). He reasoned that if the vocalization had really been Sinaitic in origin, the rabbis would not have challenged any reading whatsoever. Buxtorf naturally took the opposite position.[111] His response to Levita could just as well have been enunciated by an upstanding traditional Jew who understood the thrust and parry of the Talmudic argument. After providing the reader with various examples of the use of these key phrases, Buxtorf wrote: "There are many other such passages in the Talmud that are put forward only for the sake of the argument and for exercising the mind, but not because they entertained any doubts about the correct reading."[112]

108. It is not easy to establish which Christian sources Buxtorf made use of, or when. A note in his copy of Levita's *Massoret* (Venice, 1539), Basel, UB, FA VIII 34, rear endpaper, draws evidence from the Masorah on Numbers 12:12 to show that Ezra was a scribe and corrector of texts: "Tikkun Sopherim etiam vocatur Tikkun Ezrae Num 12. 12, quasi Ezra istorum Sopherim sive Scribarum qui Biblia inter se contulerunt et correxerunt praefectus fuerit." In the *Tiberias* (105), he makes clear that he drew this information from Génébrard's *Chronologia*, which he did not cite in his note.

109. On the rabbinic mode of conflating the identities of different biblical individuals, see de' Rossi, *Light of the Eyes*, chap. 18.

110. Buxtorf, *Tiberias*, 11: "... ut tota Scriptura tunc quasi ex nova matre renata videretur."

111. Ibid., 37–38.

112. Ibid., 38: "Similia plurima in Talmud reperiuntur quae disputandi & exercendi ingenii tantum causa proponebant, non quod de vera lectione dubitarent."

Buxtorf's description of the work done by Ezra and his colleagues has at least one striking omission. He did not discuss or refer to the tradition – widespread in Jewish texts – that a Torah written by Ezra himself had survived over the centuries. One of the most famous of these – the Bologna "rotolo d'oro" – is in fact the oldest extant complete Torah scroll, though it was written in the twelfth or early thirteenth century CE, not the fifth century BCE.[113] From the early fourteenth century on, it belonged to the Dominican convent in Bologna. As Saverio Campanini has shown, it never lacked admirers.[114] The Dominicans showed it to the French humanist François Tissard.[115] He described it as "immaculate," without any sign of damage from vermin – evidence of sanctity, perhaps.[116] Prominent visitors to Bologna, such as Queen Christina of Sweden, regularly examined the scroll.[117]

But accounts of it differed on vital details. Tissard observed that "the points have been added" to its elegant script. Fino Fini, however, claimed that "the points do not belong to the substance of the text. For in antiquity, and in the beginning, the text was not pointed. This is very clear from the Pentateuch of Moses, restored by Ezra, which is in the city of Bologna, written by Ezra himself." He also noted that Torah scrolls still lacked vowels in his own day, and that vowels were a late invention.[118]

113. See *The Ancient Sefer Torah of Bologna*, ed. Maurizio Perani (Leiden, 2019). The MS, once rotolo 2 of the University of Bologna Library, has now been restored to the Synagogue of Biella.

114. Saverio Campanini, "The 'Ezra Scroll' of Bologna: Vicissitudes of an Archetype between Memory and Oblivion," in *The Ancient Sefer Torah of Bologna*, ed. Perani, 29–50.

115. François Tissard, *Grammatica Hebraica* (Paris, 1508), sigs. hv–h2r, at hv: "Ea enim Bononiae in sacrario aedis divi Dominici tam honorifice ferme ac preciosum eiusdem Dominici caput reconditur. Quod quidem opus saepenumero vidi. Nam mihi nonnunquam committebatur, neque id tamen cum parua solemnitate."

116. Ibid.: "scripturam tam elegantem tam insignem ac immaculatam laudabam nulla nec tinea nec labe viciatam."

117. Campanini, "The 'Ezra Scroll,'" 38, according to Galeazzo Gualdo Priotato, "Sua Maestà ne lesse alcune parole" (quoted at 38, n13).

118. Fino Fini, *In Iudaeos flagellum ex sacris scripturis excerptum* (Venice, 1538), fol. 392r: "Primo, quia puncta non sunt de substantia literae: cum ab antiquo & in principio non punctaretur scriptura: quod manifeste patet ex Pentateucho Moysis restaurato per Esdram: quod est in civitate Bononiae, manu ipsius Esdrae, in monasterio fratrum de observantia sancti Dominici: & etiam ex Rotulis Hebraeorum, qui in Synagogis Hebraeorum leguntur: qui sunt similiter sine punctis: sed per magnum tempus postea ad facilius legendum adinventa fuerunt." The English Hebrew scholar Robert Wakefield dismissed with contempt the supposed evidence that the scroll was genuine – though he also argued, from other evidence, that the Jews had preserved, not corrupted, the text of the Hebrew Bible. See his *Syntagma de Hebraeorum codicum incorruptione* ([London, 1534]), sigs. Hiiv–Hiiir. Of

Sixtus of Siena described the scroll with precision, and without evident skepticism, in his *Bibliotheca sancta* of 1566: "The Pentateuch of Ezra, written without any vowel points in a unique oblong parchment in, as the Jews believe, his own hand, is preserved with the deepest veneration in our convent of the Preachers at Bologna."[119] Very likely Buxtorf had not seen the works of Tissard, Fini, or Robert Wakefield, who made fun of the scroll. But he respected Sixtus.[120] It seems odd, and striking, that Buxtorf did not discuss the manuscript. Did he dismiss it? Or did he suppress a piece of evidence that seemingly weighed against his own history of the Masorah? The evidence does not yield a verdict.

Whatever the merits of Buxtorf's thesis, one cannot help but be impressed by his familiarity with Talmudic learning. This same expertise was on show when he undertook the most novel part of his study, in which he raised a truly relevant question: "Why Tiberias?" As Yosef Hayyim Yerushalmi has taught us, Jews in the premodern era did not cultivate historical research in a systematic way.[121] But from what they recorded Buxtorf managed to scrape together a robust, if polemical, history of Jewish learning. Buxtorf the grammarian and lexicographer transformed himself here into a historian of Jewish antiquity, gathering his evidence with antiquarian zeal. And this pursuit of the evidence is on show in Buxtorf's own books. On the hunt for references to the city of Tiberias, he underlined the relevant word; trying to reconstruct the history of the Jewish academies, he noted down all references to the schools of Palestine and Babylon. Drawing on widely varied sources – they included Josephus, the Talmud, the twelfth-century travelogue of Benjamin of Tudela, and the late fifteenth-century ethical writer Judah Kalats – Buxtorf established certain facts. Tiberias was located in a felicitous position and endowed with famous hot springs. Together with cities like Sepphoris and Caesarea, it played an important role in Jewish life in Palestine after the destruction of the Temple. Rabbi, the editor of the Mishnah, the Jewish *corpus iuris*, and friend of Emperor Antoninus, lived either in Tiberias or Sepphoris. Culling information from a motley crowd of medieval and Renaissance

---

the ascription of the MS to Ezra he writes: "huic enim rei fidem meam non obstringo sed fabulae autoribus relego."

119. Sixtus of Siena, *Bibliotheca sancta* (Cologne, 1586), 60: "ESDRAE Pentateuchus, unica & oblonga membrana ipsius, ut Hebraei credunt, Esdrae manu conscriptus absque ullis vocalibus punctis, servatur summa veneratione in coenobio nostro praedicatorum Bononiae."

120. For a citation by Buxtorf, see "Ex Zohar impresso Cremonae, in fol. 1558," 3: "De punctuatione per accentus, mirum nihil notasse Sixtum Senensem."

121. Yosef Hayyim Yerushalmi, *Zakhor: Jewish History and Jewish Memory* (Seattle and London, 1996); see the critiques by Robert Bonfil, "How Golden Was the Age of the Renaissance in Jewish Historiography?," *History and Theory* 27.4 (1988): 78–102, and Moshe Idel, "Yosef H. Yerushalmi's 'Zakhor': Some Observations," *Jewish Quarterly Review* 97 (2007): 491–501.

chronologers such as Abraham ibn Daud, David Gans, Gedaliah ibn Yahya, and Abraham Zacuto – men whose works were much perused by Christian scholars, were often cited and translated into Latin, and, as we have seen, were favorites of Buxtorf's – he traced the transfer of learning from Palestine to Babylonia and eventually to Spain and beyond. Erudition in Tiberias itself, however, gradually faded. Once the Palestinian Talmud had been produced, possibly in Tiberias, Palestine's academies withered. Scholarship and leadership moved to Babylon with its academies of Nehardea, Sura, and Pumpedita.

Historical fact, according to Buxtorf, flew in the face of Elijah's arguments that Tiberian Masoretes had invented the written Masorah – the view that, as we have seen, Buxtorf himself still espoused in 1612. The truth lay elsewhere – in the much earlier age of Ezra, the "antistes" of the Jews, and his council, the so-called Men of the Great Synagogue.[122] That was the story he told in the *Tiberias*.[123] As he wrote to Louis Cappel in 1623, "All I wanted to do in the *Tiberias* was to investigate those Tiberian Masoretes of Elijah, and whether the ones who wrote the Masorah over the vowels were really the same as the inventors of the points. In my view the answer emerges clearly from it."[124] Yet he also admitted, in the same letter, that a full history of the vowels and accents – as opposed to an intellectual history of Tiberias – could not be drawn from the known sources: "we suffer from a lack of Hebrew history."[125]

When Buxtorf laid out seven pages of "rabbinic" evidence about the antiquity of the vowels, accordingly, he had little of substance to add to the materials that he and others had long since identified and discussed.[126] Out came the crucial passages from the Zohar, *Tikkunei ha-Zohar*, Zohar on Song of Songs, and several passages from the Bahir cited by the kabbalist of Saragossa, Bahya ben Asher. "The letters are like the body; the soul is like the point, for the vowel points move the letters just as the soul moves the body, as our rabbis of blessed memory expound in the book of Bahir."[127] As we have seen, Buxtorf cherished

122. This is the term used in rabbinic literature for the council of sages who worked together with Ezra in establishing the text of Scripture and to whom are attributed a number of fixed laws and rituals.

123. Buxtorf, *Tiberias*, 102–31.

124. Buxtorf to Louis Cappel, 1 January 1623, fol. 247r (copy): "Ego in Tiberiade nihil aliud volui agere quam Masorethas illos Tiberienses Elianos indagare, et num qui Masoram super vocales scripserunt iidem fuerint et Vocalium inventores. Quod ut puto non obscure inde patet."

125. Ibid., fol. 247v: "Quid multis? laboramus defectu historiae Hebraicae unde veritatem inde demonstrare."

126. Buxtorf, *Tiberias*, 74–80. See also John Prideaux, *Viginti-duae lectiones* (Oxford, 1648), lect. 12: "De punctorum hebraicorum origine," 180–98.

127. Bahya ben Asher, *Be' ur al ha-Torah* (Venice, 1546), fol. 26v, col. 2.

no friendly feelings toward this medieval exegete and kabbalist. Rather, he held Bahya to account for churning out anti-Christian sentiments on so many pages of his very popular writings. At first sight it seems that Buxtorf totally swept aside his antagonistic assessments of Bahya, and indeed of so many other Jewish authors, when he turned his mind to the crucial question of the integrity of the Hebrew Bible. A somewhat different Buxtorf appears, one whose sole aim was to gather the crucial sources for the sake of his quest to retain that integrity – even when doing so meant consulting and citing the very same Bahya whom he had readily denounced in other contexts.

Bahya, like his contemporary Ibn Caspi, interpreted the vowels and accents as crucial keys to the understanding of the biblical narrative.[128] Indeed, he espoused the view that the cantillation marks hailed back to Mosaic times. On six occasions Buxtorf saw fit to cite "Behai"; Bahya supplied the requisite testimony from the Zohar and the Bahir.[129] Essential to the argument was the rabbinic view that the verse in Nehemiah, "They read from the Book of the Law of God, making it clear and giving the meaning so that the people understood what was being read," referred to the Masoretic punctuation.[130] Bahya was clearly a useful witness for the authenticity of the Masorah.

Yet even in this context Buxtorf could not help referring to a passage in Bahya's *Kad ha-kemah* that he had cited on more than one occasion.[131] He reiterated his report on the origin of the suspended letter *ayin* in the word *ya'ar* (forest) in Psalm 80:14, which appears in the phrase "The boar from the forest (*ya'ar*) ravages it." He referred to Rashi and other commentators. As in the past, his main purpose was to emphasize Bahya's identification of the boar with Rome, and therefore also Christian Rome, when he stated that "the *ayin* is suspended because they worship the one who was hung."[132] Even as Buxtorf wove a historical and philologically

128. See Benjamin Williams, "'Some Fanciful Midrash Explanation': *Derash* on the *Ṭe'amim* in the Middle Ages and Early Modern Period," in *Studies in the Masoretic Tradition of the Hebrew Bible*, ed. Daniel Crowther, Daniel D. Hornkohl, and Geoffrey Khan (Cambridge, 2022), 329–76, at 338–43; and on mystical interpretation of the cantillation marks, see Elliot Wolfson, "Biblical Accentuation in a Mystical Key: Kabbalistic Interpretation of the Te'amim," *Journal of Jewish Music and Liturgy* 11 (1988–89): 1–16; 12 (1989–90): 1–13.

129. Buxtorf, *Tiberias*, 75, 79.

130. Ibid., 80–81.

131. See above, Chapter 1, n26, and Buxtorf, *Epitome radicum Hebraicarum et Chaldaicarum* (Basel, 1607), 369–70.

132. Buxtorf, *Tiberias*, 172: "Rabbi Bechai in libro כד הקמח, cap. 1, quod incipit גאולה, scribit: והעין התלויה כי כן הם עובדי התלוי *Litera Ain suspensa est, nam sic sunt cultores suspensi illius*, id est, Christi, quem sic solent malitiose nominare, quasi dicat litera Ain suspensa est, in symbolum, ut per חזיר מיער intelligatur Christus suspensus in cruce, quem sic suspensum Christiani colunt & adorant."

sophisticated overview of the origins, formation, and technical methods of the Masorah, he was unable to forgo his unwavering desire to detect evidence of Jewish anti-Christianity.

At the same time, virtually every page of the *Tiberias* is constructed from a careful reading of a wide range of Jewish sources: especially those that actually discussed the biblical text and its apparatus. He took account of Jewish as well as Christian debates about the vowel points and accents. What distinguished Buxtorf from his predecessors and contemporaries was above all his ability to find the relevant works. As censor he had access both to Jews and to their writings. The commentary on Song of Songs (3:8) by Elijah Loanz, the Jewish kabbalist of Hanau whom he had met in Basel and with whom he corresponded, served as evidence for the inextricable relation of Masorah to the text of Scripture. Even more important for the argument was the work of the Italian rabbi and grammarian Samuel Archivolti, a contemporary of Azariah de' Rossi. In his *Arugat ha-bosem*, Archivolti launched an attack on Levita, following in de' Rossi's footsteps but without citing him.[133] Buxtorf ignored the mystical arguments that Archivolti employed: "God created the vowel points which were transmitted to Moses at Sinai such that they were ready for application to the Torah at the propitious moment and for the sake of the 'generations of this world.'" Instead, he used one of Archivolti's counterarguments against Levita. Levita had argued that a Midrashic mode of playing with the text, "Don't read this, but rather read that," implied that the points did not exist in the Talmudic age. Buxtorf wrote אל תקרא (do not read) at the start of this passage in his copy of the book.[134] In rebuttal, Archivolti employed a Talmudic passage that alluded to the need to recognize and memorize the "signs" embedded in the text of Scripture.[135] Such mnemonic aids – "signposts for the Torah," the redundancies in Scripture – enabled the sages to expound on the traditions that had been transmitted over the ages. According to Buxtorf, these "signa" effectively undermined and erased Levita's position.[136]

133. Samuel Archivolti, *Arugat ha-bosem* (Venice, 1520), fols. 90r–91v. For an English translation of the text, see Aaron Rubin, "Samuel Archivolti and the Hebrew Vowel Point Controversy," *Jewish Quarterly Review* 101 (2011): 233–43.

134. Buxtorf's signed copy of Archivolti is Basel, UB, FA V 14: 1; his note is on fol. 91r.

135. He alludes to BT, Erubin 54b: "The Torah is acquired only by [mnemonic] signs, as it is said, 'Put it in their mouths' [Deut 31:19]; do not read '*sima*' 'put,' but rather 'its Sign' (*simana*) . . . 'set up markers for yourself' [Jer 31:21], that is, 'Make markers for the Torah.'"

136. Buxtorf, *Tiberias*, 85–86: "Alius ergo docendi scopus hujus docendi modi, *Ne legas sic, sed sic*, fuerit necesse est, nempe quod fuerit duntaxat סימן בעלמא לזכירת דבר המקבל *signum vulgare et leve pro memoriali rei a Majoribus acceptae*, ut scribit author libri ערוגת הבשם cap. 26, ubi contra Eliam in hac materia agit. Varia enim antiquorum erant docendi genera, quae איש מפי איש quisque ex ore alterius accipiebat, quorum ut non incideret oblivio tam apud ipsos quam apud posteros, constituterunt sibi סימנים *Signa & Symbola*, quaedam juxta id quod ante

Buxtorf also examined the Masorah itself, in all its forms, with a level of granular attention previously found only in Jewish scholarship. He envisioned the core of the *Tiberias* as a manual and textbook to guide students of the Hebrew Bible into the mysterious world of the Masoretic apparatus. He laid out dozens of definitions of terms, endless lists of passages, and corrections of textual errors: guide rails, in effect, for a Christian who set out to explore the vast labyrinth of the Masorah. Buxtorf knew how bewildering it could be to confront, for example, the decorative opening pages of many manuscripts and editions, and he did his best to provide orientation. As in his rabbinic bibliography, so here too he showed striking interest in the physical form as well as the textual context of a Jewish work of scholarship:

> The Masorah that is set out by the text at the very beginning of the individual books does not always have to do with the text of that leaf. For it is often too long to be concerned with that which also has its own Masorah attached to it. To prevent anyone from getting stuck at this point in an excessively detailed inquiry, we will shed light on this Masorah as well, and clear away the obstacles. The authors of this work decided to make the entrance into each book more majestic by copying its first word in capital letters and wrapping it in remarkable regalia, in the center of the page, as a sort of gate to the book. To enhance this adornment further they placed there a certain more extensive Masorah, with certain lines laid out in square form, as a sort of magnificent doorposts, so that they would inspire the visitor with this splendid decoration to approach in a more attentive and devout way.[137]

He went on to explain that the Masorah to the first word of Genesis, *Bereshit*, "in which, in ancient times, the letter Beth was written as a capital," provided "a whole alphabet of such capital letters, that appear everywhere in the Scriptures."[138] Not even Elijah had gathered the material that a reader would need in such a systematic, painstaking way: perhaps Buxtorf, the Christian, could see puzzles and obstacles that eluded the supremely expert Elijah. Never satisfied with what he had accomplished, Buxtorf was still trying to provide better guidance to the Masorah in the last weeks of his life, seven years later. He dedicated his final days to laying out a newly methodical concordance to the Bible, which would also offer guidance to the Masorah.[139]

---

fuit propositum. *Non comparatur vel possidetur Lex nisi per signa*; aut ציונים *figuras & Notas*, juxta illud הציבי לך צינים *Statue tibi signa*, Jerem. 31.21 [31:20], quod explicant, עשו ציונים לתורה *Facite signa legi*. Haec ergo signa Eliae sententiae nullum colorem affricant."

137. Buxtorf, *Commentarius Masorae*, 304–5.

138. Ibid., 305.

139. See the *praefatio* by Johann Buxtorf the Younger in Buxtorf, *Concordantiae Bibliorum Hebraicae, nova et artificiosa methodo dispositae* (Basel, 1632). He ascribes the book's

Yet Buxtorf continually went beyond this technical remit. Unlike Levita, he did not simply give the rules, or describe the problematic and complex world of spelling, vocalization, and accentuation. Rather, he often accounted for strange phenomena in the biblical text by recourse to theological explanations. Sometimes he drew inspiration from rabbinical texts: for example, Jacob ben Asher's commentary to the Pentateuch (*Ba'al ha-turim*), which appeared in the margins of Rabbinic Bibles, including Buxtorf's. With regard to the capital letter *tav* in Deuteronomy 18:13, "You shall be wholehearted (*tamim*) with the Lord your God," Buxtorf appropriated and enlarged Jacob ben Asher's marginal note: "The capital *tav* indicates that if you go wholeheartedly it is as if you have fulfilled [the Torah] from *alef* to *tav*."[140] The works of other exegetes were also put to the same goal. In explaining the large letter *mem* in the first word of Proverbs – *Mishlei* (Proverbs) – he used the same expression employed by Abraham ibn Ezra. The large *mem* indicated the "excellence of the ancient wisdom" contained in the book of Proverbs, human as well as divine and eternal, through which something was created from nothing, described in this book.[141] Though Buxtorf appropriated ibn Ezra's term "excellence of the ancient wisdom," he omitted the specific details provided by his source.[142] As he taught the Masorah, Buxtorf embedded the gigantic corpus of technical materials into a moralistic and spiritual discourse.

Sometimes, the terms of this discourse came not from Jewish tradition but from Buxtorf's Christian mother wit. Simple descriptions of the phenomena of the Masoretic text were accompanied by short exegetical notes. Thus, for example, in accounting for the phenomenon of the inverted *nun* before the verse Numbers 11:1, "And when the people complained (*mitonnenim*)," Buxtorf suggested that it was a symbol of the perversity of the Jewish people who were perpetually ungrateful for God's kindness.[143]

When de' Rossi put aside his critical scholarship to defend the antiquity of the vowel points, his purpose was not really to undermine Levita but rather to

methodology to his father. Perhaps it reflected Buxtorf's studies in Herborn, where the works of Ramus were taught and debated, or more recent projects. An unpublished letter reveals his interest in new ways of language teaching as seen in William Bathe's *Janua linguarum*, "dessen Propositum mir wohlgefallet." Buxtorf to Wolfgang Ratichius, June 13, 1617. Warm thanks to Regine Buxtorf for this text from her library.

140. Buxtorf, *Tiberias*, 161–62.

141. Ibid., 158: "non tam ob principium libri, quam ad considerandum מעלת החכמה הקדומה excellentiam Sapientiae priscae cum humanae tum divinae et aeternae, per quam ex nihilo creatum est aliquid, quae hoc libro describitur."

142. Ibn Ezra ad loc: "This is the section of admonition in which he demonstrates the excellence of ancient wisdom through which something was created from nothing; he warns that one should cling to it [i.e. wisdom] and avoid the irate woman, and he demonstrates that one should destroy folly and earthly desires and restore perfect souls."

143. Buxtorf, *Tiberias*, 169.

defend the integrity of the Hebrew text. This emerges in chapter 8 of *The Light of the Eyes,* where he set out to explain the reason for the discrepancies between the Septuagint and Hebrew Bible, "because there are some people who accuse us of sacrilege by making additions and subtractions to our texts."[144]

Mutatis mutandis, Buxtorf suffered from the same kind of scruples. In his 1623 letter to Louis Cappel, he admitted that the case for the antiquity of the vowel points and accents was not really foolproof. He confessed, however, that he would not proclaim this from the pulpit. The consequences of such a pronouncement were fraught with risk. As Stephen Burnett argued several decades ago, in this respect Buxtorf, like so many other colleagues of his time, was guided by confessional considerations. For all the depth of his Judaic scholarship, in his framing of the authority of the Hebrew text of the Bible Buxtorf followed the teachings of his own tradition.[145]

In crucial respects, though, the *Tiberias* marked both a decisive change in Buxtorf's practices as a scholar and an extraordinary achievement. As we have seen, the younger Buxtorf remarked that his father, in his last years, became increasingly conscious that the Hebrew language and Hebrew writing had evolved over time. This same awareness inspired Buxtorf to work his way through the Talmud and other sources in order to compose what he described as the missing "history of the Masoretes." Though he could not supply that, he did draw up a study of the development and decline of Jewish learning in Palestine. True, Buxtorf came to the wrong conclusions. If he had thought harder about the fact that the version of the Talmud he used was composed in Babylon, he might have thought twice about its evidence that the schools of Palestine declined – as well as about the extensive further evidence that attested to the intensity and quality of Tiberian scholarship. But "the journey, not the arrival, matters": Buxtorf's history of the Masorah offered an innovative model for many later histories of Jewish institutions and creations.

More influential, certainly, was his second achievement. With the *Tiberias* he provided something that Christian scholars had previously lacked: a key that unlocked all the mechanisms and mysteries of the Masorah. Any Christian scholar who worked his way through Buxtorf's exacting study of Masoretic terms and technique could then find his way through what had been the intricate highways and byways of the Rabbinic Bible. He could go beyond the text and its flanking commentaries to see how it had been preserved by Jewish scholarship and read in Jewish synagogues. Buxtorf disliked Jews as much, probably, as any Christian expert before or after him. Still, he had the expertise and found the patience,

144. De' Rossi, *Light of the Eyes,* 178.

145. Burnett, *From Christian Hebraism to Jewish Studies,* chap. 7.

Reformed Christian though he was, to make a great creation of Jewish scholars and scribes the object of a Benedictine labor.

In one respect at least, the reception of the *Tiberias* took an ironic turn. Buxtorf, as we have seen, described his book as a "history" of the Masoretes and their work. Half a century later, Baruch Spinoza claimed that he had created "histories" of the books of the Bible. After Spinoza, Richard Simon published what he called "critical histories" of the texts of the Hebrew Bible and the New Testament. It seems all but certain that they derived the term "history" from Buxtorf.[146] And though both men applied it formally to texts rather than people, their histories, like his, were accounts of what human industry had done to sacred texts. Though Buxtorf would have viewed their projects with horror, their cases show that his book continued to stimulate and even inspire readers for decades after it appeared.

146. See Anthony Grafton, *Inky Fingers: The Making of Books in Early Modern Europe* (Cambridge, MA, 2020), 244–46.

Conclusion

Buxtorf's life was not without incident, but it lacked the sort of arc that gladdens the hearts of biographers. His personal relationships with Jews – though vital to his sense of what made his scholarship distinctive – do not show any pattern of change over time. One of the few Jews whom Buxtorf mentioned by name was Elijah Loanz, a remarkable rabbi and kabbalist, known also as a miracle worker (*ba'al shem*).[1] Born in Frankfurt am Main in 1555, Loanz, like so many of his fellow Jews, led a peripatetic life. He had studied with some of the rabbinic luminaries of the time, including the Maharal of Prague, and served as rabbi in Fulda, Hanau, Friedberg, and Worms, where he died in 1636. For a short time he lived in Basel. There he encountered the English Hebraist Hugh Broughton, who attempted to engage him in religious debate.[2] What we know of Loanz's activities in Basel, notably the way in which he presumably earned his living, is derived almost exclusively from the various books that he published at the press of Conrad Waldkirch. His name, "Elia or Elijah son of my father, my teacher Moses Loanz," embellishes all these publications. In 1599, he brought out the *Zemirot ve-tushbehot* (Songs and praises) of his teacher Akiba Frankfurter, together with three poems of his own and a bilingual Hebrew/Yiddish dispute between wine and water, set to the melody of Dietrich von Bern.[3] Similarly entrepreneurial were Loanz's publications of the halakhic writings of Isaac ben Meir Dueren and Solomon Luria.[4]

1. It is not clear whether his name should be transcribed as Loanz or Luanz.

2. On Hugh Broughton and Loanz see Kirsten Macfarlane, *Biblical Scholarship in an Age of Controversy: The Polemical World of Hugh Broughton 1549–1612* (Oxford, 2021), 115–18, 125–28. Broughton first met Loanz in Frankfurt. He claimed that his persuasive discussion of biblical and rabbinic texts had nearly led Loanz to convert. As Macfarlane states (126), this is highly unlikely.

3. On poems of this kind, see Chava Turniansky, "The Evolution of the Poetical Contest in Ashkenaz," in *Studies in Yiddish Literature and Folklore* (Jerusalem, 1986), 60–98, esp. 74–75. Our thanks to Jean Baumgarten for this reference. In the same year, Loanz also published Frankfurter's *Tehinot be-khol yom* (Daily Supplications), which he structured and to which he added the *Viddui* (Confession) of R. Judah the pious (leader of the pietists of Ashkenaz), on fols. 15a–16b. See also Prijs, *Die Basler Hebräischen Drucke, 1492–1866*, ed. Bernhard Prijs (Olten, 1964), 263–66.

4. Isaac ben Meir, *Sha'arei dura* (Basel, 1599). Loanz printed the work together with the commentary of Solomon Luria, *Ateret Shlomo* (Basel, 1599), and Luria's *Ammudei Shlomo* (Basel, 1600), a commentary on Moses of Coucy's Semag. On Loanz's active editing of these works, see Prijs, *Die Basler Hebräischen Drucke*, 267–71.

Buxtorf referred to Loanz three times in his various publications. In his *Institutio epistolaris* he published the heading of a letter that Loanz sent to him. In good conventional form, his Jewish correspondent R. Elias heaped fulsome praise on Buxtorf, who noted Loanz's high reputation among Jews for his publications, and remarked that he "nowadays serves as rabbi of the community of Hanau."[5] On two occasions Buxtorf referred to *Rinat dodim*, a commentary on the Song of Songs that Loanz both composed and printed in Waldkirch's press in 1600.[6] Buxtorf listed the work in the *Bibliotheca Rabbinica*.[7] Again he referred to Loanz's position as rabbi of the community of Hanau, but he also proffered an additional bit of information: Elias had recently passed through Basel and had told Buxtorf that he was thinking of publishing his explanation of the difficult passages in Bahya's works.[8] As we have shown at length, Buxtorf was determined to prove the authority and authenticity of the Masorah. In the *Tiberias*, Buxtorf discussed the rabbinic saying "Masorah is the fence round the Torah." A verse from Song of Songs (3:8) provided scriptural support for this significant principle: "'Every man has his sword upon his thigh, because of fear at night.' This refers to the Masorah and the signs designed to preserve the Torah from being forgotten in exile."[9] Not content with citing this interpretation of the verse, which Levita provided, Buxtorf thought it necessary to provide additional confirmation from Loanz's *Rinat dodim*. There, he had interpreted the verse as a reference to the "war of Torah," where the "sword on the thigh" corresponded to the Masorah and its signs which would ensure that during times of trouble ("fear at night"), the correct reading of the Scriptural text would not be forgotten.[10]

5. Johann Buxtorf, *Institutio epistolaris hebraica* (Basel, 1610), 27–28: "Hoc exordio quidam R. Elias libris aliquot editis inter Judaeos clarus et hodie Rabbinum agens Synagogae Hanovianae ad me usus fuit."

6. At the end of the introduction, Loans refers to a certain Menachem ben Jacob who had read and approved his work. It may be that he is to be identified with the Menachem with whom Buxtorf had been in contact. See above, p. 38.

7. Buxtorf, *Bibliotheca Rabbinica*, in his *De abbreviaturis hebraicis liber novus & copiosus* (Basel, 1613), 257–335, at 324, s.v. "*Rinnas Dodim*." Unusually for Buxtorf, he makes a mistake, referring to Loanz as Elia ben Lama. Clearly he did not understand the abbreviation למא ("of my teacher, my father"). The *Rinat dodim* was reprinted with extracts from Loanz's unpublished manuscripts: *Rinat dodim*, ed. Moses Hillel ben David Zeri (Betar Ilit, 2002).

8. Ibid.: "Hic Basileam nuper transiens, retulit mihi, se conscripsisse Perusch in difficiliora loca R. Bechai, quem propediem praelo dare cogitet." Apart from the introduction, published in *Rinat Dodim*, ed. Zeri, 170–75, the work is still in manuscript in Oxford, Bodleian Libraries, MS Opp. 216. See Chapter 1 above.

9. Buxtorf, *Tiberias, sive commentarius Masorethicus* (Basel, 1620), 206. Buxtorf is citing Elijah Levita who is citing Rashi ad loc.

10. Elijah Loanz, *Rinat dodim* (Basel, 1600), fol. 14a.

Buxtorf was clearly in occasional communication with Elijah Loanz. He was aware of the Jew's status in his community and had carefully read his commentary on Song of Songs or had at least sought out the rabbi's interpretation of a significant verse. What kind of relationship, then, existed between Buxtorf and Loanz? As we have seen, for the most part, Jews did not get good press from Buxtorf. Usually they were not designated by name. In the rhymed prose on the title page of his commentary on Song of Songs, Loanz struck a personal note. Singing the praises of his native town, Frankfurt am Main, and its distinguished Jewish scholars, he disclosed some intimate details about his life. Driven out from Frankfurt by capricious time, "I came to a place where very few members of our people live. I reside in the shadow of fear of the grave." The phrase "in the shadow of fear of the grave" is a serious play on the name of the biblical figure Zelophahad ben Hefer (Numbers 27:1–11). The first part of the name *(Ba)-tsel* (in the shadow) connotes Basel, while "fear" (*pahad*) completes the name. Loanz's sense of alienation and insecurity living in Basel, the city without Jews, is "fearlessly" imprinted here. But it was in Basel that he was able to print Hebrew books, including one of his own, and to plan for future production. It would appear that Loanz's stay in Basel was pragmatic; indeed, not long after being there he was called to the rabbinate in Hanau. Though later he passed through Basel, he did not remain and was not allowed to settle there.

Loanz did not express any opinion of non-Jews or of Buxtorf in his writings, apart from penning the usual complaint about the gentile compositors who made mistakes when they typeset the Hebrew text on the Sabbath.[11] And yet, living as a Jew in Basel was evidently not easy. Loanz's experience was probably no different from that of Braunschweig and Gumplin. They interacted with Buxtorf on business and he with them. Buxtorf worked with Jews on an equally pragmatic footing. His attitude toward Jews as a community did not necessarily impinge – though it sometimes did – on his behavior toward individual Jews, especially those who could offer him scholarly advice and guidance. Yet he had nothing to fear. Loanz, on the other hand, was afraid, and such was the strength of his feelings that he expressed them on the title page of a book printed in Basel.

In most respects, the life of Buxtorf's mind was as consistent as his relations with Jews. This book began with a digression: anecdotes that Buxtorf told his students about two Jews and their command of the text of the Hebrew Bible. More than twenty years later, as a busy professor, author, corrector, and censor, Buxtorf

11. See Loanz's edition of Isaac ben Meir, *Sha'arei dura*, with Solomon Luria's *Ateret Shlomo*, fol. 3v. In his introduction to Bahya, Loanz referred to those "'not of our people' [i.e., Christian kabbalists] who write about the names and attributes of God." He also stated: "Our souls are endowed such that we can apprehend the wisdom of Kabbalah better than any other people" (Oxford, Bodleian Libraries, MS Opp. 216, fol. 1r).

still found the time for digressions. He inserted a substantial one into a letter to his close friend Casper Waser in Zurich, in which he confessed that he could not answer some vital questions about both the Hebrew Bible and the Gospels:

> The Gospel history shows that Christ, with his Apostles and Evangelists, spoke with the Jews in their native language, as the Syriac expressions throughout show. Why then was the Gospel history written in Greek? Why are the passages from the Old Testament cited from it [the Septuagint], even corrupt ones, and not rather according to the original Hebrew truth? Was the Greek version also used and understood then in the synagogues of the Jews? The passage where I read that does not come to mind; I remember I read it. Or did they mean to show that they had left the Jews and crossed over to the Gentiles, to whom the Greek language was then known because of its excellence and widespread use? You know that the Papists want to gain authority for the Greek version from the fact that Christ and the Apostles used it to confirm their teaching. From that they are carried along to holding that the Hebrew should be corrected from the Greek.[12]

The passage touched on topics and questions that had long mattered to Buxtorf – the Aramaic words and phrases in the New Testament, the authority of the Hebrew and Greek Old Testament texts, and the perfidy of Catholics. But it also raised a newer set of questions – questions not only about the language that Jesus and his first followers spoke, but also about their religious identity. Buxtorf described the apostles as speaking "with the Jews," a phrase that suggests that they were no longer Jews themselves. He wondered if they had already transformed their mission into a universal one. It seems that unlike such contemporaries as Scaliger and Selden but like many New Testament scholars after him, he believed that early Christianity took little from Judaism except the form of Greek in which the New Testament was written.[13] Here too, the stability of Buxtorf's views is striking.

12. Buxtorf to Waser, 15 April 1618, Zurich, ZB, MS A 52, 57: "Porro ex evangelica historia constat Christum cum Apostolis et Evangelistis vernacula lingua cum judaeis fuisse locutos, ut docent passim Syriacae locutiones. Quare ergo evangelica historia Graece scripta? quare inde loca Veteris Testamenti, etiam prava, citantur, et non potius secundum veritatem originalem Hebraicam? An etiam Graeca versio in Synagogis judaeorum tunc usurpata et intellecta? non succurrit, ubi id legerim, legisse memini. An vero indicare voluerunt, se relictis judaeis transitum facere ad gentes, quibus Graeca lingua tunc ob excellentiam et latum usum nota? Nosti Papistas velle Graecae versioni auctoritatem conciliare ex eo, quod Christus et Apostoli eam ad suae doctrinae confirmationem usurparint. Inde et eo feruntur, ex Graeca Hebraeam esse corrigendam."

13. For these contrasting positions, see, on the one hand, Jason Rosenblatt, *Renaissance England's Chief Rabbi: John Selden* (New York, 2006), and Gerald Toomer, *John Selden: A*

As Buxtorf admitted, he could not answer these questions, or even remember offhand all of the relevant texts he had gone through. When he tried to recall what he had read about the use of Greek in the synagogue, he may have remembered, a little indistinctly, a passage in which Angelo Canini, a scholar whom he admired, used the evidence of the Jerusalem Talmud to show that Jews in Palestine had said the Shema in Greek.[14] Later in that Talmudic passage, rabbis discuss the acceptability of reading the scroll of Esther on Purim "in barbarian," i.e., in Greek. Perhaps he had that passage in mind. Or he might have heard about Joseph Scaliger's argument that "Hellenistic" Jews had read the Bible in Greek in their synagogues – an argument that found only partial expression in print before Scaliger's death. What matters more, as we take leave of Buxtorf, is his closing comment to Waser: "But my immediate tasks are snatching these meditations away from me. Still, this sort of discussion refreshes me somewhat in the midst of them."[15]

Gibbon notoriously made fun of Augustine: "His learning is too often borrowed, and his arguments are too often his own."[16] Buxtorf's learning was his own, acquired by long years of exhausting work at the coal face, but his arguments were often shared with a great many of his co-religionists. This late letter reveals an assumption that Judaism and Christianity were separated, from very early times, by a chasm, and a suspicion that the arguments of those of other confessions – especially Catholics – were shaped by confessional motivations rather than historical evidence. Both are very familiar.

True, Buxtorf's scholarly ways changed to some extent, at least in private. As a senior scholar writing to a friend and colleague, Waser, he admitted – as he did in his 1623 letter to Louis Cappel – that his knowledge might be incomplete and that his understanding of key passages might be inadequate. Two decades before,

---

Life in Scholarship (Oxford, 2009); and, on the other hand, Jonathan Smith, *Drudgery Divine: On the Comparison of Early Christianities and the Religions of Late Antiquity* (Chicago, 1990).

14. Angelo Canini, *Loci aliquot Novi Testamenti cum Hebraeorum originibus collati atque historice explicati,* in his *Institutiones linguae Syriacae, Assyriacae atque Thalmudicae* (Paris, 1554), 12–13, quoting Jerusalem Talmud 7.1.21b. On Canini, see Joanna Weinberg, "A Sixteenth-Century Hebraic Approach to the New Testament," in *History of Scholarship: A Selection of Papers from the Seminar on the History of Scholarship Held Annually at the Warburg Institute,* ed. Christopher Ligota and Jean-Louis Quantin (Oxford, 2006), 231–50, and Anthony Grafton and Joanna Weinberg, with Alastair Hamilton, *"I have always loved the Holy Tongue": Isaac Casaubon, the Jews and a Forgotten Chapter in Renaissance Scholarship* (Cambridge, MA, 2011), 283–87.

15. Buxtorf to Waser, 15 April 1618, fol. 57r: "Sed haec meditatio mihi abripitur per operas praesentes. Tamen colloquium hujusmodi me inter eas nonnihil recreat."

16. Edward Gibbon, *The History of the Decline and Fall of the Roman Empire,* ed. David Womersley (London; New York, 1994), 2: 93 n79.

he assumed that Jews were Jews, and Hebrew was Hebrew. He felt comfortable drawing examples of a single stylistic point from Jewish texts that were very different in kind and very distant in time. Now he realized, and explained, that Jews had a history and that their language and religion had changed over time. The styles of writers, the practices of synagogues, and the scholarship of the Masoretes all had histories, even if too little evidence survived for him to reconstruct them in detail.

Buxtorf's assumptions never changed in any fundamental way, any more than his relations with Jews. The motivations of his scholarship were ideological from start to finish. From the 1580s to the 1620s, he steeped himself in a vast world of Jewish learning in order to serve Christian ends. He set out to locate, identify, and when possible censor anti-Christian passages in Jewish texts; to describe Jewish practices, from worship to everyday life, in accurate if selective detail, but also to reveal that superstition and sterility permeated every area of Jewish life; and to make clear that Protestant – above all, a Reformed – theology rested on an accurate vision of the history and condition of the biblical text. To attain these goals, he devoted a lifetime to hunting, gathering, and collecting information of every kind about the Jews, their languages, their books, and their way of life. Through this massive effort he hoped not to frame new hypotheses or test old ones, but to verify the assumptions and beliefs from which he started. Casaubon summed up Buxtorf's career with precision and insight in two notes on the title page of his copy of the *Jüden schul.* In one of them he pointed out that Buxtorf had conversed with rabbis and even taken part in their feasts. In the other he wrote that "Buxtorf promises a free-standing book '*On the curses of the Jews against Christians* in their prayers.'"[17]

Yet scholarship, when practiced with great energy and high technique, has a power of its own. One theme that we have traced, from Buxtorf's early notebook through the *Tiberias,* is compilation. Faced with the need to describe a world, compose a textbook, or edit the Hebrew Bible, he began by making copybooks: systematic collections of the evidence. This was a standard practice – perhaps *the* standard practice – of sixteenth-century humanists, as Ann Blair and many others have taught us. Buxtorf did not always employ it in distinctive ways. Sometimes, as we have seen, his assumptions limited his efforts. Like many contemporary experts on ecclesiastical and Jewish matters, he concentrated on collecting only evidence that supported his arguments. But he was anything but ordinary. A great

17. British Library 848.b.19, title page: "auctor consuetudinem habuit cum hodiernis Rabb. ut etiam convivijs eorum interesset"; "Librum singularem promittit *De maledictionibus Judaeorum contra Christianos* in ipsorum precibus." In the second note Casaubon refers to a remark at the end of chapter 11 of the *Jüden schul* (*Synagoga Iudaica: das ist Jüden schul* [Basel, 1603]), 392.

scholar, a master of Jewish languages, and an eager explorer of Jewish writing, he cast his net very widely. Sometimes his excerpts and his memory made him the first to see connections between texts. Occasionally he assembled materials that challenged his own core beliefs.

In some cases, Buxtorf learned to work with his evidence in new ways. He grasped something of the vast variety of Jewish culture, over time and across space. He depicted a Jewish world that was not hermetically sealed against outside languages and practices. He constructed an original history of Jewish learning, based on Jewish sources. Above all, though, he ended as he began: as the immensely energetic creator of vast bodies of useful materials for the immensely varied readers, Jews as well as Christians, who would exploit and contest his works over the next two centuries and more. He left them provocative theses that Jews and Christians alike found good to think with, even when they disagreed completely. And sometimes he let a friend or correspondent see that he knew he could not solve all of the problems with which his masses of evidence confronted him. His legacy was puzzling, and sometimes frustrating. But it was also immensely productive.

APPENDIX 1

Buxtorf's Conversations with Informants

Buxtorf recorded some of his conversations and contacts with informants in his copybook, Basel, UB, MS A XII 20. They are presented here in chronological order. He described another conversation of this kind in some detail in his study of the abbreviations in rabbinical texts, a propos of י״ש ("may his name be eradicated"), which he characterized as a curse spoken by Jews against Christians.

> A few years ago, when I discussed this matter with a certain Jew, he answered, "In fact, it is not only explained in this way, but also as 'Jesus is a lie and an abomination.'" Who would not shudder at this with all his heart? And yet these things are spoken among the circumcised, and reported, and no Christian pays attention to it. That Jew still lives, and I know where, and he is quite well known at Frankfurt. When he revealed that frightful saying to me, he was wavering in the Jewish faith and not at all hostile to the Christian faith. He also took counsel about it a number of times with me and with Dr. Amandus Polanus of blessed memory.[1]

Some of the conversations that Buxtorf reports on in the following notes may also have taken place while he and his informant discussed the possibility of the latter's conversion.

1. Buxtorf, *De abbreviaturis hebraicis* (Basel, 1613), 88: "Mihi ante paucos annos hac de re cum quodam judaeo conferenti, immo, respondit, non sic tantum explicatur, sed & י׳שו ש׳קר ו׳תועבה. Quis ad haec toto pectore non exhorreat? & tamen haec dicuntur inter recutitos istos, & feruntur, nemine Christianorum ad id advertente. Vivit iste Judaeus adhuc, & scio ubi, estque Francofurti satis notus. Cum autem mihi horrendam istam vocem proderet, in fide Judaica vacillabat, minime a fide Christiana alienus, de qua etiam mecum, ut & cum D.D. Amando Polano b.m. aliquoties contulit."

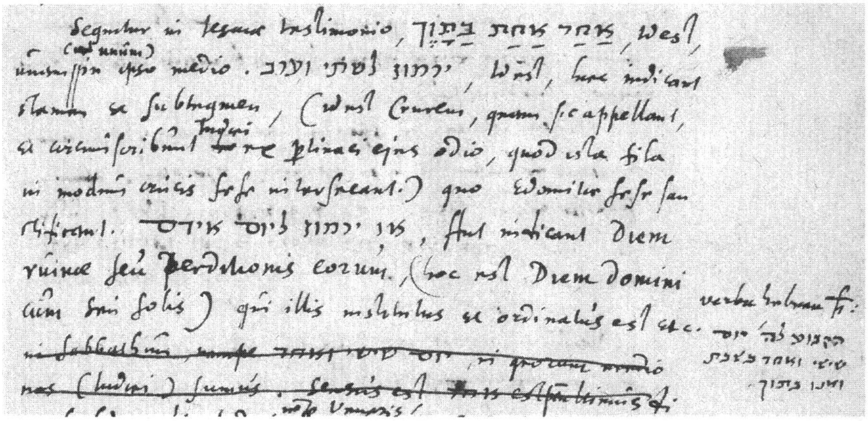

Figure A1.1a. Basel, Universitätsbibliothek, MS A XII 20, 29.

Sequitur in Iesaiae testimonio אַחַד אַחַת בַּתָּוֶך id est unam (vel unum) in ipso medio ירמוז לשתי וערב id est, haec indicant stamen et subtegmen, (id est Crucem, quem sic appellant et circumscribunt Iudaei ex pertinaci ejus odio, quod ista fila in modum crucis se intersecant) quo Edomitae sese sanctificant או ירמוז ליום אידם. Aut indicant diem ruinae seu perditionis eorum (hoc est Diem domini cum suis solis) qui illis institutus et ordinatus est etc. Verba hebraea sunt: הקבוע לה׳ יום שישי ואחד בשבת ואינו בתוך.

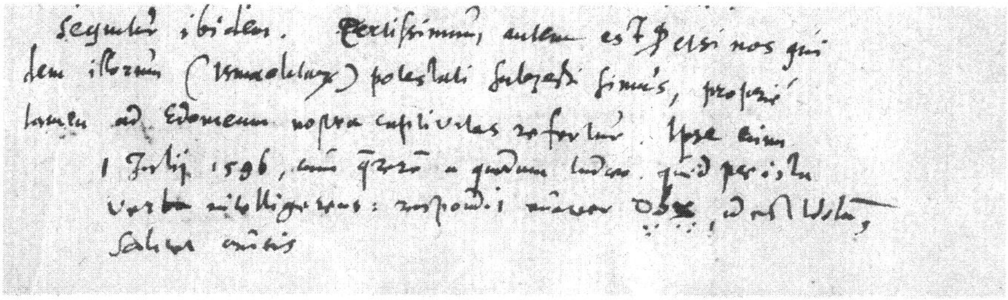

Figure A1.1b. Basel, Universitätsbibliothek, MS A XII 20, 29.

Sequitur ibidem. Rectissimum autem est quod etsi nos quidem illorum (Iesuaolatorum) potestati subjecti simus, proprie tamen ad Edomeum nostra captivitas refertur. Ipse enim 1 Julii 1596, cum quaererem a quodam Iudaeo quid per ista verba intelligerent: respondit unum צֶלֶם id est idolum, scilicet crucis.

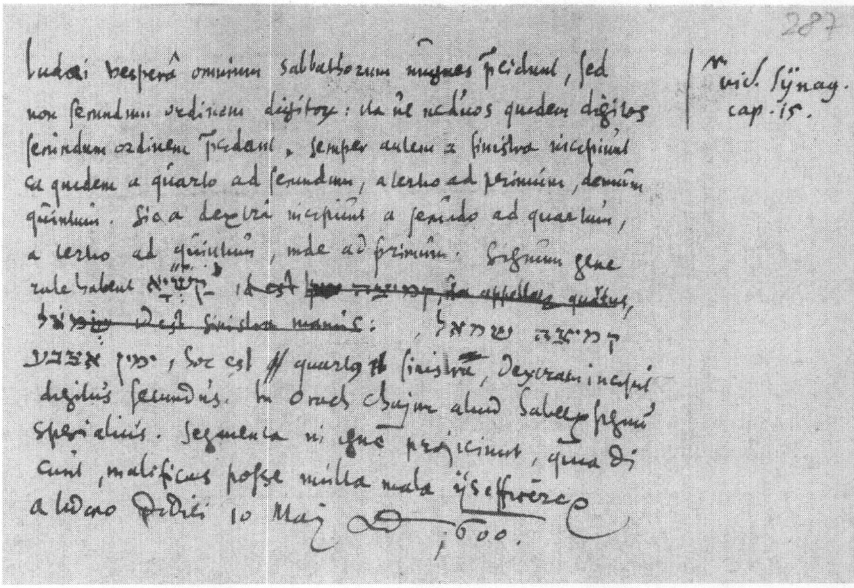

Figure A1.2. Basel, Universitätsbibliothek, MS A XII 20, 287.

Iudaei vespera omnium Sabbathorum ungues praecidunt, sed non secundum ordinem digitorum: ita ut ne duos quidem digitos secundum ordinem praecidant, semper autem a sinistra incipiunt et quidem a quarto ad secundum, a tertio ad primum, demum quintum. Sic a dextra incipiunt a secundo ad quartum, a tertio ad quintum, inde ad primum. Signum generale habent. קְשְׁיָא id est קמיצה שמאל ימין אצבע hoc est, quartus sinistram, dextram incipit digitus secundus. In Orach Chajim aliud habetur signum specialius. Segmenta in ignem projiciunt, quia dicunt, maleficus [read "maleficos"] posse multa mala illis efficere etc. A Iudaeo didici 10 Maij Anno 1600.

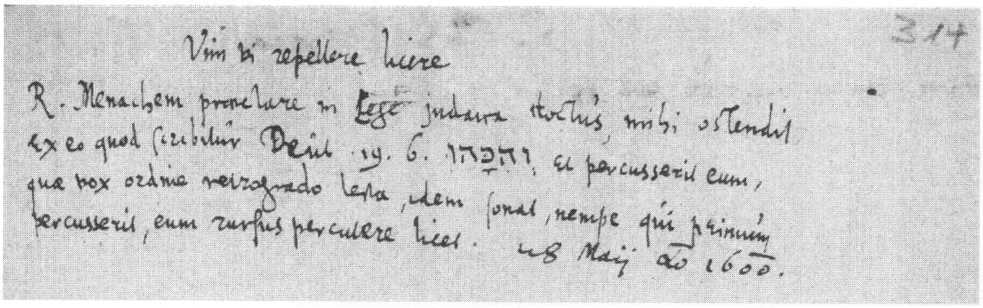

Figure A1.3. Basel, Universitätsbibliothek, MS A XII 20, 314.

Vim vi repellere licere
R. Menachem praeclare in lege Judaica doctus, mihi ostendit ex eo quod scribitur Deut. 19. 6. וְהִכָּהוּ et percusserit eum, quae vox ordine retrogrado lecta, idem sonat, nempe qui primum percusserit, eum rursus percutere licet. 28 Maij anno 1600.

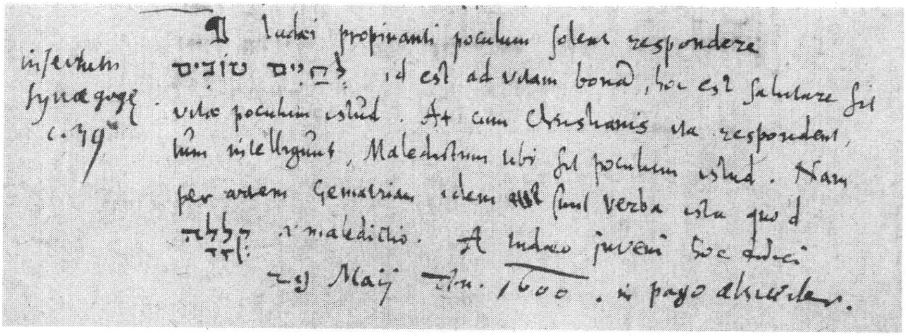

Figure A1.4. Basel, Universitätsbibliothek, MS A XII 20, 282.

Iudaei propinanti poculum solent respondere לְחַיִּים טוֹבִים, id est ad vitam bonam, hoc est salutare sit vitae poculum istud. At cum Christianis ita respondent, tum intelligunt, Maledictum tibi sit poculum istud. Nam per Artem Gematriam idem sunt verba ista quod קְלָלָה id est maledictio. A Iudaeo juveni hoc didici 29. Maij An. 1600. in pago Alswiler.

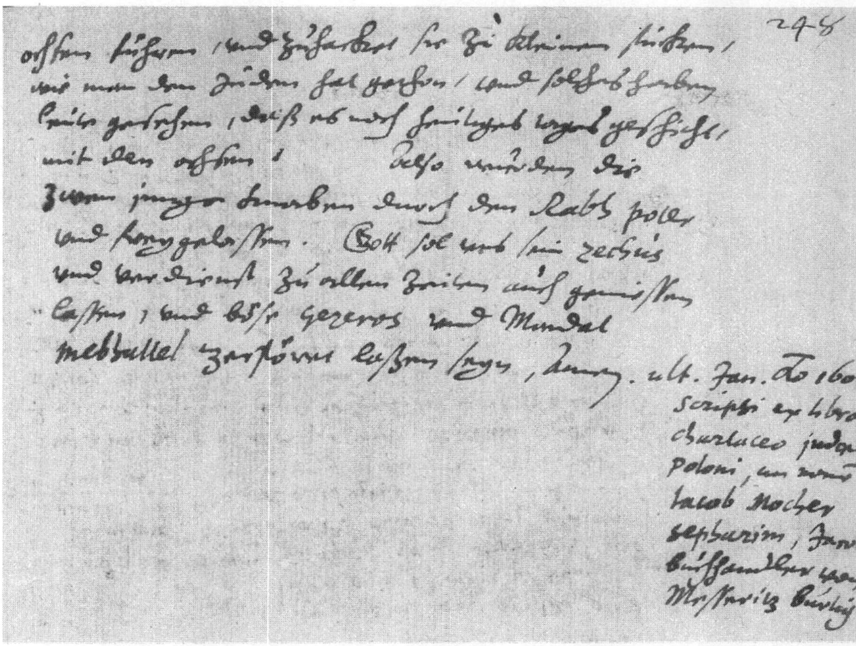

Figure A1.5. Basel, Universitätsbibliothek, MS A XII 20, 248.

Buxtorf's colophon to an extract from the *Mayse bukh*:

> ult.Jan. anno 1602 scripsi ex libro chartaceo judaei Poloni, cui nomen Iacob Mocher sephorim, Jacob Buchhandler von Messeritz burtig.

APPENDIX 2.1

From *Jüden Schul* to *Synagoga Judaica*: The Additions of the Younger Buxtorf

In an article on *The Twelfth Report of the London Society for Promoting Christianity among the Jews* (1820), the unnamed author stated: "Buxtorf pathetically exclaimed: 'Quis me pulvere aromatario aut succo malogranatorum satis confirmasset, ut fumosas et maleolentes Judaeorum caveas permeare potuissem?'[1] Few travellers indeed have been anxious to seek intimate converse with the Jews, except those whose Christian zeal has furnished a better antidote than that which Buxtorf sought."[2] He had been reading Buxtorf's *Bibliotheca Rabbinica* and had imbibed the trenchant anti-Judaism reflected in the author's introductory references to the Jews' refusal to hand over their books, though the attitude was less apparent in the main contents of the book.

When Buxtorf the Younger undertook a substantive revision of his father's *Jüden Schul* in 1661, the stridently polemical and mocking tone that permeated the work did not subside.[3] Indeed, in the new preface he added to his first edition of the text in 1641, he printed selected passages from his father's unpublished manuscript on the topic of Jewish hatred of the Christians, which buttressed the very negative assessment of Jews and Judaism that the first German edition of the work undoubtedly promoted.[4] Yet it is clear that the younger Buxtorf's mission

1. For a fuller quotation and translation of this passage, see Chapter 3 above (pp. 85–86, with n18).

2. "State of the Jews," *The British Review, and London Critical Journal* 16 (1820): 346–75, 346.

3. See, e.g., Buxtorf, *Synagoga Judaica*, ed. Johann Buxtorf the Younger (Basel, 1661), 325. Here the younger Buxtorf cited a long miracle tale about the Sabbath, concluding sarcastically, "Valete & plaudite, Calepinus recensui." This was a play on an ancient subscription, "Callopius recensui," found at the end of some Terence manuscripts and then transformed into Calepinus (the Renaissance lexicographer) in a comic passage in Rabelais's *Pantagruel*.

4. Johann Buxtorf, *Synagoga Judaica auspiciis authoris jam olim Latinitate donate, nunc primum in vulgus emissa* (Basel, 1641),):():(3v–):():((, fol. 5r. On the unpublished text by

was by no means polemical in every respect. In his first reprinting of the *Synagoga* in 1641 he made few additions, apart from providing the quotations in the original Hebrew or Greek. But in the 1661 edition and the 1680 one (which included his own son's corrections), the text expanded from thirty-six to fifty chapters. The younger Buxtorf made clear that, like his father, he had made systematic excerpts from a wide range of Jewish texts. But he used these in ways that sometimes reveal a striking intellectual independence.[5]

The additions to the text take several forms. Sometimes the son added new passages that his father had composed. In chapter 25, for example, the text adds a cross-reference to "my [*sic*] preface to the illustrious Mornay, which appears at the start of the little book on abbreviations" – the father's *De abbreviaturis Hebraicis*, first published in 1613.[6] In chapter 18, however, the son added a new passage, also in the first person, in which he referred to the discussion of the *afikomen* in "my dissertation on the Lord's Supper," first published in 1642.[7] Readers could not always identify the author of a given passage.

Many of the additions reflect the evolution of the Christian study of Hebrew – a field that had expanded as universities added instruction in Hebrew and Aramaic, whose practitioners now expected a level of precision and fullness that the elder Buxtorf had evidently not seen as necessary. The son added references to older books by Johannes Pfefferkorn, Azariah de' Rossi, and Christian Gerson and to recent works by John Selden and Leon Modena.[8] Revising the text for a more

---

Buxtorf the Elder, see Stephen Burnett, *From Christian Hebraism to Jewish Studies: Johannes Buxtorf (1564–1629) and Hebrew Learning in the Seventeenth Century* (Leiden, 1996), 91–95.

5. Buxtorf, *Synagoga* (1661), chap. 2, 9: "In libris Precum magna etiam differentia saepe conspicitur inter libros veteres, tum manuscriptos tum editos et recentiores. In illis enim passim multa acerba & virulenta reperiuntur in Christum & Christianos exprésse & aperte dicta, quae in posterioribus sunt omissa vel emollita, aut tectis & ambiguis terminis occultata, tum quod per censuram Christianorum ea exprimere ipsis ubivis non liceret, tum quod animadverterent, posteaquam Christiani linguam Hebraeam excolere, libros ipsorum legere, & haec deprehendere coeperunt, magna hinc sibi enasci odia & pericula.

"Talia inprimis occurrunt in *Machsorim* & *Selichos* manuscriptis, sicuti ex his multa ego habeo excerpta & in adversariis enotata."

6. Ibid., chap. 25, 523: "Qui plura cupit, adeat Praefationem nostram ad illustrem Virum Mornaeum, libello de Abbreviaturis praefixam." See also chap. 40, 644, where the text refers to Maimonides' quotation of a document "quae etiam extat in fine Grammaticae nostrae Chaldaicae & Syriacae." The reference is to Buxtorf, *Grammaticae chaldaicae et syriacae libri tres* (1615; repr., Basel, 1685), 432–33 (text and translation), 433–36 (notes).

7. Buxtorf, *Synagoga* (1661), chap. 18, 415: "De hac voce, ut & de re ipsa, videat Lector Lexicon Talmudicum & Exercitationem nostram, de Coena Domini [Buxtorf the Younger, *Exercitatio sacra in historiam institutionis ss. Coenae dominicae* (Basel, 1642)]."

8. Buxtorf, *Synagoga* (1661), chap. 48, 685–86; chap. 28, 551–52; chap. 35, 607; chap. 16, 371; chap. 49, 696; and chap. 1, 3 (on Modena, whose work was a response to the elder Buxtorf's polemical ethnography).

informed and demanding public, he provided original passages as well as translations. He regularly cited prayerbooks and collections of customs, some of which came from his own collections, which revealed the variety of Jewish practices. He developed this point at length at the beginning of the new edition, and in this respect at least his vision of Jewish life differed starkly from his father's dogmatic, monolithic vision and account of Jewish law and practice.[9]

Antiquarian Additions

In some cases, the new sections included substantive antiquarian investigations. Christian scholars of the seventeenth and eighteenth centuries were avid collectors and interpreters of Hebrew epitaphs.[10] The Buxtorfs were no exceptions. It is perhaps not surprising that in the much-expanded chapter on burial and mourning practices the younger Buxtorf inserted a short passage about inscriptions on tombstones. What he chose to include here seems typical of his approach to his work as revisor. He provided a new literary reference. The fourteenth-century rabbinic authority Jacob Moellin (Maharil) had referred to an inscription to a "betrothed maidservant" (*shifhah harufah*) discovered in his lifetime on one of the gravestones in the old cemetery in Mainz:

> On one occasion in his [i.e., Maharil's] lifetime, they discovered a tombstone in the cemetery of Mainz with an inscription to a betrothed maidservant. The date given from the time the stone had been erected until that particular day was 1100 years. Consequently, it is clear that the holy community of Mainz and the community of Halle [(?)] had been preeminent for a very long period of time.[11]

Maharil served as rabbi of Mainz for many years. He put together an authoritative compilation of the Customs (Minhagim) of the Ashkenazic communities. His disciple, known as Rabbi Zalman of St Goar, edited his work. Such was its importance that it was published four times in the sixteenth century and enjoyed several more printings in the seventeenth century. Buxtorf consulted the text

9. Ibid., chaps. 1–2, 1–9; chap. 10, 216–17 (on the Aleinu); chap. 23, 485; chap. 46, 669; chap. 49, 691, 694–95.

10. See Michela Andreatta, "Collecting Hebrew Epitaphs in the Early Modern Age: The Christian Hebraist as an Antiquarian," in *Jewish Books and Their Readers: Aspects of Jewish and Christian Intellectual Life in Early Modern Europe*, ed. Scott Mandelbrote and Joanna Weinberg (Leiden, 2016), 260–86.

11. The reference to Halle seems incorrect and appears to be an interpolation. The whole sentence about Mainz's importance is absent from the fifteenth-century MS of the text (Oxford, Bodleian Libraries, Mich 249).

published in Cremona in 1565.[12] The passage is not easy to decipher, but it appears to indicate that the tombstone had been erected in about 300 CE, 1100 years before Maharil's lifetime. Buxtorf had learned from his careful reading of David Gans's *Tsemah David* (Prague, 1592) that Maharil had lived at the beginning of the fifteenth century.[13] He used this information to expand the text, explaining that the tombstone had been discovered more than two hundred years previously, at the beginning of the fifteenth century. And he then paraphrased the second additional remark by stating simply that the inscription proved the community's antiquity.[14] Reading Maharil's description of the Mainz cemetery, Buxtorf must have realized that he had come across an astonishingly interesting fragment of Jewish, particularly Ashkenazic, history.[15] His discovery seems to have encouraged the epigrapher Matthias Friedrich Beck (1649–1701) to examine Maharil's work for himself and to conclude that unlike many other epitaphs published at the time, this one was genuine.[16]

12. He refers to a specific folio, 111 (v), which corresponds exactly to the Cremona edition. The same edition is held by Basel, UB (FA V 23), though it is unlikely that this copy belonged to Buxtorf.

13. David Gans, *Tsemah David* (Prague, fols. 1592), 59r–v. Buxtorf's copy is Basel, UB, V 12:1–2. In chap. 46, the younger Buxtorf's own contribution to the topic of rabbinic titles and honorific appellations, he used the information culled from Gans's work (already mentioned by his father in *De abbreviaturis hebraicis liber novus & copiosus* [Basel, 1613], 101) to discuss the honorific appellation "morenu ha-rav" (our master, the rabbi). Used in connection with Maharil, this was a relatively late (fourteenth-century) usage.

14. Buxtorf, *Synagoga* (1661), chap. 49, 715: "In Rituali Maharil f. 111 scribitur, inter sepulchra Moguntina inventam esse jam ante 200. & amplius annos, hujusmodi statuam, cui inscriptum fuerit, שפחה חרופה, erectam, juxta supputationem in ea expressam, mille & centum ante annis. Unde antiquitas illius Synagogae colligitur."

15. The old Mainz cemetery has been thoroughly excavated, but this particular tombstone – not surprisingly – has not been discovered. On the cemetery, see Nathanja Hüttenmeister and Andreas Lehnhardt, "Newly Found Gravestones from Magenza," in *Death in Jewish Life and Burial and Mourning Customs among Jews of Europe and Nearby Communities*, ed. Stefan C. Reif, Andreas Lehnardt, and Avriel Bar-Levav (Berlin, 2014), 213–23.

16. See Matthias Fridericus Beck, *Monumenta antiqua Judaica, Augustae Vindel: reperta et enarrata* (Augsburg, 1686), 9: "In לקוטים seu Collectaneis, Rituali Maharil i.e. Doct. Amplissimi, R. Jacobi, Levitae, Weilensis, annexis, pag. 111b refertur: *Dixit*, scil. R. Jacob, Levita, *quod aliquando sua aetate*, anno nim. 5160 a Creat. Mundi, i.e. A.C. 1400 *repererint Monumentum aliquod inter sepulchra Moguntina, cum hac Inscriptione*: שפחה חרופה Ancilla seu Virgo Desponsata. *Cuius calculus ab eo quo positum est tempore ad hunc usque diem, fuerat MC annorum. Praeterea,* Collector addit, *fama est, Sanctam Synagogam Moguntinam et Hallensem inde iam a longe plurimis annis celebrem esse.*" Beck's translation is more accurate than Buxtorf's. The reference to Halle in this context does not make sense, however. It may be for this reason that Buxtorf omitted it from his translation.

Deconstruction and Reconstruction: The Puzzle of the *Synagoga*

The chronologer David Gans was an authority for both Buxtorfs, father and son, and his *Tsemah David* was an indispensable source for the retrieval of biographical and bibliographical information.[17] They owed their familiarity with the great names of rabbinic Judaism in no small part to their reading of Gans's easily digestible chronicle. The very last entry of the first part of *Tsemah David* is devoted to Mordechai Jaffe (1525/30–1612), the renowned rabbi of Prague. Bestowing abundant praise on his rabbinic contemporary, Gans wrote: "He is the foremost leader of the heads of the academies and judges of the Three Lands."[18] Gans's assessment of Jaffe's paramount position among rabbinic figures of the time must surely have had some impact on Buxtorf the Younger. For throughout his revisions to his father's text he constantly made reference to Jaffe's great legal compilation, the *Levush*.[19] Though his father also referred to the work, Buxtorf not only inserted all sorts of information culled from Jaffe's work, including long swathes of calendrical material, but also used Jaffe in order to modify or qualify statements about regulations or offer reasons for certain types of behavior.[20]

In the chapter devoted to morning prayer rituals, Buxtorf the Elder had reported how prostration during the standing prayer would be transformed into "a prostration of the heart" should a Christian wearing a cross come into the presence of the person praying.[21] Buxtorf the Younger deemed it important to convey the reason for this rule. In his revised edition, he referred to Jaffe's statement in the *Levush* that a person should prostrate himself in his heart so that "it should not appear as though he was bowing to an idol."[22] By including the reason for Jewish practices the son made them comprehensible, and simultaneously provided evidence for Jewish anxiety about Christians and their crosses.

17. The family copy of *Tsemah David* is Basel, UB, FA V 12 (1–2).

18. Gans, *Tsemah David*, fol. 64v.

19. The family copy of Mordehai Jaffe's *Sefer levush yalkut* (Prague, 1623) is Basel, UB, FA IV 10:1, 2.

20. Jaffe wrote a commentary on Maimonides' Laws of the New Moon, but also interspersed his legal compilation with relevant calendrical information and tables. Buxtorf used this material in the chapters on the new moon and New Year. He noted Jaffe's reputation as an expert in calendrical matters. See his comments in *Synagoga* (1661), 472: "Qui accurate cupit cognoscere Judaeorum calculum de supputatione Mensium, Annorum, Revolutionum etc. consulere potest R. Mardochaeum Japhe in *Levush malkhus* [hur] par. 1. Num. 427. & 428. Quo, Judaeorum judicio, nemo accuratius ea de re scripsit."

21. Buxtorf, *Jüden Schul* (Basel, 1603), chap. 5, 225.

22. Buxtorf, *Synagoga* (1661), chap. 10, 225 = *Levush malkhut, Levush ha-tehelet* 113: 8. The younger Buxtorf has underlined this passage and marked it with a small cross in Basel, UB, FA IV 10a.

Mordechai Jaffe was a prominent halakhist, known for the clarity and conciseness of his rulings, which were grounded in Ashkenazic legislative tradition. He often proffered reasons for legal decisions, and occasionally criticized popular customs. Buxtorf the Younger seems to have valued Jaffe's approach to Jewish law. Otherwise, it would be difficult to account for his constant reports on what "R Mardochaeus Japhe wrote in his Levush." In his greatly expanded description of the customs for the New Year, for example, Buxtorf cited a long passage from Jaffe. The rabbi explained that in certain communities, prior to the reciting of the petitionary prayer (Amidah) at the Musaf service, the precentor would enunciate a range of adjurations, calling upon a whole host of angels to help him lead the congregation, to ensure that he did not falter or err in any utterance. Jaffe dismissed this custom. Adjurations had no place in the context of petitionary prayer. In fact, they smacked of blasphemy.[23] By documenting Jaffe's disapproval of this particular practice, Buxtorf conveyed to his readers that rabbinic Judaism could not be straitjacketed or reduced to a single uniform system of rituals and customs. In this respect, he undermined his father's appraisal of Judaism as a superstitious religion designed and governed by overbearing rabbis.

Several examples enable us to follow this process of adding to the elder Buxtorf's narrative. Consider, for example, the two men's treatment of *Kol Nidrei*, the declaration of annulment of vows recited on the eve of the Day of Atonement. Buxtorf the Elder was not the first Christian to point to the prayer as evidence that the oath of a Jew was worthless:

> The meaning of the prayer is that all vows, oaths and promises, truth and faith, which a Jew has not kept throughout the year, shall be dissolved and annulled, not considered a sin, but held to be nothing, dissolved, and forgiven. Therefore, the pious and the truthful, along with the perjurers and the breakers of vows and promises, are joined together into a holy congregation, which may pray and praise God together. (Observe how much a Christian can rely on the oath of a Jew.)[24]

This somewhat abbreviated and oddly non-Ashkenazic rendition of the *Kol Nidrei* was not left untouched in Buxtorf the Younger's revision.[25] In addition, he challenged the premises underlying his father's description by citing Jaffe's

23. Mordehai Jaffe, *Levush malkhut, Levush ha-hur* 584,1 = Buxtorf, *Synagoga* (1661), chap. 24, 500–502. Buxtorf also refers to Joseph Albo's *Sefer ikkarim* (Lublin, 1597). There Albo discusses in a philosophical context whether angels can be invoked in prayer and by whom.

24. Buxtorf, *Jüden Schul*, chap. 26, 528.

25. The elder Buxtorf held that the annulment of vows refers to vows taken from the previous Day of Atonement to the present Day of Atonement. This was the formulation used in Geonic times, which had been rescinded by the twelfth-century Tosafists. Standard

clear overview of Ashkenazic practice. Jaffe had explained that the wording of *Kol Nidrei* referred not to unfulfilled obligations of the past year but to vows that might remain unfulfilled or forgotten during the year to come. Moreover, as Jaffe wrote and Buxtorf the younger reported, the dispensation from vows in the *Kol Nidrei* referred only to those which an individual voluntarily assumed for him or herself.[26]

Additional information that the son culled from Jaffe created even greater havoc with the original narrative. The elder Buxtorf claimed to have seen Jews stand in prayer for twenty-seven hours throughout the Day of Atonement. The younger Buxtorf retained this anecdote. But he also quoted, seemingly with approval, a passage from Jaffe, who denounced the practice of not sleeping on the Day of Atonement because it prevented proper devotion in prayer.[27] As usual, the son intercalated this material into his father's account without breaks, leaving the reader unable to know which Buxtorf was speaking at which point.[28]

When Shemaiah the Jew wrote to Buxtorf the Younger in 1660, he expressed bewilderment about the fiercely anti-Jewish nature of the *Jüden Schul*; at the same time he seemed to imply that the work did not represent Buxtorf's true feelings.[29] Reading the extensive revisions that Buxtorf the Younger made to his father's original text, we might go some way in sharing Shemaiah's assessment – and, perhaps, his puzzlement as to the purpose of the Buxtorfs' enterprise in its new form.

Ashkenazic practice, documented in all the many prayer books that Buxtorf had examined, reflected the reading "from this Day of Atonement until the next Day of Atonement."

26. Buxtorf, *Synagoga* (1661), chap. 26, 530–32 = Jaffe, *Levush malkhut, Levush ha-hur* 619, 1. Buxtorf does make one intervention, claiming that this dispensation has given Jews a bad reputation, because it seems one should place no trust in vows that they make with Christians: "Ob hanc abolitionem et juramentorum solutionem male audiunt Judaei, quasi Juramentis illorum quae Christianis faciunt, nulla fides sit habenda [531]."

27. Buxtorf, *Synagoga* (1661), chap. 26, 532–33: "hunc tamen morem multis modis improbat R. Mordechai Japhe in *Levusch hachor* Num. 619. §. 4. et judicat illos melius facere qui post Precationes et Deprecationes solennes domum dormitum se conferunt. Quidam, qui pietate sunt admirabili, & gravem poenitentiam agere volunt, per totum festum, die noctuque stant in pedes, canunt & orant indesinenter, ut ego quosdam vidi qui per septem & viginti horas in loco uno steterunt."

28. The younger Buxtorf's appropriation of his father's work and legacy deserves a comprehensive study. Note, for example, the two men's inscriptions on the title page of the father's *Album amicorum*, Basel, UB, MS AN VI 26[z]. The father wrote: "Ioannes Buxtorfius Camensis, Westphalus / בֵּאלֹהִים יִשְׁעִי / In Deo salus mea. / Fidentem nescit deseruisse Deus." The son added: "Nunc haeres est filius / Johannes Buxtorfius, / Basileensis."

29. See Appendix 2.2, below.

APPENDIX 2.2

The Letter of Shemaiah to Buxtorf the Younger

This letter, dated 1660 and addressed to Buxtorf the Younger, provides precious information about the strategies that Jews had to devise in order to negotiate and cooperate with Christians. Written in a highly ornate and rhetorical style, the letter conforms to the conventions of Hebrew epistolography so familiar to Buxtorf, and showers rabbinic praise on the recipient of the letter. Yet the author, Shemaiah, who was apparently a member of the renowned Segal Ish Horovitz family, did not refrain from expressing his dismay at the vehemently anti-Jewish sentiments of the *Jüden Schul*. It is not clear whether Shemaiah realized that the bulk of the work had been written by Buxtorf's father, and that the son had reedited the Latin version of the book in 1641. Whatever the case, this outspoken critique appears all the more courageous given that Shemaiah was apparently hoping to be employed by Buxtorf as a corrector or typesetter.

The letter also encapsulates some of the characteristic elements of seventeenth-century Ashkenazic culture. For the author, who is peddling his literary wares, clearly belongs to the group that Elchanan Reiner defines as the secondary elite, who were reading and popularizing Lurianic and other kabbalistic or mystical works.[1] This is obvious not only in the case of the majority of books that Shemaiah claims to be writing or completing but also from his discussion of a sixteenth-century kabbalistic work, the *Galya razaya*.

1. Elchanan Reiner, "The Ashkenazi Élite at the Beginning of the Modern Era: Manuscript versus Printed Book," in *Jews in Early Modern Poland*, ed. Gershon David Hundert (London, 1997), 85–98, at 87–89.

Basel, UB, MS A XII 19, Nr. 28

With God's help

May the Lord who rides on a swift cloud[2] support him[3] and deliver him from any accident or misfortune – he in question is the distinguished and great authority, a cedar in Lebanon.[4] His reputation will last as long as the sun – our teacher the rabbi Johanan Buxdorfer. May he continue to flourish until those who sleep in the earth rise again.

I received his letters, which like his name pour out kindness, grace, and compassion.[5] I was overjoyed by the letters just like Moses when the Lord spoke to him in the tent of meeting from a cloud. I took note of his [i.e., Buxtorf's] orders. Now in response to my task I have taken it upon myself to speak, in this case, about the pleasant writings of his honor and his success in writing a commentary on the *Kuzari*, a book about which it is said among the Jews "you shall not forsake the Levite in your midst."[6] It is a fine work to view and pleasant to delve into this precious document every day. For he [Halevi] was the most outstanding writer in all disciplines, both in minor matters such as the arguments of Abaye and Rabba and in major matters such as the story of [Ezekiel's vision of the] chariot.[7] Indeed, God's spirit inspired him and his Torah became great and glorious throughout the world.

All nations say that his honor knows both the revealed and esoteric. Thus, being of Jacob's root, I, Shemaiah, have also come, to listen to his wisdom and to get to know his greatness in which his humility resides. But I have a small question to ask of him and he should give me a response. What is the purpose and why is it, his honor, that he thought to explain the laws of the Jews to the nations?

2. The phrase "May the Lord who rides on a swift cloud" is an adaptation of Isaiah 19:1.

3. In Chapter 2 above, references in the third person to Buxtorf in quotations from Shemaiah were changed to the second person, to make the text more readable. In this appendix the translation is done strictly. Shemaiah almost always refers to Buxtorf in the third person.

4. This is one of the conventional phrases used in formal letter writing. Interestingly, Shemaiah uses one of the expressions for "viro Honorato" discussed in Buxtorf's *Institutio epistolaris hebraica* (Basel, 1610), 21. Here, the ending *non* in *Lebanon* rhymes with *non* in *yanon* in the next phrase.

5. The words "grace, compassion," allude to the name Johanan, which in turn relates to the verb *hnn*, meaning "to be gracious or compassionate."

6. I.e., Judah Halevi, author of the *Kuzari*. Shemaiah is referring to Buxtorf's publication of a bilingual edition of this text, *Liber Cosri continens colloquium seu disputationem de religione*, in Basel in 1660, the year in which he was writing this letter. On the reception of this work, see Adam Shear, *The Kuzari and the Shaping of Jewish Identity, 1167–1900* (Cambridge, 2008).

7. This is a well-known passage, given, e.g., in BT, Baba Batra 134a, according to which Johanan ben Zakkai is claimed to have been expert in both the minor and major matters.

I noticed that when he entered the domain of Torah with his studies in his *Jüden Schul* he spoke perversely about them [i.e., the Jews]. For even by his words he has confirmed his power over them. So now I would judge him in accordance with the saying of the sages: It is permitted for a scholar in a place where he is not known to make himself known and claim that he is a worshipper of fire.[8] Does this not explain the praises that the sages and rabbis of Frankfurt and their yeshivah declared at the time of writing the document inserted in the copy that I am sending to him and setting before his eyes?[9] When he sees it his eyelids will move and his eyes will confront the truth as to who you are and what I am and what I say and how I justify myself, and furthermore, that the Lord has spoken.[10]

Moreover, I am unable to hide from his honor all the books that I have written. I know that I am a blacksmith.[11] Therefore, indeed, his honor should make a decree and it will be fulfilled for his honor. I would say to him that the book *Galya razaya* is a very ancient book. It is very rare and hardly found in the possession of the rabbis, and these do not contain the commentary of 300 pages according to the kabbalistic interpretation that I composed.[12] It was written by dictation from the angel Gabriel and it consists completely of secrets, the holy names.[13] For if you read my commentary, to which I gave the title *Razi li* (My Secret Is Mine), you

8. This is a conflation of two passages in BT, Nedarim 62a and 62b: "A man may reveal his identity when he is unknown"; "Raba said: A rabbinical scholar may declare, 'I will not pay toll tax... [a reference to Ezra 7:24].' Raba also said: A rabbinical scholar may assert, 'I am a servant of fire, and will not pay poll tax.'" That is, he was claiming that he was a worshipper of fire, a Zoroastrian priest, and would therefore be exempted from paying the tax.

9. In other words, he seems to be insinuating that Buxtorf is somewhat hypocritical in the *Jüden Schul*, an anti-Jewish work that does not properly represent his true views or behavior toward the Jews.

10. We do not know what was in this document.

11. This is an allusion to BT, Berakhot 28a, where Rabbi Joshua is said to have an arduous job, that of blacksmith. In other words, Buxtorf should take Shemaiah's circumstances into account.

12. This is a fascinating use of the work *Galya razaya* (*Raza*); Rachel Elior published a critical edition of part of the work from the complete manuscript (Oxford, Bodleian Libraries, MS Oppenheimer 104): *Revealing Secrets–Galia Raza: Critical edition of Manuscript Bodleian Library 104* (Jerusalem, 1981). The authorship of this work has been disputed, but it is agreed that it was written by a sixteenth-century kabbalist. It is possible that Shemaiah is referring to that same Bodleian Libraries MS, which contains 344 folios, or MS Opp. 526, which contains 250 folios and belonged to Eleazar Altschul of Prague. The work was cited in many contemporary works, such as the *Yalkut hadash* on which Shemaiah purports to have written a commentary.

13. This statement is probably based on the statement on MS Opp. 104, fol. 52r, referring to Gabriel's involvement in the castration of Potiphar: "For Gabriel is in charge of dreams and his name is *Galya razaya* and he is Gabriel according to Gematria [of adding up the letters of the name Gabriel and the words *Galya razaya*] and therefore I called my work 'This is the revealed secret,' for by the power of dreams in nightly visions and sometimes

will be amazed and the hairs of your head will stand on end on account of the incredible wonder that you will encounter: how great is the power of the Lord, great and greatly exalted who is called "praise befits You" (Ps. 65:2), God of law,[14] the *Ein sof*, blessed be He and blessed is His name.

You should know, my lord, and honestly trust and completely believe that if he would strive [to lend] me substantial powers,[15] exalted honor, I would be there for you and thank you. I would serve you in perpetuity like the slave of the bored ear[16] and serve you in the holy work, the task of correcting, and I will do all you command me to do and we will be united and delight in our friendship. Thus I pray to the Lord, who bestows dew and rain that his silver and gold will multiply and become as numerous as the stars and illuminate his life with every good, joyousness, and illumination.[17]

Thus I beseech you, I, Shemaiah the lowly, a dry tree full of blemishes, son of the honorable Rabbi Meir Segal Horovitz, writing on the eve of Purim, on the fast of Esther, 25 February 1660 according to your era [CE] here in Pfersee near Augsburg.

These are the titles of the books I have written and they are not completed. I hope, however, that soon, with God's help, they will be completed and finished apart from those that our distinguished teacher knows. Here is the list.

1. *Ma'amarot tehorot* (Pure sayings), a third commentary on the Ten sayings about Kabbalah.[18]
2. A work on truth and exposition about the truth of free will and absolute knowledge of God, regarding the general and the particular with regard to every action and speech of human beings.
3. A commentary on various passages of the Zohar.
4. A commentary on the entire *Sefer Bahya* and its meaning with an esoteric commentary and *Naftulei*.[19] I called it *Nishmat kol hai* and it is totally finished, praise be to God.

---

when awake I have built all the structures that I have constructed." On the various readings of this passage, see *Galya raza*, ed. Elior, 10.

14. The phrase "God of law" should be "fiery God," as in Deuteronomy 33:3.

15. This is in Yiddish "ein fass mi-sheroroh."

16. "Slave of the bored ear" is another conventional phrase used in the formal part of letter writing. Buxtorf the Elder often used this expression in his Hebrew letters.

17. With "illumination," Shemaiah appears to be using a kabbalistic phrase relating to emanation.

18. This is a reference to *Asarah ma'amarot* by Menahem Azariah da Fano (Venice, 1597). It is striking how many times Shemaiah refers to the work of this author (1548–1620), an Italian Jewish kabbalist influenced by Lurianic Kabbalah, whose works were widely distributed in sixteenth- and seventeenth-century Ashkenaz.

19. This is a reference to the mystical commentary titled *Naftulei elohim* (Ferrara, 1556) by Naftali Hirts Treves on Bahya ben Asher's very popular commentary on the Pentateuch. The 1556 first edition was followed by others.

5. Commentary on *Yalkut hadash*. I've called it *Man hu* (What is it?).[20]
6. A commentary on *Imrei sefer* on the Pentateuch written by the honorable Rabbi Nathan Spira, blessed be his memory. It is called *Pesher davar* (Interpretation of the matter).[21]
7. A commentary on the *Ma'arekhet ha-elohut*, i.e., animadversions on the words of Hayyat of blessed memory.[22]
8. *Iggerot shelomim* on writing letters to each other in verse according to the??.[23]
9. *Ma'asi katon ve-ha-talmud gadol* (My action is little but learning great). Indeed it is virtually finished.
10. A discourse on the decree of the evil inclination, that is, the Satan, the angel of death over scholars (*talmidei hakhamim*) and above all over Israel.
11. A discourse *ashrunu navah tehilla* on *Ashrei yoshevei veitekha* and on three Psalms.[24]
12. A discourse *The Tower Flying in the Air* – it is about the Otiyyot of Rabbi Akiva.[25]
13. A very old book on all the words in the Torah according to Michael good and evil; the right side, the left side "and understand."
14. Commentary on *Kaftor va-ferah*, the sayings of the sages of blessed memory.[26]
15. Midrashim of the Yalkut according to the Kabbalah, and I have called it *Sefer Izen ve-tikken* since it harmonizes one [interpretation] with the other,

20. *Yalkut hadash* was a very popular homiletical work by Israel ben Benjamin of Belzec (Amsterdam, 1657). The title *Man hu* is a play on Exodos 15:15 about the manna.

21. Nathan ben Samson Spira (d. 1577) wrote a commentary on Rashi and on the supercommentary on Rashi by Elijah Mizrachi. For a discussion of interest in Mizrahi's supercommentary on Rashi among Ashkenazic Jews, see Eric Lawee, *Rashi's Commentary on the Torah: Canonization and Resistance in the Reception of a Jewish Classic* (Oxford, 2019), 261.

22. This is a reference to the anonymous kabbalistic work published in Ferrara in 1556 with the commentary of Judah Hayyat.

23. The last words are unclear. He seems to be referring to "literal meanings of Torah," but the word פשטי is spelled פטשי. This title is a plagiarism. The *Iggerot shelomim* on Hebrew epistolography was published in Augsburg in 1536, and again in Basel in 1603, and of course Shemaiah knew that the Buxtorfs, father and son, printed works on Hebrew epistolography. It was probably through Jacob Buchländler's mediation that Buxtorf acquired a copy of *Iggerot shelomim* and, taking his cue from the Jewish entrepreneur, proceeded to make excerpts from it with Latin translations in order to advance his pedagogic project to instruct Christians in the art of Hebrew epistolography.

24. *Ashrei* is Psalm 145, read in the Sabbath liturgy.

25. "Tower flying in the air" is a Talmudic expression, used in a variety of ways in kabbalistic literature. His title is taken from BT, Hagigah 15b and BT, Sanhedrin 106a and is also perhaps a reference to the letter *lamed*, which, in the Zohar, is said to have the structure of a "tower flying in the air."

26. *Kaftor va-ferah*, on the Aggadot of the Talmud, was written by Jacob Luzzatto and printed in Basel in 1581.

according to the hermeneutical principle of using a third verse to reconcile two seemingly contradictory verses.

16. On *Pirkei d'Rabbi Eliezer ben Hyrcanus*. I have titled it *The Lord's Help*.[27]
18. *Asof asifem* (I will gather them up; Jer. 8:13) on the strange texts cited in the *Be'er ha-golah* of R. Lev (Judah) of Prague (i.e., Maharal of Prague) on him be peace.
19. *Sefer ha-magen* (Book of the shield) on the sayings of Rabbah bar bar Hana and the like.[28]
20. Commentary on Ruth according to the Kabbalah. I have called it *Toren ha-sefinah* (Ship's mast) to allude to the transmigration of the human soul and the secret of the levirate marriage and *ha[lizah]*.
21. *Kisse kavod* (Throne of Glory). This is an important discourse on the Throne of Glory according to the interpretation of the kabbalists of blessed memory.
22. Commentary on *Sefer ha-aggudah*.[29] It is not finished.
23. Commentary on the old and new *Arukh*, published again in Amsterdam according to the Kabbalah. It is called *Shemurah ve-arukhah* (Kept and ordered).[30]
24. Discourse called *Orhei ve-farhei* (Guests and wayfarers).[31]
25. Commentary on Ecclesiastes called *Kahalu lishmoa*, "Assemble to hear."[32]
26. *Zer zahav* (Golden crown) on the secrets of the Pentateuch, Prophets, and Hagiographa.
27. Commentary of *Olam katan* (Small world) and it is called *Tsel Elohim hayyim* (Shadow of the living God).[33]
28. Commentary on *Yonat elem*, part one of *Asarah ma'amarot* and it is called *Yonati tamati* (*Song of Songs* 5:2).
29. Some commentaries on the *Kavvanot ha-Ari* and it's called *Kavvanot la-shemayim* (directions to heaven).[34]

27. No seventeenth book appears: the list goes directly from 16 to 18.

28. Rabbah bar bar Hana's tall tales are related in BT, Baba Batra 73a–77b and became the subject of much discussion and analysis by later Jewish commentators.

29. The *Book of the Collection* by Alexander Suslin ha-Cohen (d. 1391), a German halakhic authority whose work was very influential among German Jews.

30. This is a reference to Benjamin Musaphia's *Sefer musaf he-arukh* (Amsterdam, 1655), an addition to Nathan ben Yehiel's talmudic dictionary *Arukh*.

31. The phrase is taken from BT, Ketubot 64b.

32. This is a play on Kohelet and on the author's name, Shemaiah.

33. This is probably a reference to one of the sections of Menahem Azariah da Fano's *Asarah Ma'am rot*.

34. *Sefer ha-kavvanot* is the work attributed to Isaac ben Solomon Luria (the Ari) on the intentions of and meditations on prayer. It is not clear what the title of Shemaiah's work means.

30. *Ori ve-yishi* (My light and salvation). It is a discourse on the life of vanity and learning Torah by illumination and by means of any teacher.

I am also contemplating and indeed my soul tells me to produce an infinite number of books since I am proficient in 10 languages and in the tongue in which I know "I will hear the Lord."[35]

35. A pun on his name *Shemaiah*, which in Hebrew means "hearing God."

APPENDIX 3

Abraham Braunschweig's Apology

What follows is a full translation of the apology that Braunschweig placed at the start of the errata list of Buxtorf's Rabbinic Bible, to introduce the printed errata list that he drew up. Copies of the original list – as opposed to those of a type facsimile of the copy in Copenhagen – are rare.[1] This translation rests on the original text, preserved in Princeton University Library's copy of the edition.[2]

An apology to the intelligent reader for the errors that occur throughout this publication. He should read this entire document and then be in a position to correct the accents, vowels, and each and every letter.

The human intellect struggles to apprehend or achieve anything in full, perfectly and flawlessly. Furthermore, human nature is limited. This is true either of the one who performs an action, since his knowledge and intellect are incapable of embracing and understanding any matter with complete clarity and precision, or from the perspective of the undertaking that is so wondrous in its greatness, profundity, and importance that it is impossible to correct it in a complete and absolute manner. It so happened that God disposed me to meet an intelligent and wise man, generous and wealthy, the lord Ludwig König – may God protect him. The Lord had inspired him to undertake a work, holy work, the publication of our holy Torah and the prophets by whose words we live; their paths are too profound for the inhabitants of the earth to discover.[3] The task was to repair the fractured texts and to compile the irregular words, *plene* or defective, however many or few they may be, with the letters that take a strong

1. Copies of the original appear in Basel, UB, FG II 19 (Buxtorf's working copy), and Princeton, Princeton University Library, 2019–0001F.

2. This apology was translated into French by Joseph Derenbourg, in "L'édition de la Bible rabbinique de Jean Buxtorf," *Revue des études juives* 30 (1895): 70–78.

3. Cf. Ecclesiastes 7:24.

or weak *dagesh*, with the variants in vowels and the accents and the markers of the ends of the verses.[4] All these indications came from heaven and were given by one shepherd.[5] There is no human being who can perform an action like his Maker – can a man be pure?[6] The intelligent reader may come across a text that is not clear and declare it to be defective. This might be his reaction on seeing, for example, a letter *bet* instead of a *kaf*, and sometimes, he would notice that the vowels or accents are hardly visible such that it was impossible to make out their correct and complete form.[7] Now I will implore you to pay respect to my lord, the great eagle with plush plumage whose influence has spread abroad;[8] his seat is in the city's high place. He specializes in the grammar of our holy tongue and Targum as well as Masorah. I am referring to the man of Torah, the respected JOHANN BUXTORF – may God protect him – who resides in Basel. He was put in charge of the work of correction. Now one who prays on behalf of another for something that he too desires, he will be answered.[9] And so I articulated my prayer: "If the intelligent reader were to regard this work as defective he should judge us to be blameless. He should not impute blemishes on the sacred nor spread calumnies about that which is holy; that either deliberately and willfully or on account of the absence of proper supervision this and other such things have happened, God forbid. Does he not realize that this work is heavier and more protracted than the earth and more expansive than the sea?"[10] A vowel point like a hair twisted in a grain of sand which is found in the paint can appear distorted.[11] Or can one throw at a hair and not miss?[12] In all the

4. He is referring here to the complex matters of Hebrew grammar and the forms of the letters, vowels, and accents that adorn the Masoretic text. The "markers of the ends of the verses" refers to a cantillation mark that is put on the last word of every verse; the *dagesh* is a point that is inserted in the letter. The weak *dagesh* is inserted in certain letters to indicate plosive pronunciation of the letter, whereas the strong *dagesh* generally indicates a gemination of the consonant.

5. This seems to be a deliberate attempt to assert the divine origin of the Masoretic text.

6. This is an adaptation of a line in Elijah ben Shemaiah's penitential poem (*selihah*) for the eve of Rosh Hashanah, אדון בשפטך אנוש רימה, which itself is a play on Job 4:14.

7. Throughout this passage, Braunschweig uses the terminology of Leviticus 14 about the examination of the leper in order to express the problems of detecting the flaws in the biblical text.

8. Literally, "his fountains flow out"; cf. Proverbs 5:16.

9. Cf. BT, Baba Kama 92a.

10. Job 11:9.

11. The images Braunschweig employs throughout this passage are very obscure. Here, he uses a word that usually refers to women's eye makeup (i.e., paint), and therefore perhaps also ink. Yakov Mayer kindly advised us on the meaning of this difficult passage.

12. A play on Judges 20:16.

sections [of the Bible], each one takes its own delegated place with their own sections.[13] Their words are pure, wonderful, and esoteric; this is particularly the case for the Great and Small Masorah, whose words are sometimes contradictory in a way that seems irrational. Sometimes the Great Masorah speaks of three and then demonstrates four or two instances, as in Genesis 4:15 where in the Small Masorah it is written that there are seven cases of the word שבעתים and the Great Masorah displays six. Similarly in Genesis 6:18 the small Masorah indicates that the word ובאת occurs ten times with the accent on the last syllable, but the Great Masorah indicates only seven instances – and there are many other such examples.

Therefore, their words need very close scrutiny and investigation. Truly, it has been executed correctly in the many places where there was clear proof and exact evidence. Therefore, wherever we were able we worked in honor of God and for the glorification of our holy Torah. Out of devotion to the task we have tried to bring light to our search in an effort to establish the text for the fundamental part but have as it were eradicated the rest – for we cannot rely on our limited knowledge and on a fundamentally damaged and broken support.[14] This occurred during the reading of the proofs. It is impossible to carry out two tasks properly at the same time.[15] So if some errors do remain through oversight, one should surely recall the words of Solomon, on him be peace, that in love you will always go astray.[16] It must have come to the attention of a wise reader that sometimes in a Torah scroll that has been written out and gone through and read with great scrutiny as far as human powers permit, and read out a hundred times in the synagogue, suddenly there is a discovery of either a defective or redundant letter or one letter exchanged for another. In the case of work of such magnitude what can one say! After all, with the art of printing as we have it nowadays it is as impossible to print anything without errors as it is for the world to be without air. In particular, this came about on account of the non-Jewish workers who had not performed God's service from childhood and who were completely incapable of reading or recognizing one letter of our holy tongue. Thus, when they embarked on the work, a holy undertaking, and set out to accomplish and bring it to completion, they were like aliens in a foreign land; and though they were professionals in their own language they seemed beginners in the art of printing. As a consequence, the designated work could

13. Here Braunschweig adopts the terminology of Numbers 33 describing the ranks of the Israelite encampments.

14. Literally, "staff of a bruised reed"; cf. Isaiah 36:6. This highly rhetorical statement, in keeping with the tenor of the entire apology, is probably not meant to be taken too literally.

15. This is a play on the noted passage in Hullin 60b about the domination of the sun and moon.

16. Braunschweig alludes to Proverbs 5:19.

not be brought to completion as originally conceived without enormous effort and trouble. When they brought a page to the corrector for editing, refining, and purging of errors he would sit and read it and become overcome with despair until he stopped counting [the errors] – there were so many that he was unable to examine them at one sitting. And when they returned to correct the mistakes, they were so numerous, and being under pressure and impatient to complete their daily task, that sometimes the defects became duplicated and even reduplicated.[17] For example, if a letter was incorrect and he would indicate how to correct it properly, they would produce another suitable letter but then insert another one that was not correct. In that way the original error remained. A similar situation occurred with regard to the vowel points. If he had indicated that there should be a *patah* instead of a *kamets*, and they struggled to find the *patah*, they produced another appropriate vowel as they saw fit Thus while working for me they have committed two wrongs for me, that is, against me. For they retained the error and have violated the fount of living water. And this foolish way of proceeding was simply because they did not know or understand our holy tongue, and such occurrences came about on the eves of Sabbaths and festivals, and in particular, on Sabbaths and the Feast Days when no Jew was present. Now in every matter, two are better than one; for if the one person omits something, his colleague can reinstate it. The workers were also particularly under pressure on the Sabbaths, anxious to ensure that they could take their rest on their feast day [i.e., Sunday]. And so it could happen that the work got adulterated. Moreover, on occasion, the corrector had to bear down on them to ensure that the work was produced absolutely on time so that while one was speaking [about one page] another page was produced for him – for three pages had to be printed every day.[18] The corrector would speak to them severely, "Finish your work according to the indications in the corrected text." But when he came to inspect the holy text, had he ever come across anything like it? For they had not heeded his instruction and had simply done what they felt like. So he had to inspect it again and had to examine in fine detail that which he had already gone over.[19] The printed pages were therefore irreparably damaged – for who can challenge the king [i.e., König

17. The description of the duplication of errors is constructed from a combination of several rabbinic passages.

18. This passage is not entirely clear. He seems to mean that they printed three sheets a day, which would have meant working very rapidly indeed. Alternatively, he could be referring to the practice of pulling three proofs for every sheet, standard in good Basel shops. See Johan Gerritsen, "Printing at Froben's: An Eye-Witness Account," *Studies in Bibliography* 44 (1991): 144–63; Theodor Zwinger, "Artis typographicae operum et operarum consideratio," in Theodor Zwinger, *Methodus apodemica* (Basel, 1577), 398–400.

19. As Derenbourg states in his notes, this whole passage is an elaborate play on the description of the priest's examination of the leper in Leviticus 13:6.

the printer] once it has been done? Thus, if an error is found in any volume,[20] it is probably their fault, but not in all cases.

Now ABRAHAM [i.e., Braunschweig], who is still involved in the correction of books, would make the following plea to the wise reader. Should any error remain he should correct it but judge us favorably and let him realize how enormous is the undertaking and how in respect of our limited knowledge, which is the human condition, it is impossible to produce a fundamental work without flaws. I myself, a lowly servant working for the noble and respected lord mentioned above on this holy task, had to read the text out to him [to Buxtorf] and to help in the process of correction. So many troubles and miseries did I experience (may you not encounter them); how much did I endure. I undertook a self-examination so that I should reach the gates of repentance. I said: Perhaps I am responsible for the erroneous text and the remaining errors are due to my failings. And yet, I did everything in my power to execute this work. What other means can I bring in order to justify myself before the respected king and gain expiation from him? I made a silent prayer in the presence of the noble lord LUDWIG KÖNIG mentioned before.[21] I then said: May I proffer advice and may God be with us. Let a copy of the new Bible be put before scholars and let them go through the entire Bible from beginning to end and mark the inaccuracies in print so that the reader can easily correct the errors. Thereby, the Lord's Torah will be made perfect and later generations will realize that we did our utmost to ensure that there would be no possibility of superseding what we have accomplished. And God procured my favor with the king [i.e., König] – he accepted my proposition and was generous. For he paid for the revision of the entire twenty-four books, which was to be undertaken by the wise and eminent, honorable scholar of Torah, Rabbi Gumpel of Posen, bookseller,[22] which he did indeed accomplish with intellectual probity.

Much time has elapsed. So this letter is to provide information about this matter, and as to whether it was my responsibility that I have not brought it to you and set it before you. And if either deliberately or by mistake I have failed in anything, I would be forever in a state of sin in regard to my Father in heaven. However, as far as I understand, the purpose was to add and not to detract and to produce a revision of the *haftarot*;[23] but time was short. So once having read through this letter, every Jew can do it for himself and correct the errors.[24]

20. The text reads קרח (frost) but he may have meant כרך (i.e., volume).

21. He uses the biblical representation of Hannah making a silent prayer to God in the presence of Eli the priest in 1 Samuel 1:13.

22. Or perhaps, as Derenbourg states, "corrector of books." The abbreviation מ"ג could signify either term.

23. This is a reference to the liturgical practice of reading out selected passages from the prophetic books of the Bible.

24. In other words, every reader can enter the corrections from the errata list: for this as a practice, see Ann Blair, "Errata Lists and the Reader as Corrector," in *Agent of Change:*

Now you see how precious the word of the Lord is to him [i.e., König], for out of love for our Torah he has been generous with his financial aid and not desisted from the work. Therefore, you, too, who are blessed by the Lord, His holy seed, do not spare your own goods, despised silver, but with your possessions set up wisdom, for you are a city that contains all.[25] Do not consider or think an improper thought, nor act in excess and in an improper place lest God's anger be inflamed and jealousy directed against you.[26] For that which is hidden is known to our God, but we are aware of that which is ours and that of our children. We should therefore correct every place as the king [König] has said, and his law extends as far as correction and each person should abandon his evil thoughts. Thus we will return to the Lord our God and He will have compassion on us, have mercy on us and allow us to deserve Isaiah's prophecy "In place of brass I will produce gold and I will appoint your custodians in peace" [60:17], "and your sun will no longer go down for the Lord will be your everlasting light" "they will inherit the earth forever, the branches of my planting..." "the smallest shall become a thousand."

The humble Abraham son of his teacher and father Rabbi Eliezer, of blessed memory, Braunschweig.

Print Culture Studies after Elizabeth L. Eisenstein, ed. Sabrina Alcorn Baron, Eric N. Lindquist, and Eleanor F. Shevlin (Amherst, 2007), 21–41

25. This is a quotation from BT, Hullin 56b with a play on Zechariah 10:4, which suggests that the Jewish people are totally self-sufficient.

26. Once again, this is an unduly ornate passage with a variety of biblical allusions.

Bibliography

Works in Manuscript

Avkat rohel. Basel, UB, MS R III 3.

Bahya ben Asher. Commentary on Genesis and Exodus. Trans. Konrad Pellikan. Zurich, ZB, MS Car C 63.

Bernegger, Matthias, notes. In Wolfenbüttel, Herzog August Bibliothek, MS 81.30 Aug. 20.

—, notes. In Wolfenbüttel, Herzog August Bibliothek, MS Gram. 73.

Blotius, Hugo. *Odeporicon Italiae*. Vienna, Österreichische Nationalbibliothek, MS 6070.

Buxtorf, Johann, the Elder. *Album amicorum*, Basel, UB, MS AN VI 26z.

—. "Alia praefatiuncula, quibus necessitas linguae Hebraeae ex Novo Testamento potissimum evincitur. An. 1596. mense Novemb. recitata." Basel, UB, Ki Ar 189, pt. 4.

—. *Babylonia*. Basel, UB, F IX 41.

—. Commentary on Proverbs. Basel, UB, Fr Gr III 22, a.

—. Copybook. Basel, UB, A XII 20.

—. *De necessitate linguae hebraeae*, 28 October 1595. Basel, UB, Ki Ar 189, pt. 4.

—. *De quibusdam qui multas linguas calluerunt*. Basel, UB, Ki Ar 189, pt. 4.

—. *Encomion Hebraeae Linguae*. Basel, UB, Ki Ar 189, pt. 4.

—. "Ex Zohar impresso Cremonae, in fol. 1558." Basel, UB, MS A XII 7.

—. *Praelectiones Grammaticae in Proverbia Salomonis*. Basel, UB, MS Fr Gr III 22, a.

Even shetiyyah. Vatican City, BAV, MS Vat. Ebr. 250.

Galya razaya (Raza). Oxford, Bodleian Libraries, MS Opp. 104.

Galya razaya (Raza). Oxford, Bodleian Libraries, MS Opp. 526.

Levita, Elijah. *Liber Masoreth Hammasoreth*, tr. Johann Buxtorf sr. Copenhagen, Det Kongelige Bibliotek, MS E Don. var. 13 8º.

Loanz, Elijah. *Ma'aglei tsedek*. Oxford, Bodleian Libraries, MS Opp. 216.

Mahzor. Bern, Burgerbibliothek, MS 409.

Psalter (Hebrew-Latin). Ed. and trans. Matthaeus Adriani. Basel, UB, A VII 2.

Selihot. Leiden, UB, MS Or. 4721 (Scal. 4).

Solomon ben Isaac (Rashi). *Commentary* on Bible. Leiden, UB, MS Or. 4718 (Scal. 1).

Talmud, Palestinian. Leiden, UB, 4720 (Scal. 3).

Waser, Caspar. *De abbreviaturis Hebraeorum*. Zurich, ZB, MS F 172c, 863r–891v.

Zohar. Rome, Biblioteca Casanatense, MS 2971.

Manuscript Collections of Correspondence

Basel, UB, MS A XII 19
Basel, UB, MS Fr Gr I 11
Basel, UB, MS Frey-Gryn I 15
Basel, UB, MS Fr Gr II 9
Basel, UB, MS Fr Gr II 14
Basel, UB, MS G I 59
Basel, UB, MS G I 60
Basel, UB, MS G II 12
Basel, UB, MS G II 24
Basel, UB, MS Ki Ar 189
London, British Library, Burney MS 363
Marburg, Staatsarchiv Marburg, SA 81 B 1Rubr 81 Nr 21
Moscow, Russian State Library, MS Guenzburg 1213/2
Munich, BSB, clm 103359
Zurich, ZB, MS A 52
Zurich, ZB, MS F 45
Zurich, ZB, MS F 51
Zurich, ZB, MS F 167
Zurich, ZB, MS S 149
Zurich, ZB, MS S-150
Zurich, ZB, MS S-151
Zurich, ZB, MS S-154
Zurich, ZB, MS S-155
Zurich, ZB, MS S-160

Printed Books with Manuscript Notes by Johann Buxtorf the Elder

Abraham ben Samuel Zacuto: *see* Zacuto
Abravanel, Isaac. *Zevah Pesah*. Cremona: Vincenzo Conti, 1557. Basel, UB, FA V 3.
Alfasi, Isaac. *Sefer ha-halakhot*. Cracow: Isaac ben Aaron Prostitz, 1597. Basel, UB, FA III 3a–c.
Archivolti, Samuel. *Arugat ha-bosem*. Venice: Giovanni di Gara, 1520. Basel, UB, FA V 14: 1.
Avkat rohel. Venice: Giorgio di Cabali, 1566. Basel, UB, FA VIII 28: 4.
Azariah de' Rossi: *see* Rossi
Ben Sira (pseud.). *Alphabet*. Constantinople: Hayim Cesarini/Astruk of Toulon, 1519. Basel, UB, FA VI 10.
—. *Alphabet*. Isny: Paulus Fagius, 1542. Basel, UB, FB III 10/2.

Bible (second Rabbinic Bible). Venice: Daniel Bomberg, 1525. Vol. 2. Basel, UB, A 1 17.
Biblia sacra. Basel: König, 1618–19. Basel, UB, FG 19–20.
Gans, David. *Tsemah David*. Prague: Solomon and Moses Katz, 1592. Basel, UB, V 12:1–2.
Halevi, Judah. *Kuzari*. Venice: Giovanni di Gara, 1591. Basel, UB, FA V ll.
Halevi, Yeshuah. *Halikhot olam*. Mantua: Moses Elishama ben Zifroni, 1593. Basel, UB, A V 32 (3).
Hayyei olam. Freiburg i. Breisgau: n.p., 1583. Basel, UB, VI 24:2.
Hirsch, Salman Zvi. *Yudisher Theriak*. Hanau: Henricus Meyer, 1615. Basel, UB, FA VI 26: 3.
Humash (Pentateuch). Venice: Daniel Bomberg, 1525. Basel, UB, FA VI 109.
Jaffe, Mordehai ben Abraham. *Sefer levush malkhut*. Prague: Moses ben Betsalel Katz, 1623. Basel, UB, FA IV 10:1, 2.
Jonah Gerondi (ascribed). *Sefer ha-yirah*. Zurich: n.p., 1546. Basel, UB, VI 24: 1.
Joseph ibn Verga. *She'erit Joseph*. Mantua: Moses Elishama ben Zifroni, 1593. Basel, UB, FA VI 18 (1).
Judah ben Samuel he-hasid. *Sefer hasidim*. Basel: Froben, 1581. Basel, UB, FA V 1:2.
Karo, Joseph. *Shulhan arukh* (with *Hiddushim* of Moses Isserles). Venice: Giovanni di Gara, 1593–94. Basel, UB, FA IV 13.
Katz, Issahar Moses ben Betsalel. *Imrei binah*. Prague: Moses ben Betsalel Katz, 1610. Basel, UB, FA V 37.
Levita, Elijah. *Massoret ha-massoret*. Basel: Henricus Petri, 1539. Basel, UB, FA VIII 34.
Luzzatto, Jacob ben Isaac. *Kaftor va-ferah*. Basel: Froben, 1581. Basel, UB, FA V 1:1.
Menahem Azariah of Fano. *Asarah ma'amarot*. Venice: Giovanni di Gara, 1597.
Menahem Recanati: *see* Recanati
Minhagim (in Yiddish). Venice: Giovanni di Gara, 1601. Basel, UB, FA V 47.
Moses Maimonides. *Moreh nevukhim*. Sabbonieta: T. Foa, 1553. Basel, UB, FA IV 17 1.
Moses of Coucy. *Sefer mitsvot*. Ed. and trans. Sebastian Münster. Basel: Henricus Petri, 1533. Basel, UB, FA VIII 41a.
Nathan ben Yehiel. *Arukh*. Ed. Isaac Eckendorf. Basel: Waldkirch, 1599. Basel, UB, FA III 24.
Poggetto, Jacob. *Reshit hokhmah ha-katsar*. Basel: Waldkirch, 1603. Basel, UB, FA VIII 28:1.
Pomis, David de. *Tsemah David*. Venice: Giovanni di Gara, 1587. Basel, UB, FA IV 23.
Recanati, Menahem. *Sefer ta'amei ha-mitsvot*. Basel: Froben, 1591. Basel, UB, FA V 1:3.
Rossi, Azariah de'. *Me'or einayim*. Mantua: n.p., 1574. Basel, UB, FA V 4.
Sefer amana. Basel, UB, R IV 3.
Selihot. Prague: Mordechai bar Gershom haCohen, 1587. Basel, UB, FA IV 3.
Yedaiah ben Abraham Bedersi. *Behinat ha-olam*. Prague: Salman Setzer b. David, 1598. Basel, UB, FA VI 17.
Yosippon. Basel: Henricus Petri, 1541. Basel, UB, Frey-Gryn II 52 (with notes by Johann Buxtorf the Younger).

Zacuto, Abraham ben Samuel. *Sefer yuhasin*. Cracow: Isaac ben Aaron Prostitz, 1580–81. Basel, UB, FA VI 28:2.
Zohar. Cremona: Vincenzo Conti, 1559–60. Basel, UB, FA III 13.

Printed Books with Manuscript Notes by Scholars other than Buxtorf

Bernegger, Matthias. In copy of Joannes Meursius, *Glossarium Graeco-Barbarum* (Leiden: Elzevir, 1614). Wolfenbüttel, Herzog August Bibliothek, Gram. 7.3.
Buxtorf, Johannes, the Younger. In copy of Johannes Buxtorf the Elder, *De abbreviaturis* (Basel: Waldkirch, 1613). Basel, UB, FA II 24b.
Casaubon, Isaac. In copy of Benjamin of Tudela, *Itinerarium*, trans. Benito Arias Montano (Antwerp: Plantin, 1575). London, British Library, 1046.b.2.
—. In copy of Herodotus, *Historiae* (Geneva: Henri Estienne, 1566). Cambridge, Cambridge University Library, Adv.a.3.2.(1.).
—. In copy of Johannes Buxtorf the Elder, *Synagoga Iudaica* (Hanau: Wilhelm Antonius, 1604). London, British Library, 848.b.19.
—. In copy of Leo Africanus, *De totius Africae descriptione, libri ix* (Antwerp: Lattius, 1556). London, British Library, 793.d.2.
Crusius, Martin. In copy of *Thesaurus Cornucopiae* (Venice: Aldus Manutius, 1596). New Haven, Beinecke Library, Zi + 5551 Copy 3.
Gessner, Conrad. Working copy of his *Bibliotheca universalis* (Zurich: Froschauter, 1545). Zurich, ZB, DrM 3.
Scaliger, Joseph. In copy of Herodotus, *Historiae* (Geneva: Henri Estienne, 1566). Cambridge, Cambridge University Library, Adv.a.19.2.
Tremellius, Immanuel. In copy of Samson ben Isaac of Chinon, *Sefer keritot* (Cremona: Conti, 1558). London, British Library, 1956.b.13.

Printed Primary Sources

Albo, Joseph. *Sefer ikkarim*. Lublin: Kalonymus ben Mordehai Jaffe, 1597.
Altschuler, Elkana ben Gadur. *Keneh binah*. Prague: Gershon ben Mordehai Katz, 1609.
Archivolti, Samuel. *Arugat ha-bosem*. Venice: Giovanni di Gara, 1520.
Azariah de' Rossi: *see* Rossi
Bacon, Francis. *The Advancement of Learning*. Ed. Michael Kiernan. Oxford: Clarendon Press, 2000.
—. "Transportata ex comentario vetere." In *The Letters and the Life of Francis Bacon*, ed. James Spedding. 7 vols. London: Longmans, Green, Reader and Dyer, 1861–74.
Bahya ben Asher. *Be' ur 'al ha-Torah*. Venice: Bomberg, 1546.
—. *Kad ha-kemah*. Venice: Giustiniani, 1545.

Beck, Matthias Fridericus. *Monumenta antiqua Judaica, Augustae Vindel: reperta et enarrata*. Augsburg: Widow of Goebel, 1686.

Bedersi: *see* Yedaiah ben Abraham Bedersi

Bellièvre, Claude. *Souvenirs de voyages en Italie et en Orient, notes historiques, pièces de vers*. Ed. Charles Perrat. Geneva: Droz, 1956.

Bibliotheca historico-philologico-theologica 4, no. 3. Bremen: Grimm, 1721.

Bracciolini, Poggius, and Nicolaus de Niccolis. *Two Renaissance Book Hunters*. Ed. Phyllis Walter Goodhart Gordan. New York: Columbia University Press, 1974.

Buber, Salomon, ed. *Midrash Tanhuma*. Vilna: Witte and Gebrüder, 1885.

Busbecq, Ogier Ghiselin de. *Legationis Turcicae epistolae quatuor*. Paris: Plantin-Beys, 1595.

—. *The Turkish Letters of Ogier Ghiselin de Busbecq, Imperial Ambassador at Constantinople, 1554–1562*. Trans. Edward Seymour Forster. Oxford: Clarendon Press, 1927; repr., Baton Rouge: Louisiana State University Press, 2005.

Buxtorf, Johann, the Elder, ed. *Biblia Sacra Hebraica & Chaldaica*. 4 pts. in 2 vols. Basel: König, 1618–19.

—. *Bibliotheca Rabbinica*. In his *De abbreviaturis hebraicis liber novus & copiosus*, 257–335. Basel: Waldkirch, 1613.

—. *Concordantiae Bibliorum Hebraicae, nova et artificiosa method dispositae*. Basel: König, 1632.

—. *De abbreviaturis hebraicis liber novus & copiosus*. Basel: Waldkirch, 1613.

—. *De abbreviaturis Hebraicis liber novus & copiosus: cui accesserunt operis Talmudici brevis recensio: Item Bibliotheca rabbinica nova*. Ed. Johann Buxtorf the Younger. 2nd ed. Basel: König, 1640.

—. *Epitome radicum Hebraicarum et Chaldaicarum*. Basel: Waldkirch, 1607.

—. *Grammaticae chaldaicae et syriacae libri tres*. 1615; repr., Basel: König, 1685.

—. *Institutio epistolaris hebraica*. Basel: Waldkirch, 1610.

—. *Institutio epistolaris hebraica*. Basel: König, 1629.

—. *Lexicon chaldaicum talmudicum et rabbinicum*. Basel: König, 1639.

—. *Lexicon Hebraicum et Chaldaicum*. Basel: König, 1631.

—, ed. *Sylvula epistolarum Hebraicarum familiarium*. Basel: Waldkirch, 1603.

—. *Synagoga judaica: Das ist, Jüden Schul*. Basel: Sebastian Henricpetri, 1603.

—. *Synagoga Judaica auspiciis authoris jam olim Latinitate donate, nunc primum in vulgus emissa*. Ed. Johann Buxtorf the Younger. Basel: König, 1641.

—. *Synagoga Judaica*. Ed. Johann Buxtorf the Younger. Basel: Buxtorf and Decker, 1661.

—. *The Jewish Synagogue or an historical narration of the State of the Jews at this day dispersed over the face of the whole Earth, in which their religion, education, manners, sects, death and burial are fully delivered, and that out of their own writers*. Trans. "A.B." London: H.R. and Thomas Young, 1663.

—. *Synagoga Judaica*. Ed. Johann Buxtorf the Younger. 4th ed. Basel: König, 1680.

—. *Synagoga Judaica (Juden-schül)*. Trans. Alan D. Corré. 2001. https://minds.wisconsin.edu/bitstream/handle/1793/7382/index.html. Repr. as *The Jews' Synagogue*, published by the author, Kindle, 2008.

—. *Thesaurus Grammaticus linguae sanctae Hebraeae*. Basel: Waldkirch, 1609.
—. *Thesaurus grammaticus linguae sanctae Hebraeae*. Basel: König, 1651.
—. *Tiberias, sive commentarius Masorethicus*. Basel: König, 1620.
Buxtorf, Johann, the Younger. *Exercitatio sacra in historiam institutionis ss. Coenae dominicae*. Basel: König, 1642.
Calvin, John. "Ad quaestiones et obiecta Iudaei cuiusdam." In John Calvin, *Epistolae et Responsa*, 306–14. Geneva: Santandreanus, 1575.
Canini, Angelo. *Loci aliquot Novi Testamenti cum Hebraeorum originibus collati atque historice explicati*. In his *Institutiones linguae Syriacae, Assyriacae atque Thalmudicae*. Paris: Charles Estienne, 1554.
Castellio, Sébastien. *Annotationes in Pauli Epistulam ad Romanos, ex cap. ix*. Ed. and trans. Michiel Op de Coul and Mirjam van Veen. Geneva: Droz, 2020.
Chevalier, Antoine Rodolphe. *Rudimenta hebraicae linguae*. Geneva: Le Preux, 1590.
Chevalier, Pierre. "In Caput IIII Annot." In Antoine Rodolphe Chevalier, *Rudimenta hebraicae linguae*. Geneva: Le Preux, 1590.
Crusius, Martin. *Diarium*. Ed. Wilhelm Göz et al. 4 vols. Tübingen: Laupp, 1927–61.
—. *Turcograeciae libri octo*. Basel: Henricpetri, 1584.
Cyprianus, Ernst Salomon, ed. *Clarorum virorum epistolae cxvii: E bibliothecae Gothanae authographis*. Leizpig: Gleditsch, 1714.
Drusius, Ioannes. *Annotationum in totum Jesu Christi Testamentum, sive praeteritorum libri decem*. Franeker, 1612.
Fini, Fino. *In Iudaeos flagellum ex sacris scripturis excerptum*. Venice: Torresani, 1538.
Flacius Illyricus, Matthias. *Regulae et tractatus quidam de sermone sacrarum literarum*. Magdeburg: Lotter, 1551.
Galatino, Pietro. *De arcanis Catholicae veritatis*. Frankfurt: heirs of Marne, 1612.
Gans, David. *Tsemah David*. Prague: Solomon and Moses ha-Cohen, 1592.
Gedaliah ibn Yahya. *Shalshelet ha-kabbalah*. Venice: Giovanni di Gara, 1587.
Génébrard, Gilbert. *Eisagoge ad legenda Rabbinorum commentaria*. Paris: Martin le Jeune, 1559.
Gessner, Conrad. *Bibliotheca universalis*. Zurich: Froschauer, 1545.
Gikatilla, Joseph. *Ginnat egoz*. Hanau: Eliezer ben Hayyim and Elijah ben Seligman, 1614.
Halevi, Yeshuah. *Halikhot olam*. Mantua: Moses Elishama ben Zifroni, 1593.
Halperin, Israel. *The Records of the Council of the Four Lands*, 1: *1580–1792*. Rev. and ed. Israel Bartal. Jerusalem: The Bialik Institute, 1975.
Hayyim ben Bezalel. *Ets hayyim* (Tree of Life). Ed. Shraga Schneebalg. Jerusalem: Alfa, 1973.
—. *Sefer ha-hayyim*. Cracow: Aaron Prostitz, 1593.
Heller, Yom Tov Lipmann. *Megillat Eivah*. Breslau: Sulzbach, 1836.
Hess, Ernest Ferdinand. *Flagellum iudeorum: Juden Geissel*. Erfurt: Martin Wittel, 1600.
Hirsch, Salman Zvi. *Yudisher Theriak*. Hanau: Henricus Meyer, 1615.
—. *Yudisher Theriak: An Early Modern Jewish Defense of Judaism*. Ed. and trans. Morris Faierstein. Detroit: Wayne State University Press, 2018.
Isaac ben Meir. *Sha'arei dura*. Basel: Waldkirch, 1599.

Israel ben Benjamin of Belzec. *Yalkut hadash*. Amsterdam: sons of Yaakov Bak, 1657.

Jacob ben Asher, comp. *Sefer arba'ah turim*. Hanau: Hans Jakoba Henne, 1610.

Jaffe, Mordechai. *Sefer levush malkhut*. Prague: Moses ben Joseph Bezalel Katz, 1623.

Jerome. *Sancti Eusebii Hieronymi Epistulae, Pars 1: Epistulae I–LXX*. Ed. Isidor Hilberg. Corpus scriptorum ecclesiasticorum Latinorum 54. Vienna: F. Tempsky, 1910.

Joseph ibn Verga. *She'erit Joseph*. Mantua: T. Rufinelli, 1593.

Kalats, Judah. *Sefer ha-mussar*. Constantinople: Soncino, 1536; Mantua: Rufinelli, 1560; Cracow: Isaac ben Aaron Prostitz, 1598.

Kimhi, David. *Commentary on Amos*. Ed. Elijah Levita. Basel: Henricus Petri, 1531.

L'Empereur, Constantine. *Clavis Talmudica*. Leiden: Elzevir, 1634.

Léry, Jean de. *History of a Voyage to the Land of Brazil, Otherwise Called America*. Trans. Janet Whatley. Berkeley: University of California Press, 1990.

—. *Voyage au Brésil*. Ed. Michel Contat. Lausanne: Bibliothèque romande, 1972.

Levita, Elijah. *Grammatica Hebraica absolutissima*. Ed. Sebastian Münster. Basel: Io. Froben, 1525.

—. *Lexicon Hebraicum utilissimum*. Ed. and trans. Paul Fagius. Basel: no publisher, but mostly made up of sheets from the first edition [Isny: Fagius, 1541], 1557.

—. *Massoret ha-massoret*. Venice: Bomberg, 1538.

—. *Massoret ha-massoret*. Ed. and partially trans. Sebastian Münster. Basel: Henricus Petri, 1539.

—. *Meturgeman*. Isny: Fagius, 1541.

—. *Tuv ta'am*. Venice: Bomberg, 1538.

Loanz, Elijah. *Rinat dodim*. Basel: Waldkirch, 1600.

Lombardus, Markus. *Gründlicher Bericht und Erklärung von der Juden Handlungen und Ceremonien*. Basel: Apiarius, 1573.

—. *Tractatus de imposturis et ceremoniis Iudaeorum nostri temporis*. Trans. Conrad Huser. Basel: Perna. 1575.

Lorenzo of Brindisi. *Opera Omnia*. 9 vols. Padua: Ex officina typographica seminarii, 1928–56.

Luria, Solomon. *Ammudei Shlomo*. Basel: Waldkirch, 1600.

—. *Ateret Shlomo*. Basel: Waldkirch, 1599.

Luther, Martin. *Vom Schem Hamphoras* (1543). Ed. F. Cohrs and O. Brenner. In *D. Martin Luthers Werke: Kritische Gesamtausgabe*, 53: 573–648. Weimar: Böhlau, 1920.

—. *Von den Juden und ihren Lügen*. Wittemberg: Lufft, 1543.

Margaritha, Anthonius. *Der gantz Jüdisch glaub*. Augsburg: n.p., 1530.

Marti, Ramon. *Pugio fidei*. Ed. Benedict Carpzov. Leipzig: heirs of Lanckisch, 1687.

Meursius, Joannes. *Glossarium Graeco-Barbarum*. Leiden: Elzevir, 1614.

Moses ben Abraham Mat. *Ho'il Mosheh*. Prague: Moses Prostitz, 1611.

Moses Mendelssohn. *Or li-netivah*. Berlin: no publisher, 1783.

Moses of Coucy. *Sefer mitsvot gadol* (Semag). Ed. Sebastian Münster. Basel: Henricus Petri, 1533.

Münster, Sebastian. *Accentuum Hebraicorum... liber: Item liber traditionum*. Basel: Henricus Petri, 1530.

Musaphia, Benjamin. *Sefer musaf he-arukh*. Amsterdam: Benveniste, 1655.
Nathan ben Yehiel. *Arukh*. Ed. Isaac Eckendorf. Basel: Waldkirch, 1599.
Obadiah ben Jacob, Isaac. *Iggeret dofi ha-zeman*. Venice: Giovanni di Gara, 1598.
Platter, Thomas. *Beschreibung der Reisen durch Frankreich, Spanien, England und die Niederlande 1595–1600*. Ed. Rut Keiser. Basel and Stuttgart: Schwabe, 1968.
—. *Journal of a Younger Brother: The Life of Thomas Platter as a Medical Student in Montpellier at the Close of the Sixteenth Century*. Trans. Sean Jennett. London: F. Muller, 1963.
Poggetto, Jacob. *Reshit hokhmah ha-katsar*. Basel: Waldkirch, 1603.
Porchetus Salvagus. *Victoria adversus impios Hebraeos*. Paris: Guillaume Des Palins for Gilles de Gourmont and Francois Regnault, 1520.
Prideaux, John. *Viginti-duae lectiones*. Oxford: Cripps, 1648.
Reuchlin, Johannes. *Defensio . . . contra calumniatores suos Colonienses*. Tübingen: Anshelm, 1514.
—. *Johannes Reuchlins Sämtliche Werke*, Band 4: *Schriften zur Bücherstreit*, Teil 1: *Reuchlins Schriften*. Ed. Widu Wolfgang Ehlers et al. Stuttgart-Bad Cannstatt: Frommann-Holzboog, 1999.
—. *Recommendation Whether to Confiscate, Destroy, and Burn All Jewish Books*. Trans. Peter Worstman. New York: Paulist Press, 2000.
Rossi, Azariah de'. *The Light of the Eyes: An English Translation with Introduction and Notes*. Ed. and trans. Joanna Weinberg. Yale Judaica Series 31. New Haven: Yale University Press, 2001.
Samson ben Isaac of Chinon. *Sefer keritot*. Ed. Meir b. Moses. Cremona: Vincenzo Conti, 1558.
Samuel Archivolti: *see* Archivolti
Scaliger, Joseph. *The Correspondence of Joseph Justus Scaliger*. Ed. Paul Botley and Dirk van Miert. 8 vols. Geneva: Droz, 2012.
—. *Opuscula*. Paris: Beys, 1610. Repr. Frankfurt: Fischer, 1612.
—. *Opus novum de emendatione temporum*. Paris: Pattison, 1583.
—. *Scaligerana, Thuana, Perroniana, et Colomesiana*. Ed. Pierre des Maizeaux. 2 vols. Amsterdam: Covens & Mortier, 1740.
Seder tefillot. Basel: Ambrosius Froben, 1579.
Sefer ha-Zohar. Ed. Reuven Margaliot. 3 vols. 5th ed. Jerusalem: Mosad ha-Rav Ḳuḳ, 1970.
Selden, John. *The Historie of Tithes*. London: n.p., 1618.
Selihot. Augsburg: Hayyim ben David [Shahor], 1536.
Selihot mi-kol ha-shanah ke-minhag ha-ashkenazim. . . . Venice: Cornelio Adelkind, 1548.
Selihot. Cracow: Tsevi Hirsch Sundeles, 1584.
Selihot. Prague: Mordehai bar Gershon, 1587.
Selihot. Basel: Waldkirch, 1599–1600.
Selihot. Venice: Gian Bragadin, 1600.
Selihot. Lublin: no publisher, 1603.
Selihot. Hanau: no publisher, 1616.
Selihot. Amsterdam: Uri Phoebus, 1664.

Selihot. Metz: Ephraim Habemard, 1822.

Sixtus of Siena. *Bibliotheca sancta*. 3rd ed. Cologne: Cholin, 1586.

"State of the Jews." *The British Review, and London Critical Journal* 16 (1820): 346–75.

Talmud, Babylonian. Basel: Ambrosius Froben, 1579.

Tissard, François. *Grammatica Hebraica*. Paris: Gourmont, 1508.

Torah Nevi'im Ketuvim. Ed. Jacob ben Hayyim ibn Adonijah. Venice: Bomberg, 1525.

Tossanus, Daniel. *Johannis Buxtorfii senioris .. Vita et mors*. Basel: König, 1630.

Treves, Naftali Hirts. *Naftulei elohim*. Ferrara: Abraham Usque, 1556.

Un beau livre d'histoires = Eyn schön mayse bukh. Ed. and trans. Astrid Starck. Basel: Schwabe, 2004.

Ussher, James. *Answer to a Jesuit*. Cambridge: Deighton and Parker, 1835.

—. *The Correspondence of James Ussher, 1600–1656*. Ed. Elisabethanne Boran. Trans. David Money. 3 vols. Dublin: Irish Manuscripts Commission, 2015.

Wagenseil, Johann. *Belehrung der Jüdisch-Teutschen Red- und Schreibart*. Königsberg: Rhode, 1699.

Wakefield, Robert. *Syntagma de Hebraeorum codicum incorruptione*. London: Wynkyn de Worde, 1534.

Waser, Caspar. *Archetypus grammaticae Hebraeae duabus praecipuis partibus, etymologia & syntaxi, absolutus*. Basel: Waldkirch, 1600.

Weil, Jacob. *Hilkhot shehitah u-vedikah*. Cracow: Isaac ben Aaron Prostitz, 1600.

Yalkut ha-Torah (Shimoni). 2 vols. Cracow: Isaac ben Aaron Prostitz, 1595.

Yedaiah ben Abraham Bedersi. *Behinat ha-olam*. Mantua: Estellina Conat, 1474; reprints incl. Cracow: Isaac ben Aaron Prostitz, 1591, and Prague: Yekutiel Zalman, 1598.

Zacuto, Abraham ben Samuel. *Sefer yuhasin*. Constantinople: Samuel Sullam, 1566.

—. *Sefer yuhasin*. Cracow: Isaac ben Aaron Prostitz, 1580–81.

Zwinger, Theodor. *Methodus apodemica,in eorum gratiam, qui cum fructu in quocunq[ue] tandem vitae genere peregrinari cupiunt*. Basel: Episcopius, 1577.

Secondary Sources

Abate, Emma. "Elias Levita the Lexicographer and the Legacy of *Sefer ha-Shorashim*." *Sefarad* 76 (2016): 289–312.

—. "Filologia e qabalah: La collezione ebraica di Egidio da Viterbo alla biblioteca Angelica di Roma." *Archivio italiano per la storia della pietà* 26 (2013): 413–51.

Adams, Jonathan, and Cordelia Heß, eds. *Revealing the Secrets of the Jews: Johannes Pfefferkorn and Christian Writings about Jewish Life and Literature in Early Modern Europe*. Berlin: De Gruyter, 2017.

Almási, Gábor. "*Experientia* and the Machiavellian Turn in Religio-Political and Scientific Thinking: Basel in 1580." *History of European Ideas* 42 (2016): 857–81.

Andreatta, Michela. "Collecting Hebrew Epitaphs in the Early Modern Age: The Christian Hebraist as Antiquarian." In *Jewish Books and Their Readers: Aspects of the*

Intellectual Life of Christians and Jews in Early Modern Europe, ed. Scott Mandelbrote and Joanna Weinberg, 260–86. Leiden: Brill, 2016.

Barbier, Frederic, et al., eds. *Le livre et l'historien: Études offertes en l'honneur du Professor Henri-Jean Martin.* Histoire et civilisation du livre 24. Geneva: Droz, 1997.

Becker, Peter, and William Clark, eds. *Little Tools of Knowledge: Historical Essays on Academic and Bureaucratic Practices.* Ann Arbor: University of Michigan Press, 2001.

Beit-Arié, Malachi. "Eliyahu Levita as a Scribe, Author-Scribe and Codicologist." Paper presented at Between Jews and Christians: The Many Faces of the Jewish Polymath Elijah (Bocher) Levita (1469–1549), Oxford, 2007.

Bell, Dean Phillip. *Jewish Identity in Early Modern Germany: Memory, Power and Community.* Aldershot: Ashgate, 2007.

Bell, Dean Phillip, and Stephen G. Burnett, eds. *Jews, Judaism, and the Reformation in Sixteenth-Century Germany.* Studies in Central European Histories 37. Leiden: Brill, 2006.

Ben-Sasson, Haim Hillel. "The Reformation in Contemporary Jewish Eyes." *Proceedings of the Israel Academy of Sciences and Humanities* 4 (1971): 239–326.

Ben-Zaken, Avner. *Cross-Cultural Scientific Exchanges in the Eastern Mediterranean, 1560–1660.* Baltimore: Johns Hopkins University Press, 2010.

Berger, Andreas. *Stadt ohne Juden? Präsenz und Absenz in der Frühen Neuzeit.* Basler Beiträge zur Geschichtswissenschaft. Basel: Schwabe, 2024.

Berkovitz, Jay R. "Crisis and Authority in Early Modern Ashkenaz." *Jewish History* 26 (2012): 179–99.

Blair, Ann. "Errata Lists and the Reader as Corrector." In *Agent of Change: Print Culture Studies after Elizabeth L. Eisenstein*, ed. Sabrina Alcorn Baron, Eric N. Lindquist, and Eleanor F. Shevlin, 21–41. Amherst: University of Massachusetts Press, in association with the Center for the Book, Library of Congress, Washington, DC, 2007.

—. "Humanist Methods in Natural Philosophy: The Commonplace Book." *Journal of the History of Ideas* 53 (1992): 541–51.

—. *The Theater of Nature: Jean Bodin and Renaissance Science.* Princeton: Princeton University Press, 1997.

—. *Too Much to Know: Managing Scholarly Information before the Modern Age.* New Haven: Yale University Press, 2010.

Boehmer, Edward, ed. *Bibliotheca Wiffeniana: Spanish Reformers of Two Centuries from 1520.* 3 vols. Strasbourg: Karl Trübner; London: Trübner, 1874–1904.

Bollbuck, Harald. "Searching for the True Religion: The Church History of the *Magdeburg Centuries* between Critical Methods and Confessional Polemics." *Renaissance Studies* 35 (2021): 100–117.

—. *Wahrheitszeugnis, Gottes Auftrag und Zeitkritik: Die Kirchengeschichte der Magdeburger Zenturien und ihre Arbeitstechniken.* Wolfenbütteler Forschungen 138. Wiesbaden: Harrassowitz, 2014.

Bonfil, Robert. "How Golden Was the Age of the Renaissance in Jewish Historiography?" *History and Theory* 27.4 (1988): 78–102.

Brisman, Shimeon. *A History and Guide to Judaic Bibliography*. Bibliographica Judaica 7. Cincinnati: Hebrew Union College; New York: Ktav, 1977.

Burman, Thomas. *Reading the Qur'an in Latin Christendom, 1140–1560*. Material Texts. Philadelphia: University of Pennsylvania Press, 2007.

Burnett, Stephen. *Christian Hebraism in the Reformation Era (1500–1600): Authors, Books and the Transmission of Jewish Learning*. Library of the Written Word 19. Leiden: Brill, 2012.

—. "The Christian Humanism of Johann Buxtorf (1564–1629)." PhD diss. University of Wisconsin, Madison, 1990.

—. "Distorted Mirrors: Antonius Margaritha, Johann Buxtorf and Christian Ethnographies of the Jews." *Sixteenth Century Journal* 25 (1994): 275–87.

—. *From Christian Hebraism to Jewish Studies: Johannes Buxtorf (1564–1629) and Hebrew Learning in the Seventeenth Century*. Studies in the History of Christian Thought. Leiden: Brill, 1996.

—. "German Jewish Printing in the Reformation Era (1530–1633)." In *Jews, Judaism, and the Reformation in Sixteenth-Century Germany*, ed. Dean Phillip Bell and Stephen G. Burnett, 503–27. Studies in Central European Histories 37. Leiden: Brill, 2006.

—. "Hebrew Censorship in Hanau: A Mirror of Jewish-Christian Coexistence in Seventeenth-Century Germany." In *The Expulsion of the Jews: 1492 and After*, ed. Raymond Waddington and Arthur Williamson, 199–222. New York: Garland, 1994.

—. "Johannes Buxtorf I and the Circumcision Incident of 1618." *Basler Zeitschrift für Geschichte und Altertumskunde* 89 (1989): 135–44.

—. "Luther's Chief Witness: Anthonius Margarita's *Der gantz jüdisch glaub* (1530/1531)." In *Revealing the Secrets of the Jews: Johannes Pfefferkorn and Christian Writings about Jewish Life and Literature in Early Modern Europe*, ed. Jonathan Adams and Cordelia Heß, 183–200. Berlin: De Gruyter, 2017.

Buxtorf-Falkeisen, Carl. *Johannes Buxtorf Vater, Prof. ling. hebr. 1564–1629, erkannt aus seinem Briefwechsel*. Basel: Detloff, 1860.

Calis, Richard. "The Impossible Reformation: Protestant Europe and the Greek Orthodox Church." *Past & Present* 259 (May 2023): 43–76.

—. "Martin Crusius and the Discovery of Ottoman Greece." PhD diss. Princeton University, 2020. Revised version forthcoming as *The Discovery of Ottoman Greece: Knowledge, Encounter, and Belief in the Mediterranean World of Martin Crusius (1526–1607)*, Cambridge: Harvard University Press, 2025.

Campanini, Saverio. "The 'Ezra Scroll' of Bologna: Vicissitudes of an Archetype between Memory and Oblivion." In *The Ancient Sefer Torah of Bologna*, ed. Mauro Perani, 29–50. Studies in Jewish History and Culture 59. Leiden: Brill, 2019.

Cancik, Hubert. "Crusius contra Frischlinum: Geschichte einer Feindschaft." In *Nicodemus Frischlin (1547–1590): Poetische und prosaische Praxis under den Bedingungen des konfessionellen Zeitalters*, ed. Sabine Holtz and Dieter Mertens, 261–95. Arbeiten und Editionen zur mittleren deutschen Literatur 1. Stuttgart: Frommann-Holzboog, 1999.

Carlebach, Elisheva. *Divided Souls: Converts from Judaism in Germany, 1500–1750*. New Haven: Yale University Press, 2001.

Chatelain, Jean-Marc. “Les recueils d’*Adversaria* aux xvie et xviie siècles: Des pratiques de la lecture savante au style de l’érudition.” In *Le livre et l’historien: Études offertes en l’honneur du Professor Henri-Jean Martin*, ed. Frederic Barbier et al., 169–86. Geneva: Droz, 1997.

Clark, Frederic. *The First Pagan Historian: The Fortunes of a Fraud from Antiquity to the Enlightenment*. Oxford: Oxford University Press, 2020.

—. “Reading the Life Cycle: History, Antiquity, and *fides* in Lambarde’s *Perambulation* and Beyond.” *Journal of the Warburg and Courtauld Institutes* 81 (2018): 191–208.

—. “Writing about Reading: On Early Modern Annotation Practice and the Future of Book History.” In *Gabriel Harvey and the History of Reading: Essays by Lisa Jardine and Others*, ed. Anthony Grafton, Nicholas Popper, and William Sherman, 307–24. London: UCL Press, 2024.

Cohen, Mark. “Leone da Modena’s *Riti*: A Seventeenth-Century Plea for Social Toleration of Jews.” *Jewish Social Studies* 34 (1972): 287–321.

Crowther, Daniel, Daniel D. Hornkohl, and Geoffrey Khan, eds. *Studies in the Masoretic Tradition of the Hebrew Bible*. Semitic Languages and Cultures. Cambridge: Open Book Publishers, 2022.

Daston, Lorraine. “Perchè i fatti sono brevi?” *Quaderni storici* 108 (2001): 745–70.

Daston, Lorraine, and Elizabeth Lunbeck, eds. *Histories of Scientific Observation*. Chicago: University of Chicago Press, 2011.

Davies, Surekha. *Renaissance Ethnography and the Invention of the Human: New Worlds, Maps and Monsters*. Cambridge Social and Cultural Histories 24. Cambridge: Cambridge University Press, 2016.

de Beer, Esmond Samuel. Introduction to John Evelyn, *The Diary of John Evelyn*, 1: 1–131. 6 vols. Oxford: Clarendon Press, 1955.

de Lange, Nicholas. “Jews in the Age of Justinian.” In *The Cambridge Companion to the Age of Justinian*, ed. Michael Maas, 401–26. Cambridge Companions to the Ancient World. Cambridge: Cambridge University Press, 2005.

Delaruelle, Louis. “Un recueil d’adversaria autographes de Girolamo Aleandro.” *Mélanges d’archéologie et d’histoire publiés par l’École française de Rome* 20 (1900): 3–21.

Della Rocca de Candal, Geri, Anthony Grafton, and Paolo Sachet, eds. *Printing and Misprinting: A Companion to Mistakes and In-House Corrections in Renaissance Europe, 1450–1650*. Oxford: Oxford University Press, 2022.

Derenbourg, Joseph. “L’édition de la Bible rabbinique de Jean Buxtorf.” *Revue des Études Juives* 30 (1895): 70–78.

Deutsch, Yaacov. *Judaism in Christian Eyes: Ethnographic Descriptions of Jews and Judaism in Early Modern Europe*. Trans. Avi Aronsky. Oxford: Oxford University Press, 2012.

DeVore, David. “Character and Convention in the Letters of Eusebius’s *Ecclesiastical History*.” *Journal of Late Antiquity* 7 (2014): 223–52.

Diemling, Maria. “Anthonius Margaritha and his ‘Der Gantz Jüdisch Glaub.’” In *Jews, Judaism and the Reformation in Sixteenth-Century Germany*, ed. Dean Philip Bell and Stephen Burnett, 303–33. Studies in Central European Histories. Leiden: Brill, 2006.

—. "Patronage, Representation and Conversion: Victor von Carben (1423–1515) and His Social Networks." In *Revealing the Secrets of the Jews: Johannes Pfefferkorn and Christian Writings about Jewish Life and Literature in Early Modern Europe*, ed. Jonathan Adams and Cordelia Heß, 157–81. Berlin: de Gruyter, 2017.

Dunkelgrün, Theodor. "The Hebrew Library of a Renaissance Humanist: Andreas Masius and the Bibliography to his *Iosuae Imperatoris Historia* (1574), with a Latin Edition and an Annotated English Translation." *Studia Rosenthaliana* 42–43 (2010–11): 197–252.

—. "The Humanist Discovery of Hebrew Epistolography." In *Jewish Books and Their Readers: Aspects of the Intellectual Life of Christians and Jews in Early Modern Europe*, ed. Scott Mandelbrote and Joanna Weinberg, 211–59. Leiden: Brill, 2016.

Dweck, Yaacob. *The Scandal of Kabbalah: Leon Modena, Jewish Mysticism, Early Modern Venice*. Princeton: Princeton University Press, 2011.

Eddy, Matthew. *Media and the Mind: Art, Science, and Notebooks as Paper Machines, 1700–1830*. Chicago: University of Chicago Press, 2023.

Eden, Kathy. *Hermeneutics and the Rhetorical Tradition: Chapters in the Ancient Legacy and Its Humanist Reception*. Yale Studies in Hermeneutics. New Haven: Yale University Press, 1997.

Eideneier, Hans, ed. *Graeca recentiora in Germania: Deutsch-griechische Kulturbeziehungen vom 15. bis 19. Jahrhundert*. Wolfenbütteler Forschungen 59. Wiesbaden: Harrassowitz, 1994.

—. "Von der Handschrift zum Druck: Martinus Crusius und David Höschel als Sammler griechischer Venezianer Volksdrucke des 16. Jahrhunderts." In *Graeca recentiora in Germania: Deutsch-griechische Kulturbeziehungen vom 15. bis 19. Jahrhundert*, ed. Hans Eideneier, 93–111. Wolfenbütteler Forschungen 59. Wiesbaden: Harrassowitz, 1994.

Elyada, Aya. *A Goy Who Speaks Yiddish: Christians and the Jewish Language in Early Modern Germany*. Stanford Studies in Jewish History and Culture. Stanford: Stanford University Press, 2012.

—. "Protestant Scholars and Yiddish Studies in Early Modern Europe." *Past & Present* 203 (2009): 69–98.

Epstein, Isidore, Ephraim Levine, and Cecil Roth, eds. *Essays in Honour of the Very Rev. Dr. J.H. Hertz, Chief Rabbi of the United Hebrew Congregations of the British Empire, on the Occasion of his Seventieth Birthday, September 25, 1942*. London: Edward Goldston, 1942.

Forster, Leonard. *The Icy Fire: Five Studies in European Petrarchism*. Cambridge: Cambridge University Press, 1969.

Fränkel Goldschmidt, Chava. "Jüdische Religion und Kultur in Frankfurt am Main am 16 und 17 Jahrhundert: Yuspa Hahn und sein 'Yosif omez.'" In *Jüdische Kultur in Frankfurt am Main von den Anfängen bis zur Gegenwart*, ed. Karl E. Grőzinger, 101–21. Jüdische Kultur 1. Wiesbaden: Otto Harrassowitz, 1997.

Fragnito, Gigliola. *Rinascimento perduto: La letteratura italiana sotto gli occhi dei censori (secoli XV–XVII)*. Collezione di testi e di studi. Bologna: il Mulino, 2019.

Frakes, Jerrold. *The Cultural Study of Yiddish in Early Modern Europe*. New York: Palgrave Macmillan, 2007.

Franz-Klauser, Olivia. *Die hebräischen Handschriften: Verzeichnis mit Einleitung*. Bern: Burgerbibliothek Bern, 1999.

Gaisser, Julia Haig. "Poggio and Other Book Hunters." In *Poggio Bracciolini and the Re(dis)covery of Antiquity: Textual and Material Conditions: Proceedings of the Symposium held at Bryn Mawr College on April 8–9, 2016*, ed. Roberta Ricci, 173–88. Florence: Firenze University Press, 2020.

Gerritsen, Johan. "Printing at Froben's: An Eye-Witness Account." *Studies in Bibliography* 44 (1991): 144–63.

Gierl, Martin. *Pietismus und Aufklärung: theologische Polemik und die Kommunikationsreform der Wissenschaft am Ende des 17. Jahrhunderts*. Veröffentlichungen des Max-Planck-Instituts für Geschichte 129. Göttingen: Vandenhoeck & Ruprecht, 1997.

Gilly, Carlos. "Die Censur von Castellios Dialogi quattuor durch die Basler Theologen (1578)." In *Freiheitsstufen der Literaturverbreitung: Zensurfragen, verbotene und verfolgte Bücher*, ed. József Jankovics and S. Katalin Németh, 147–76. Wiesbaden: Harrassowitz, 1998.

—. *Spanien und der Basler Buchdruck bis 1600: Ein Querschnitt durch die spanische Geistesgeschichte aus der Sicht einer europäischen Buchdruckerstadt*. Basel: Helbing & Lichtenhahn, 1985.

Ginzberg, Lewis. *Legends of the Jews*. Trans. Henrietta Szold. 7 vols. Philadelphia: Jewish Publication Society of America, 1909–38.

Ginzburg, Carlo. *Clues, Myths, and the Historical Method*. Trans. John Tedeschi and Anne Tedeschi. Baltimore: Johns Hopkins University Press, 1992.

—. *Threads and Traces: True, False, Fictive*. Trans. Anne Tedeschi and John Tedeschi. Berkeley: University of California Press, 2012.

Goldin, Simha. *Apostasy and Jewish Identity in High Middle Ages Northern Europe: "Are you still my brother?"* Manchester: Manchester University Press, 2014.

Goldschmidt, Ernst D., ed. *Seder ha-selihot ke-minhag Polin ve-rov ha-kehillot be-Erets Yisra'el*. Jerusalem: Mosad harav Kook, 1965.

Goshen-Gottstein, Moses. "Foundations of Biblical Philology in the Seventeenth Century: Christian and Jewish Dimensions." In *Jewish Thought in the Seventeenth Century*, ed. Isadore Twersky and Bernard Septimus, 77–94. Harvard Judaic Texts and Studies. Cambridge, MA: Harvard University Press, 1987.

Gottlieb, Ephraim. *The Kabbalah in the writings of R. Baḥya ben Asher ibn Halawa* (in Hebrew). Jerusalem: Kiryat sefer, 1970.

Goyet, Francis. *Le sublime du "lieu commun": L'invention rhétorique dans l'Antiquité et à la Renaissance*. Bibliothèque littéraire de la Renaissance Série 3 32. Paris: Champion; Geneva: Slatkine, 1996.

Grafton, Anthony. *The Culture of Correction in Renaissance Europe*. Panizzi Lectures. London: British Library, 2011.

—. "Humanism and the Mishnah: Paul Fagius Edits Avot." In *The Mishnaic Moment: Jewish Law among Jews and Christians in Early Modern Europe*, ed. Piet van Boxel,

Kirsten Macfarlane, and Joanna Weinberg, 47–67. Oxford: Oxford University Press, 2022.

—. *Inky Fingers: The Making of Books in Early Modern Europe*. Cambridge, MA: Belknap Press of Harvard University Press, 2020.

—. *Joseph Scaliger: A Study in the History of Classical Scholarship*. 2 vols. Oxford: Clarendon Press, 1983–93.

—. "'Pandects of the Jews': A French, Swiss and Italian Prelude to John Selden." In *Jewish Books and Their Readers: Aspects of the Intellectual Life of Christians and Jews in Early Modern Europe*, ed. Scott Mandelbrote and Joanna Weinberg, 169–88. Leiden: Brill, 2016.

—. "Some Early Citizens of the *Respublica Litterarum Sacrarum*: Christian Scholars and the Masorah Before 1550." *Reformation* 23 (2018): 6–28.

Grafton, Anthony, and Glenn W. Most, eds. *Canonical Texts and Scholarly Practices: A Global Comparative Approach*. Cambridge: Cambridge University Press, 2016.

Grafton, Anthony, Nicholas Popper, and William Sherman, eds. *Gabriel Harvey and the History of Reading: Essays by Lisa Jardine and Others*. London: UCL Press, 2024.

Grafton, Anthony, and Joanna Weinberg. "Johann Buxtorf Makes a Notebook." In *Canonical Texts and Scholarly Practices: A Global Comparative Approach*, ed. Anthony Grafton and Glenn W. Most, 275–98. Cambridge: Cambridge University Press, 2016.

Grafton, Anthony, and Joanna Weinberg, with Alastair Hamilton. *"I have always loved the Holy Tongue": Isaac Casaubon, the Jews and a Forgotten Chapter in Renaissance Scholarship*. Cambridge, MA: Belknap Press of Harvard University Press, 2011.

Grendler, Paul. *The Roman Inquisition and the Venetian Press, 1540–1605*. 1977. Repr. Princeton: Princeton University Press, 2015.

Grőzinger, Karl E., ed. *Jüdische Kultur in Frankfurt am Main von den Anfängen bis zur Gegenwart*. Wiesbaden: Otto Harrassowitz, 1997.

Habermann, Abraham M. "The Printer Israel Zifroni and His Son Elishma and Their Books." In *Studies in the History of Hebrew Printers and Books* (in Hebrew), ed. Abraham M. Habermann, 215–92. Jerusalem: Rubin Mas, 1978.

de Hamel, Christopher. *The Posthumous Papers of the Manuscripts Club*. London: Allen Lane, 2022.

Hardy, Nicholas. *Criticism and Confession: The Bible in the Seventeenth-Century Republic of Letters*. Oxford-Warburg Studies. Oxford: Oxford University Press, 2017.

Hardy, Nicholas, and Dmitri Levitin, eds. *Confessionalism and Erudition in Early Modern Europe: An Episode in the History of the Humanities*. Oxford: Published for the British Academy by Oxford University Press, 2019.

Hartmann, Martina. *Humanismus als Kirchenkritik: Matthias Flacius Illyricus als Erforscher des Mittelalters*. Beiträge zur Geschichte und Quellenkunde des Mittelalters 19. Stuttgart: Thorbecke, 2001.

Heller, Marvin. "Ambrosius Froben, Israel Zifroni and Hebrew Printing in Freiburg im Breisgau." *Gutenberg-Jahrbuch* 80 (2005): 137–48.

—. *Printing the Talmud: A History of the Earliest Printed Editions of the Talmud*. Brooklyn, NY: Im Hasefer, 1992.

—. *The Seventeenth-Century Hebrew Book: An Abridged Thesaurus*. Brill's Series in Jewish Studies 41. Leiden: Brill, 2011.

—. *The Sixteenth-Century Hebrew Book: An Abridged Thesaurus*. Brill's Series in Jewish Studies 33. Leiden: Brill, 2004.

Heszer, Catherine. "The Codification of Legal Knowledge in Late Antiquity: The Talmud Yerushalmi and Roman Law Codes." In *The Talmud Yerushalmi and Graeco-Roman Culture*, ed. Peter Schäfer, 1: 581–641. Tübingen: Mohr Siebeck, 1998.

Hieronymus, Frank. "Gewissen und Staatskirchenraison: Basler Theologie und Zensur um 1578." *Archiv für Reformationsgeschichte* 82 (1991): 209–38.

Holford-Strevens, Leofranc and Amiel Vardi, eds. *The Worlds of Aulus Gellius*. Oxford: Oxford University Press, 2004.

Hollerich, Michael. *Making Christian History: Eusebius of Caesarea and His Readers*. Christianity in Late Antiquity 11. Oakland: University of California Press, 2021.

Holtz, Sabine, and Dieter Mertens, eds. *Nicodemus Frischlin (1547–1590): Poetische und prosaische Praxis under den Bedingungen des konfessionellen Zeitalters*. Arbeiten und Editionen zur mittleren deutschen Literatur 1. Stuttgart: Frommann-Holzboog, 1999.

Horovitz, Marcus. *Frankfurter Rabbinen: Ein Betrag zur Geschichte der Israelitischen Gemeinde in Frankfurt am M.* 4 vols. Frankfurt a. M.: Commissions-Verlag der Jaeger'schen Buchhandlung, 1882–85.

Hotson, Howard. "Irenicism in the Confessional Age: The Holy Roman Empire, 1563–1648." In *Conciliation and Confession: Struggling for Unity in the Age of Reform, 1415–1648*, ed. Howard Louthan and Randall Zachman, 228–85. Notre Dame, IN: University of Notre Dame Press, 2004.

Hunter, Michael, ed. *Archives of the Scientific Revolution: The Formation and Exchange of Ideas in Seventeenth-Century Europe*. Woodbridge: Boydell, 1998.

—. "Mapping the Mind of Robert Boyle: The Evidence of the Boyle Papers." In *Archives of the Scientific Revolution: The Formation and Exchange of Ideas in Seventeenth-Century Europe*, ed. Michael Hunter, 121–36. Woodbridge: Boydell, 1998.

Hüttenmeister, Nathanja, and Andreas Lehnhardt. "Newly Found Gravestones from Magenza." In *Death in Jewish Life and Burial and Mourning Customs among Jews of Europe and Nearby Communities*, ed. Stefan C. Reif, Andreas Lehnardt, and Avriel Bar-Levav, 213–23. Berlin: de Gruyter, 2014.

Idel, Moshe. "Yosef H. Yerusalmi's 'Zakhor': Some Observations." *Jewish Quarterly Review* 97 (2007): 491–501.

Jankovics, József, and S. Katalin Németh, eds. *Freiheitsstufen der Literaturverbreitung – Zensurfragen, verbotene und verfolgte Bücher*. Wolfenbütteler Abhandlungen zur Renaissanceforschung 18. Wiesbaden: Harrassowitz, 1998.

Jardine, Lisa and Anthony Grafton. "'Studied for Action': How Gabriel *Harvey* Read His Livy." *Past & Present* 129 (1990): 30–78.

Johnson, Carina. *Cultural Hierarchy in Sixteenth-Century Europe: The Ottomans and Mexicans*. Cambridge: Cambridge University Press, 2011.

Jütte, Daniel. *Das Zeitalter des Geheimnisses: Juden, Christen und die Ökonomie des Geheimen (1400–1800)*. Göttingen: Vandenhoeck & Ruprecht, 2011.

Kaegi, Werner. "Machiavelli in Basel." *Basler Zeitschrift für Geschichte und Altertumskunde* 39 (1940): 5–51.

Katritzky, M.A. *Healing, Performance and Ceremony in the Writings of Three Early Modern Physicians: Hippolytus Guarinonius and the Brothers Felix and Thomas Platter*. History of Medicine in Context. Farnham: Ashgate, 2012.

Keller, Marcus, and Javier Irigoyen-García, eds. *The Dialectics of Orientalism in Early Modern Europe*. London: Palgrave Macmillan, 2018.

Knoch-Mund, Gaby. *Disputationsliteratur als Instrument antijüdischer Polemik: Leben und Werk des Marcus Lombardus, eines Grenzgängers zwischen Judentum und Christentum im Zeitalter des deutschen Humanismus*. Bibliotheca Germanica 33. Tübingen: Francke, 1997.

Kühlmann, Wilhelm. *Gelehrtenrepublik und Fürstenstaat: Entwicklung und Kritik des deutschen Späthumanismus in der Literatur des Barockzeitalters*. Studien und Texte Zur Sozialgeschichte der Literatur 3. Tübingen: Max Niemeyer, 1982.

Langer, Ruth. *Cursing the Christians?: A History of the Birkat HaMinim*. Oxford: Oxford University Press, 2011.

Lawee, Eric. *Rashi's Commentary on the Torah: Canonization and Resistance in the Reception of a Jewish Classic*. Oxford: Oxford University Press, 2019.

Lehmann, Daniel. "'Such an Illumination Cannot Occur': Anthonius Margaritha, the Reformation, and the Polemic against the Jews." *Archiv für Reformationsgeschichte – Archive for Reformation History* 111 (2020): 55–77.

Letteney, Mark. *The Christianization of Knowledge in Late Antiquity: Intellectual and Material Transformations*. Cambridge: Cambridge University Press, 2023.

Lipschitz, Avi. "A Natural Yet Providential Tongue: Moses Mendelssohn on Hebrew as a Language of Action." In *Language as Bridge and Border: Linguistic, Cultural and Political Constellations in Eighteenth to Twentieth Century German-Jewish Thought*, ed. Sabine Sander, 1–15. Berlin: Hentrich & Hentrich, 2015.

Loop, Jan. *Johann Heinrich Hottinger: Arabic and Islamic Studies in the Seventeenth Century*. Oxford-Warburg Studies. Oxford: Oxford University Press, 2013.

Loop, Jan, and Jill Kraye, eds. *Scholarship between Europe and the Levant: Essays in Honour of Alastair Hamilton*. The History of Oriental Studies 8. Leiden: Brill, 2020.

Lüber, Alban Norbert. "*Die Basler Zensurpolitik* in der zweiten Hälfte des 16. Jahrhunderts." *Basler Zeitschrift für Geschichte und Altertumskunde* 97 (1997): 77–141.

Lupher, David. *Romans in a New World: Classical Models in Sixteenth-Century Spanish America*. History, Languages, and Cultures of the Spanish and Portuguese Worlds. Ann Arbor: University of Michigan Press, 2003.

Lyon, Gregory. "Flacius, Baudouin, and the Plan for the Magdeburg Centuries." *Journal of the History of Ideas* 64 (2003): 253–72.

Maas, Michael, ed. *The Cambridge Companion to the Age of Justinian*. Cambridge Companions to the Ancient World. Cambridge: Cambridge University Press, 2005.

Macfarlane, Kirsten. *Biblical Scholarship in an Age of Controversy: The Polemical World of Hugh Broughton 1549–1612*. Oxford: Oxford University Press, 2021.

Maclean, Ian. *Scholarship, Commerce, Religion: The Learned Book in the Age of Confessions, 1560–1630*. Cambridge, MA: Harvard University Press, 2012.

Maïer, Ida. *Les manuscrits d'Ange Politien: Catalogue descriptif.* Travaux d'humanisme et Renaissance 70. Geneva: Droz, 1965.

Malcolm, Noel. "Thomas Harrison and His 'Ark of studies': An Episode in the History of the Organisation of Knowledge." *The Seventeenth Century* 19 (2004): 196–232.

Mandelbrote, Scott, and Joanna Weinberg, eds. *Jewish Books and Their Readers: Aspects of Jewish and Christian Intellectual Life in Early Modern Europe*. Brill's Series in Church History 75. Leiden: Brill, 2016.

Mantel, Hilary. *Wolf Hall*. London: Fourth Estate, 2010.

Marcus, Hannah. *Forbidden Knowledge: Medicine, Science and Censorship in Early Modern Italy*. Chicago: University of Chicago Press, 2020.

Mayer, Yakov Z. *Editio Princeps: The 1525 Venice Edition of the Palestinian Talmud and the Beginning of Hebrew Printing* (in Hebrew). Jerusalem: Magnes, 2022.

McKenzie, Donald. "Printers of the Mind: Some Notes on Bibliographical Theories and Printing-House Practices." *Studies in Bibliography* 22 (1969): 1–75. Repr. in McKenzie, *Making Meaning: "Printers of the Mind" and Other Essays*, ed. Peter McDonald and Michael Suarez, 13–85. Amherst: University of Massachusetts Press, 2002.

Menzel-Reuthers, Arno, and Marina Hartmann, eds. *Catalogus und Centurien: Interdisziplinäre Studien zu Matthias Flacius und den Magdeburger Centurien*. Spätmittelalter, Humanismus, Reformation 45. Tübingen: Mohr Siebeck, 2008.

Miller, D. Gary. *The Oxford Gothic Grammar*. Oxford: Oxford University Press, 2019.

Moennig, Ulrich. "Matthias Berneggers Handexemplar des Glossarium graecobarbarum des Ioannes Meursius mit Korrekturen des Metrophanes Kritopoulos." In *Graeca recentiora in Germania: Deutsch-griechische Kulturbeziehungen vom 15. bis 19. Jahrhundert*, ed. Hans Eideneier, 161–98. Wolfenbütteler Forschungen 59. Wiesbaden: Harrassowitz, 1994.

Molino, Paolo. "Istruzioni per un viaggio in Italia: 'Hodoeporicum Hugoni Blotii earum rerum quas in Italia vidit et observavit.'" *Biblos* (2006): 115–37.

Momigliano, Arnaldo, ed. *The Conflict Between Paganism and Christianity in the Fourth Century*. Oxford-Warburg Studies. Oxford: Clarendon Press, 1963.

—. "Pagan and Christian Historiography in the Fourth Century A.D." In *The Conflict Between Paganism and Christianity in the Fourth Century*, ed. Momigliano, 79–99. Oxford: Clarendon Press, 1963.

—. "Pagan and Christian Historiography in the Fourth Century A.D." In *Terzo contributo alla storia degli studi classici e del mondo antico*, 1: 87–109. 2 vols. Rome: Storia e Letteratura, 1966.

—. "The Place of Herodotus in the History of Historiography." In *Studies in Historiography*, ed. Momigliano, 127–42. London: Weidenfeld and Nicolson, 1966.

—. *Studies in Historiography*. Harper Torchbooks: Academic Library TB1288. London: Weidenfeld and Nicolson, 1966.

—. *Terzo contributo alla storia degli studi classici e del mondo antico*. 2 vols. Rome: Storia e Letteratura, 1966.

Moss, Ann. *Printed Commonplace-Books and the Structuring of Renaissance Thought*. Oxford: Clarendon Press, 1996.

Mulsow, Martin. *Knowledge Lost: A New View of Early Modern Intellectual History*. Trans. Hans Christian Erik Midelfort. Princeton: Princeton University Press, 2022.

Nelson, Eric. *The Hebrew Republic: Jewish Sources and the Transformation of European Political Thought*. Cambridge, MA: Harvard University Press, 2010.

Neveu, Valerie. "De Guillaume Postel à Richard Simon: Zohar et autres sources hébraïques dans les collections de la Bibliothèque municipale de Rouen." *Revue des études juives* 155 (1996): 75–105.

Olds, Katrina. *Forging the Past: Invented Histories in Counter-Reformation Spain*. New Haven: Yale University Press, 2015.

Parenty, Hélène. *Isaac Casaubon, helléniste: Des studia humanitatis à la philologie*. Travaux d'humanisme et Renaissance 454. Geneva: Droz, 2009.

Penkower, Jordan. "A Renewed Inquiry into the *Sefer Masoret ha-Masoret* of Elijah Bahur." In Hebrew. *Italia* 8 (1989): 7–73

Perani, Maurizio, ed. *The Ancient Sefer Torah of Bologna*. Studies in Jewish History and Culture 59. Leiden: Brill, 2019.

Pohlig, Matthias. *Zwischen Gelehrsamkeit und konfessioneller Identitätsstiftung: Lutherische Kirchen- und Universalgeschichtsschreibung, 1546–1617*. Spätmittelalter und Reformation neue Reihe 37. Tübingen: Mohr Siebeck, 2007.

Popper, Nicholas. *Walter Raleigh's "History of the World" and the Historical Culture of the Late Renaissance*. Chicago: University of Chicago Press, 2012.

Popper, William. *The Censorship of Hebrew Books*. New York: Knickerbocker Press, 1899.

Prijs, Joseph. *Die Basler hebräischen Drucke, 1492–1866: Im Auftrag der Öffentlichen Bibliothek der Universität Basel bearb. von Joseph Prijs. Ergänzt und hrsg. von Bernhard Prijs*. Ed. Bernhard Prijs. Olten: Urs Graf, 1964.

—. *Die Handschriften der Universitätsbibliothek Basel: Die Hebräischen Handschriften: Katalog auf Grund der Beschreibungen von Joseph Prijs*. Ed. Bernhard Prijs and David Prijs, with Stephen G. Burnett and Thomas Willi. Basel: Universitätsbibliothek, 1994.

Principe, Lawrence. *The Aspiring Adept: Robert Boyle and His Alchemical Quest*. Princeton: Princeton University Press, 1998.

Quantin, Jean-Louis. *The Church of England and Christian Antiquity: The Construction of a Confessional Identity in the 17th Century*. Oxford-Warburg Studies. Oxford: Oxford University Press, 2009.

Raspe, Lucia. "Three Sixteenth-Century Translations of Penitential Prayers into Yiddish." *Jewish Studies Quarterly* 26 (2019): 224–42.

—. "Tradition, Migration, and the Impact of Print: Local Rites of Seliḥot Recitation in Early Modern Ashkenaz." *Jewish Quarterly Review* 113 (2023): 83–10.

Rassem, Mohammed, and Justin Stagl, eds. *Geschichte der Staatsbeschreibung: Ausgewählte Quellentexte, 1456–1813*. Berlin: Akademie, 1994.

Raz-Krakotzkin, Amnon. *The Censor, the Editor and the Text: The Catholic Church and the Shaping of the Jewish Canon in the Sixteenth Century*. Jewish Culture and Contexts. Philadelphia: University of Pennsylvania Press, 2007.

Reif, Stefan. *Shabbetai Sofer and His Prayer-Book*. Cambridge: Cambridge University Press, 1979.

Reif, Stefan, Andreas Lehnardt, and Avriel Bar-Levav, eds. *Death in Jewish Life and Burial and Mourning Customs among Jews of Europe and Nearby Communities*. Berlin: de Gruyter, 2014.

Reiner, Elchanan. "The Ashkenazi Élite at the Beginning of the Modern Era: Manuscript versus Printed Book." In *Jews in Early Modern Poland*, ed. Gershon David Hundert, 85–98. Polin: Studies in Polish Jewry 10. London: Littman Library of Jewish Civilization, 1997.

Reynolds, L.D., and N.G. Wilson. *Scribes and Scholars: A Guide to the Transmission of Greek and Latin Literature*. 4th ed. Oxford: Oxford University Press, 2013.

Richler, Benjamin, ed. *Hebrew Manuscripts in the Vatican Library: Catalogue*. Palaeographical and codicological descriptions by Malachi Beit-Arié in collaboration with Nurit Pasternak. Città del Vaticano: Biblioteca Apostolica Vaticana, 2008.

Rosenblatt, Jason. *Renaissance England's Chief Rabbi: John Selden*. New York: Oxford University Press, 2006.

Rowe, John Howland. "Ethnography and Ethnology in the Sixteenth Century." *Kroeber Anthropological Society Papers* 30 (1964): 1–19.

Rubiés, Joan-Pau. *Travel and Ethnology in the Renaissance: South Asia Through European Eyes, 1250–1625*. Cambridge: Cambridge University Press, 2000.

—. *Travellers and Cosmographers: Studies in the History of Early Modern Travel and Ethnology*. Collected Studies 888. Aldershot: Ashgate, 2007.

Rubin, Aaron. "Samuel Archivolti and the Hebrew Vowel Point Controversy." *Jewish Quarterly Review* 101 (2011): 233–43.

Şahin, Kaya. "Political Pragmatism, Humanist Ideals, and Early Modern Orientalism in Busbecq's *Turkish Letters*." In *The Dialectics of Orientalism in Early Modern Europe*, ed. Marcus Keller and Javier Irigoyen-García, 33–49. London: Palgrave Macmillan, 2018.

Sander, Sabine, ed. *Language as Bridge and Border: Linguistic, Cultural and Political Constellations in Eighteenth to Twentieth Century German-Jewish Thought*. Berlin: Hentrich & Hentrich, 2015.

Schäfer, Peter. *Jesus in the Talmud*. Princeton: Princeton University Press, 2007.

—, ed. *The Talmud Yerushalmi and Graeco-Roman Culture*. 3 vols. Tübingen: Mohr Siebeck, 1998–2002.

Scheible, Heinz. *Der Entstehung der Magdeburger Zenturien: Ein Beitrag zur Geschichte der historiographischen Methode*. Schriften des Vereins für Reformationsgeschichte 183. Gütersloh: Mohn, 1960.

Scheidegger, Christian. "Die Bücherzensur in der Schweiz von der Reformation bis 1600." *Zwingliana* 45 (2018): 365–435.

Scholem, Gershom. *On the Kabbalah and Its Symbolism*. New York: Schocken, 1965.

Schwartz, Eduard. "Über Kirchengeschichte." *Nachrichten der Königlichen Gesellschaft der Wissenschaften zu Göttingen, Geschäftliche Mitteilungen* (1908), 106–22. Repr. in Eduard Schwartz, *Gesammelte Schriften*, 1: 110–30. 5 vols. Berlin: de Gruyter, 1938–63.

Sharpe, Kevin. *Reading Revolutions: The Politics of Reading in Early Modern England.* New Haven: Yale University Press, 2000

Shear, Adam. *The Kuzari and the Shaping of Jewish Identity, 1167–1900.* Cambridge: Cambridge University Press, 2008.

Sheehan, Jonathan. "Sacred and Profane: Idolatry, Antiquarianism and the Polemics of Distinction in the Seventeenth Century." *Past & Present* 192 (2006): 37–66.

Sherman, William. *Used Books: Marking Readers in Renaissance England.* Philadelphia: University of Pennsylvania Press, 2008.

Sherwin, Byron L. "In the Shadow of Greatness: Rabbi Hayyim Ben Betsalel of Friedberg." *Jewish Social Studies* 37 (1975): 35–60.

Sidorko, Clemens. *Basel und der jiddische Buchdruck (1557–1612): Kulturexport in der frühen Neuzeit.* Schriften der Universitätsbibliothek Basel 8. Basel: Schwabe, 2014.

Simpson, Percy. *Proof-Reading in the Sixteenth, Seventeenth and Eighteenth Centuries.* London: Oxford University Press, 1935.

Skloot, Joseph. *First Impressions: Sefer Hasidim and Early Modern Hebrew Printing.* Waltham, MA: Brandeis University Press, 2023.

Sládek, Pavel. "'Before the Law': Jewish Correctors of Early Printed Books." In *Printing and Misprinting: A Companion to Mistakes and In-House Corrections in Renaissance Europe (1450–1650),* ed. Geri Della Rocca de Candal, Anthony Grafton, and Paolo Sachet, 261–77. Oxford: Oxford University Press, 2023.

Slatkine, Menahem. *Reishit bikkurei ha-bibliyografyah ba-sifrut ha-'Ivrit: ha-sefer Siftei yeshenim me-R. Shabbetai Meshorer Bass.* Tel Aviv: Malan, 1957/8.

Smend, Rudolf. *Vier Epitaphe: Die Basler Hebraistenfamilie Buxtorf.* Litterae et Theologia. Berlin: de Gruyter, 2010.

Smith, Jonathan. *Drudgery Divine: On the Comparison of Early Christianities and the Religions of Late Antiquity.* Jordan Lectures in Comparative Religion 14. Chicago: University of Chicago Press, 1990.

Smyth, Adam. *Material Texts in Early Modern England.* Cambridge: Cambridge University Press, 2018.

Sonne, Isaiah. "Expurgation of Hebrew Books: The Work of Jewish Scholars. A Contribution to the History of the Censorship of Hebrew Books in Italy in the Sixteenth Century." *Bulletin of the New York Public Library* 46 (1942): 975–1015.

Sorkin, David. *The Religious Enlightenment: Protestants, Jews, and Catholics from London to Vienna.* Jews, Christians and Muslims from the Ancient to the Modern World. Princeton: Princeton University Press, 2008.

Staehelin, Ernst. "Des Basler Buchdruckers Ambrosius Froben Talmudausgabe und Handel mit Rom." *Basler Zeitschrift für Geschichte und Altertumskunde* 30 (1931): 7–37.

Stagl, Justin. *A History of Curiosity: The Theory of Travel, 1550–1800.* Studies in Anthropology and History 13. Chur: Harwood Academic, 1995.

Stagl, Justin, et al. *Apodemiken: Eine räsonnierte Bibliographie der reisetheoretischen Literatur des 16., 17. und 18. Jahrhunderts.* Quellen und Abhandlungen zur Geschichte der Staatsbeschreibung und Statistik 2. Paderborn: F. Schöningh, 1983.

Stein, Siegfried. "Philippus Fernandus Polonus, A Sixteenth-Century Hebraist in England." In *Essays in Honour of the Very Rev. J.H. Hertz*, ed. I. Epstein, E. Levine and C. Roth, 397–412. London: Edward Goldston, 1942.

Steinmann, Martin. *Johannes Oporinus: Ein Basler Buchdrucker um die Mitte des 16. Jahrhunderts*. Basler Beiträge zur Geschichtswissenschaft 105. Basel: Helbing & Lichtenhahn, 1967.

Stroumsa, Guy. *A New Science: The Discovery of Religion in the Age of Reason*. Cambridge, MA: Harvard University Press, 2010.

Suchland, Klaus-Henning. *Das Byzanzbild des Tübinger Philhellenen Martin Crusius (1526–1607)*. PhD diss., Würzburg, 2001.

Tanner, Regula. "Sefer ha-Yira und Sefer Hayye Olam: Zwei jiddische Drucke aus Freiburg i. Br. von 1583." *Judaica* 68 (2012): 105–28.

Taylor, Kathryn. "Ancients and Moderns in Sixteenth-Century Ethnography." *History of European Ideas* 46 (2020): 113–30.

—. "Making Statesmen, Writing Culture: Ethnography, Observation and Diplomatic Travel in Early Modern Venice." *Journal of Early Modern History* 22 (2018): 279–98.

Timm, Erika. "Abraham ibn Ezra und das *Maiśebuch*." In *Leket: Yidishe shtudyes haynt: Jiddistik heute: Yiddish Studies Today*, ed. Marion Aptroot et al, 1: 281–308. Düsseldorf: Düsseldorf University Press, 2012.

Toomer, Gerald. *John Selden: A Life in Scholarship*. 2 vols. Oxford: Oxford University Press, 2009.

Toufexis, Panagiotis. *Das Alphabetum vulgaris linguae graecae des deutschen Humanisten Martin Crusius (1526–1607): Ein Beitrag zur Erforschung der gesprochenen griechischen Sprache im 16. Jh*. Neograeca Medii Aevi 8. Cologne: Romiosini, 2005.

Turniansky, Chava. "The Events in Frankfurt am Main (1612–1616) in *Megillas Vints* and in an Unknown Yiddish 'Historical' Song." In *Schöpferische Momente des europäischen Judentums in der frühen Neuzeit*, ed. Michael Graetz, 121–37. Heidelberg: Universitätsverlag C. Winter, 2000.

—. "The Evolution of the Poetical Contest in Ashkenaz." In *Studies in Yiddish Literature and Folklore*, 60–98. Jerusalem: Hebrew University of Jerusalem, 1986.

Twersky, Isadore, and Bernard Septimus, eds. *Jewish Thought in the Seventeenth Century*. Cambridge, MA: Harvard University Press, 1987.

Twining, Timothy. "The Early Modern Debate over the Age of the Hebrew Vowel Points: Biblical Criticism and Hebrew Scholarship in the Confessional Republic of Letters." *Journal of the History of Ideas* 81 (2020): 337–58.

—. *The Limits of Erudition: The Old Testament in Post-Reformation Europe*. Cambridge: Cambridge University Press, forthcoming.

van Boxel, Piet. "Hebrew Books and Censorship in Sixteenth-Century Italy." In *Jewish Books and Their Readers: Aspects of the Intellectual Life of Christians and Jews in Early Modern Europe*, ed. Scott Mandelbrote and Joanna Weinberg, 75–99. Brill's Series in Church History 75. Leiden: Brill, 2016.

van Boxel, Piet, Kirsten Macfarlane, and Joanna Weinberg, eds. *The Mishnaic Moment: Jewish Law among Jews and Christians in Early Modern Europe*. Oxford: Oxford University Press, 2022.

van Liere, Katherine, Simon Ditchfield, and Howard Louthan, eds. *Sacred History: Uses of the Christian Past in the Renaissance World*. Oxford: Oxford University Press, 2012.

Vine, Angus. *Miscellaneous Order: Manuscript Culture and the Early Modern Organization of Knowledge*. Oxford: Oxford University Press, 2019.

Vrolijk, Arnoud, and Joanna Weinberg. "Thomas Erpenius, Oriental Scholarship and the Art of Persuasion." In *Scholarship between Europe and the Levant: Essays in Honour of Alastair Hamilton*, ed. Jan Loop and Jill Kraye, 34–59. The History of Oriental Studies 8. Leiden: Brill, 2020.

Waddington, Raymond, and Arthur Williamson, eds. *The Expulsion of the Jews: 1492 and After*. Garland Reference Library of the Humanities 1844. New York: Garland, 1994.

Walton, Michael. *Anthonius Margaritha and the Jewish Faith: Jewish Life and Conversion in Sixteenth-Century Germany*. Detroit: Wayne State University Press, 2012.

—. "Anthonius Margaritha – Honest Reporter?." *Sixteenth Century Journal* 36 (2005): 129–41.

Wandel, Lee Palmer. *Reading Catechisms, Teaching Religion*. Leiden: Brill, 2016.

Weinberg, Joanna. "Crossroads in Hebraism: Johann Buxtorf Gives a Hebrew Lesson to Philippe Duplessis-Mornay." In *Jewish Books and Their Readers: Aspects of the Intellectual Life of Christians and Jews in Early Modern Europe*, ed. Scott Mandelbrote and Joanna Weinberg, 151–68. Leiden: Brill, 2016.

—. "A Sixteenth-Century Hebraic Approach to the New Testament." In *History of Scholarship: A Selection of Papers from the Seminar on the History of Scholarship Held Annually at the Warburg Institute*, ed. Christopher Ligota and Jean-Louis Quantin, 231–50. Oxford: Oxford University Press, 2006.

Wendebourg, Dorothea. *Reformation und Orthodoxie: Der ökumenische Briefwechsel zwischen der Leitung der Württembergischen Kirche und Patriarch Jeremias II. von Konstantinopel in den Jahren 1573–1581*. Forschungen zur Kirchen- und Dogmengeschichte 37. Göttingen: Vandenhoeck & Ruprecht, 1986.

White, Paul. *Jodocus Badius Ascensius: Commentary, Commerce, and Print in the Renaissance*. Oxford: Oxford University Press, 2013.

Wilhelmi, Thomas. *Die griechischen Handschriften der Universitätsbibliothek Tübingen: Sonderband Martin Crusius: Handschriftenverzeichnis und Bibliographie*. Handschriftenkataloge der Universitätsbibliothek Tübingen 2. Wiesbaden: Harrassowitz, 2002.

Wilke, Carsten. "Splendeurs et infortunes du talmudisme académique en Allemagne." In *Les Textes judéophobes et judéophiles dans l'Europe chrétienne à l'époque moderne, XVIe–XVIIIe siècles*, ed. Daniel Tollet, 97–134. Paris: Presses Universitaires de France, 2000.

Williams, Benjamin. "'Some Fanciful Midrash Explanation': *Derash* on the *Ṭeʿamim* in the Middle Ages and Early Modern Period." In *Studies in the Masoretic Tradition of the Hebrew Bible*, ed. Daniel Crowther, Daniel D. Hornkohl, and Geoffrey Khan, 329–76. Semitic Languages and Cultures. Cambridge: Open Book Publishers, 2022.

Wolfson, Elliot. "Biblical Accentuation in a Mystical Key: Kabbalistic Interpretation of the Te'amim." 2 parts. *Journal of Jewish Music and Liturgy* 11 (1988–89): 1–16; 12 (1989–90): 1–13.

Yaari, Abraham. "Correctors' Complaints about Printing on Shabbat by Non-Jews" (in Hebrew). *Kiryat Sefer* 14 (1937): 524–28; addenda, ibid., 16 (1939): 500–501.

Yale, Elizabeth. *Sociable Knowledge: Natural History and the Nation in Early Modern England.* Philadelphia: University of Pennsylvania Press, 2016.

—. "With Slips and Scraps: How Early Modern Naturalists Invented the Archive." *Book History* 12 (2009): 1–36.

Yeo, Richard. "Between Memory and Paperbooks: Baconianism and Natural History in Seventeenth-Century England." *History of Science* 45 (2007): 1–46.

—. *Notebooks and English Virtuosi: Memory, Information, and Early Modern Science.* Chicago: University of Chicago Press, 2014.

Yerushalmi, Yosef Hayyim. *Zakhor: Jewish History and Jewish Memory*. Samuel and Althea Stroum Lectures in Jewish Studies. Seattle and London: University of Washington Press, 1996.

Zimmer, Eric. *Rabbi Hayyim b. R. Betsalel me-Friedberg, aḥi ha-Maharal mi-Prag.* Jerusalem: Mosad harav Kook, 1987.

Index